Remote Sensing of the European Seas

Remote Sensing
of the European Seas

Edited by

Vittorio Barale

Joint Research Centre
European Commission
Ispra, Italy

and

Martin Gade

Institute of Oceanography
University of Hamburg
Germany

 Springer

Vittorio Barale
Joint Research Centre
European Commission
Ispra, Italy

Martin Gade
Institute of Oceanography
University of Hamburg
Germany

ISBN: 978-1-4020-6771-6   e-ISBN: 978-1-4020-6772-3

Library of Congress Control Number: 2007942178

Cover illustrations: Figures from this book, pp. 82, 99, 328, 390, 465

Printed on acid-free paper.

9 8 7 6 5 4 3 2 1

springer.com

To Enheduanna
Daughter of Sargon of Akkad
High Priestess of the Moon God Nanna

Contents

Preface

Princess Enheduanna, daughter of king Sargon of Akkad, lived around 2300 BC. She was a high priestess of the moon god Nanna in the ancient city of Ur. And an accomplished poet too. In fact, she is the author of a number of Sumerian hymns, and is generally considered to be the earliest author known by name. When she came to honor Inanna – the goddess of sexual love, fertility, and warfare, daughter of Nanna and often associated with the planet Venus (the one that the Akkadians called Ishtar) – above all the other gods of the Sumerian pantheon, she mentioned for the very first time, in her Hymn number 8, nothing less than the "Seven Seas"...

Septem Maria, would call them the Romans centuries later, after inheriting the concept from the Greeks (for whom seven probably just meant several), but perhaps applying it to the wrong place – *i.e.* the extensive system of coastal lagoons, which at the time dotted the northern Adriatic Sea – at least in the description of Pliny the Elder, Roman fleet commander and scholarly author of *Historia Naturalis*. Indeed, which seven seas are intended depends on the context. According to the historians, there are at least nine bodies of water in the medieval European and Arabic literature that can aspire to qualify as one of the famous seven. And, as suggested in poetry by British author Rudyard Kipling, there are seven oceans on planet Earth, if one counts after separating into North and South both the Pacific Ocean and the Atlantic Ocean. But here, in this Volume devoted to the European Seas, the seven mythical basins are those that surround the European continental landmass. Maybe they are not even seven, at a careful account, but they are certainly several and, just as certainly, they display a wide range of environmental traits.

Studying and understanding the natural history of these modern Seven Seas requires integrated observation systems, which must include up-to-date remote sensing techniques. This Volume reviews the current potential of Earth Observations, while devoting particular attention to those applications that deal with the issues, peculiarities and special challenges posed by the European marginal, semi-enclosed and enclosed seas. The assessment of surface parameters by means of both passive and active techniques – measuring reflected visible and near-infrared sunlight, or surface emissions at thermal infrared or microwave frequencies, or again the return of transmitted impulses of visible or microwave radiation – is addressed in a collection of topical papers. Satellite remote sensing from Earth's orbit is the focus of most of them, but selected, promising examples of airborne measurements and ground-based applications are also covered. The review of the most recent results achieved by each of these techniques, and of the new scientific ground broken by them, provides an unprecedented insight into the inimitable mix of dynamical and bio-geo-chemical features that characterize the European Seas.

The peer-reviewed papers collected in this Volume – which targets researchers working in the field of Earth and Marine Sciences, but also teachers, as well as (graduate) students, in the same field – are organized into four main sections. The first provides a brief overview of the European Seas, followed by an historical outlook on the development of the remote sensing approach to marine environmental issues in Europe. The second part is devoted to Visible & Thermal Infrared (passive and active) remote sensing, and the third to Microwave (passive and active) remote sensing. The fourth offers a few examples of multi-sensor techniques, which are becoming increasingly useful to exploit the synergies of complementary sensors and to enhance the value of their combined views.

The breath of the environmental themes covered, and of the diverse techniques dealt with, called for the contribution of the entire European marine remote sensing community. Not surprisingly, then, very many scientists helped, either as authors or as reviewers, sometimes as both, in the realization of this Volume. Their names and affiliations are recalled in the "contributors" list that follows. Sincere thanks are due to all.

June 2007   Vittorio Barale & Martin Gade

Contributors

Allan T.D.
Satellite Observing Systems
Godalming, UK

Alpers W.
Institute of Oceanography
Universiy of Hamburg
Hamburg, Germany

Ambar I.
Institute of Oceanography
University of Lisbon
Lisboa, Portugal

Andersen S.
Danish Meteorological Institute
Copenhagen, Denmark

Askne J.
Radio & Space Science Department
Chalmers University of Technology
Gothenburg, Sweden

Babichenko S.
Laser Diagnostic Instruments AS
Tallinn, Estonia

Bakhanov V.
Institute of Applied Physics
Russian Academy of Sciences
Nizhny Novgorod, Russia

Ballabrera-Poy J.
Institute of Marine Sciences
Spanish Research Council
Barcelona, Spain

Barale V.
Joint Research Centre
European Commission
Ispra, Italy

Bell P.S.
Proudman Oceanographic
    Laboratory
Liverpool, UK

Bergamasco A.
Institute of Marine Sciences
National Research Council
Venezia, Italy

Berger M.
European Space Research and
    Technology Centre
European Space Agency
Nordwijk, The Netherlands

Berthon J.F.
Joint Research Centre
European Commission
Ispra, Italy

Bocharova T.
Space Research Institute (IKI)
Russian Academy of Sciences
Moscow, Russia

Bouzinac C.
European Space Research and
    Technology Centre
European Space Agency
Nordwijk, The Netherlands

Brandt P.
Leibniz Institute of Marine Sciences
    at the University of Kiel
Kiel, Germany

Brekke C.
Department of Informatics
University of Oslo
Oslo, Norway

Bricaud A.
Laboratory of Oceanography
National Center for Scientific
    Research (CNRS)
Villefranche-sur-Mer, France

Brockmann C.
Brockmann Consult
Geesthacht, Germany

Bulgarelli B.
Joint Research Centre
European Commission
Ispra, Italy

Buongiorno Nardelli B.
Institute for Atmospheric and
    Climate Sciences
National Research Council (CNR)
Roma, Italy

Burenkov V.I.
Shirshov Institute of Oceanology
Russian Academy of Sciences
Moscow, Russia

Byfield V.
Ocean Observing & Climate Group
National Oceanography Centre
Southampton, UK

Camps A.
Department of Signal Theory and
    Communications
Polytechnic University of Catalonia
Barcelona, Spain

Christiansen M.B.
Wind Energy Department
Risø National Laboratory
Roskilde, Denmark

Cipollini P.
Ocean Observing & Climate Group
National Oceanography Centre
Southampton, UK

Da Silva J.C.B.
Institute of Oceanography
University of Lisbon
Lisbon, Portugal

Dias J.
Institute of Oceanography
University of Lisbon
Lisboa, Portugal

Dierking W.
Alfred Wegener Institute for Polar
    and Marine Research
Bremerhaven, Germany

Doerffer R.
Institute for Coastal Research
GKSS Research Center
Geesthacht, Germany

Donlon C.
Met Office Hadley Centre
Exeter, UK

Dowell M.
Joint Research Centre
European Commission
Ispra, Italy

Dransfeld S.
Institute of Oceanography
Universiy of Hamburg
Hamburg, Germany

Drinkwater M.
European Space Research and
    Technology Centre
European Space Agency
Noordwijk, The Netherlands

Durand D.
Institute for Water Research
Bergen, Norway

Ermakov S.
Institute of Applied Physics
Russian Academy of Sciences
Nizhny Novgorod, Russia

Ezraty R.
Space Oceanography Laboratory
French Research Institute for Ex-
    ploitation of the Sea (IFREMER)
Plouzane, France

Fenoglio-Marc L.
Institute of Physical Geodesy
Darmstadt University of Technology
Darmstadt, Germany.

Ferraro G.
Joint Research Centre
European Commission
Ispra, Italy

Font J.
Department Physical Oceanography
Institute of Marine Sciences
Barcelona, Spain

Gade M.
Institute of Oceanography
Universiy of Hamburg
Hamburg, Germany

Garello R.
GET-ENST Bretagne
National Center for Scientific
    Research (CNRS)
Brest, France

Gasparini G.P.
Institute of Marine Sciences
National Research Council (CNR)
Pozzuolo di Lerici, Italy

Gerth M.
Baltic Sea Research Institute (IOW)
Rostock-Warnemünde, Germany

Girard-Ardhuin F.
Space Oceanography Laboratory
French Research Institute for Ex-
    ploitation of the Sea (IFREMER)
Plouzane, France

Gommenginger C.
Laboratory for Satellite Oceano-
    graphy
National Oceanography Centre
Southampton, UK

Greidanus H.
Joint Research Centre
European Commission
Ispra, Italy

Gücü A.C.
Institute of Marine Sciences
Middle East Technical University
Erdemli-İçel, Turkey

Gudmandsen P.
Danish National Space Center
Technical University of Denmark
Lyngby, Denmark

Gurgel K.W.
Institute of Oceanography
Universiy of Hamburg
Hamburg, Germany

Haarpainter J.
Norut Information Technology Ltd
Tromsø, Norway

Hainbucher D.
Institute of Oceanography
Universiy of Hamburg
Hamburg, Germany

Harms I.
Institute of Oceanography
Universiy of Hamburg
Hamburg, Germany

Hasager C.
Wind Energy Department
Risø National Laboratory
Roskilde, Denmark

Hessner K.G.
OceanWaveS GmbH
Lüneburg, Germany

Heygster G.
Institute of Environmental Physics
University of Bremen
Bremen, Germany

Hoepffner N.
Joint Research Centre
European Commission
Ispra, Italy

Hoogeboom P.
International Research Centre for
    Telecommunications and Radar
Delft University of Technology
Delft, The Netherlands

Horstmann J.
Institute of Coastal Research
GKSS Research Center
Geesthacht, Germany

Hühnerfuss H.
Institute of Organic Chemistry
University of Hamburg
Hamburg, Germany

Johannessen J.
Nansen Environmental and Remote
    Sensing Centre
Bergen, Norway

Kaitala S.
Department of Biological Oceano-
    graphy
Finnish Institute of Marine Research
Helsinki, Finland

Kaleschke L.
Institute of Oceanography
Universiy of Hamburg
Hamburg, Germany

Kern S.
Institute of Oceanography
University of Hamburg
Hamburg, Germany

Klein B.
Federal Maritime and Hydrographic
    Agency (BSH)
Hamburg, Germany

Koch W.
Institute of Coastal Research
GKSS Research Center
Geesthacht, Germany

Kopelevich O.V.
Shirshov Institute of Oceanology
Russian Academy of Sciences
Moscow, Russia

Korosov A.
Nansen International Environmental
    and Remote Sensing Centre
St. Petersburg, Russia

Krężel A.
Institute of Oceanography
University of Gdansk
Gdynia, Poland

Kudryavtsev V.
Nansen International Environmental
    and Remote Sensing Centre
St. Petersburg, Russia

Lacroix G.
Royal Institute for Natural Sciences
Brussels, Belgium

Lancelot C.
Ecology of Aquatic Sytrems (ESA)
Free University of Brussels
Brussels, Belgium

Lange P.A.
Institute of Oceanography
Universiy of Hamburg
Hamburg, Germany

Larnicol G.
Space Oceanography Division
Satellite Data Center (CLS)
Ramonville Saint-Agne, France

Lavender S.J.
Marine Institute
University of Plymouth
Plymouth, UK

Lavrova O.Y.
Space Research Institute (IKI)
Russian Academy of Sciences
Moscow, Russia

Lidicky L.
Delft University of Technology
Delft, The Netherlands

Litovchenko K.
Russian Institute of Space Device
    Engineering
Moscow, Russia

Lyard F.
Laboratory for Space Studies in
    Geophysics and Oceanography
    (LEGOS)
Toulouse, France

Marcos M.
Ocean Observing & Climate Group
National Oceanography Centre
Southampton, UK

Marullo S.
National Agency for New Technolo-
    gies, Energy and the Environment
    (ENEA)
Frascati, Italy

Mauri E.
National Institute of Oceanography
    and Experimental Geophysics
    (OGS)
Trieste, Italy

Mélin F.
Joint Research Centre
European Commission
Ispra, Italy

Melsheimer C.
Institute of Environmental Physics
University of Bremen
Bremen, Germany

Meyer-Roux S.
Joint Research Centre
European Commission
Ispra, Italy

Mitchelson-Jacob E.G.
Centre for Applied Marine Sciences
Marine Science Laboratories
Anglesey, UK

Mityagina M.
Space Research Institute (IKI)
Russian Academy of Sciences
Moscow, Russia

Morales J.
IFAPA, Junta de Andalucia
Huelva, Spain

Morovic M.
Institute of Oceanography and
    Fisheries
Split, Croatia

Muellenhoff O.
Joint Research Centre
European Commission
Ispra, Italy

Nechad B.
Royal Institute for Natural Sciences
Brussels, Belgium

Nieto-Borge J.C.
Department of Signal Theory and
    Communications
Escuela Politécnica Superior
Madrid, Spain

Neumann A.
Remote Sensing Technology
    Institute (IMF)
German Aerospace Center (DLR)
Berlin, Germany

Nykjaer L.
Joint Research Centre
European Commission
Ispra, Italy

Olsen R.
Norwegian Defence Research
    Establishment (FFI)
Kjeller, Norway

Park Y.
Royal Institute for Natural Sciences
Brussels, Belgium

Pasaric M.
Andrija Mohorovicic Geophysical
    Institute, University of Zagreb
Zagreb, Croatia

Pascual A.
Institut Mediterrani d'Estuidis
    Advançats, IMEDEA
Palma de Mallorca, Spain

Peters S.
Faculty of Earth and Life Sciences
Free University of Amsterdam
Amsterdam, The Netherlands

Pettersson L.
Nansen Environmental and Remote
Sensing Centre
Bergen, Norway

Pozdnyakov D.
Nansen International Environmental
    and Remote Sensing Centre
St. Petersburg, Russia

Poulain P.M.
National Institute of Oceanography and
    Experimental Geophysics (OGS)
Trieste, Italy

Pradhan Y.
University of Plymouth
Plymouth, UK

Raitsos D.E.
Marine Institute
University of Plymouth
Plymouth, UK

Redondo J.
Department of Applied Physics
Polytechnic University of Catalonia
Barcelona, Spain

Reuter R.
Institute of Physics
University of Oldenburg
Oldenburg, Germany

Roblou L.
Noveltis
Ramonville-Saint-Agne, France

Romeiser R.
Institute of Oceanography
Universiy of Hamburg
Hamburg, Germany

Rubino A.
Department of Environmental
   Sciences
Ca' Foscari University
Venezia, Italy

Ruddick K.
Royal Institute for Natural Sciences
Brussels, Belgium

Runge H.
Remote Sensing Technology Institute
   (IMF)
German Aerospace Center (DLR)
Oberpfaffenhofen, Germany

Salusti E.
National Institute of Nuclear Physics
   (INFN)
Frascati, Italy

Sandven S.
Nansen Environmental and Remote
Sensing Centre
Bergen, Norway

Santoleri R.
Institute for Atmospheric and
   Climate Sciences
National Research Council (CNR)
Roma, Italy

Schlick T.
Institute of Oceanography
Universiy of Hamburg
Hamburg, Germany

Schulz-Stellenfleth J.
Remote Sensing Technology
   Institute (IMF)
German Aerospace Center (DLR)
Oberpfaffenhofen, Germany

Shaw A.G.P.
Ocean Observing & Climate Group
National Oceanography Centre
Southampton, UK

Sheberstov S.V.
Shirshov Institute of Oceanology
Russian Academy of Sciences
Moscow, Russia

Siegel H.
Baltic Sea Research Institute (IOW)
Rostock-Warnemünde, Germany

Skou N.
Danish National Space Center
Technical University of Denmark
Lyngby, Denmark

Smyth T.
Plymouth Marine Laboratory
Plymouth, UK

Snaith H.M.
Scientific Data Group
National Oceanography Centre
Southampton, UK

Solberg A.H.S.
Department of Informatics
University of Oslo
Oslo, Norway

Spreen G.
Institute of Oceanography
University of Hamburg
Hamburg, Germany

Stelzer K.
Brockmann Consult
Geesthacht, Germany

Stips A.
Joint Research Centre
European Commission
Ispra, Italy

Stoffelen A.
Royal Netherlands Meteorological
    Institute (KNMI)
De Bilt, The Netherlands

Stanichny S.
Marine Hydrophysical Institute
National Academy of Sciences
Sevastopol, Ukraine

Tarchi D.
Joint Research Centre
European Commission
Ispra, Italy

Taupier-Letage I.
Laboratory of Oceanography and
    Biogeochemistry
University of the Mediterranean Sea
La Seyne sur Mer, France

Tonboe R.T.
Danish Meteorological Institute
Copenhagen, Denmark

Topouzelis K.
Joint Research Centre
European Commission
Ispra, Italy

Tournadre J.
Space Oceanography Laboratory
French Research Institute for Ex-
    ploitation of the Sea (IFREMER)
Plouzané, France

Tsimplis M.N.
Ocean Observing & Climate Group
National Oceanography Centre
Southampton, UK

Van Der Wal D.
Centre Estuarine Marine Ecology
Netherlands Institute of Ecology
Yerseke, The Netherlands

Van Mol B.
Royal Institute for Natural Sciences
Brussels, Belgium

Vignudelli S.
Institute of Bio-Physics
National Research Council (CNR)
Pisa, Italy

Vogelzang J.
Royal Netherlands Meteorological
    Institute (KNMI)
De Bilt, The Netherlands

Volpe G.
Institute for Atmospheric and
    Climate Sciences
National Research Council (CNR)
Roma, Italy

Waldteufel Ph.
Institute Pierre-Simon-Laplace
National Center for Scientific
    Research (CNRS)
Verrieres Le Buisson, France

Wensink H.
ARGOSS
Vollenhove, The Netherlands

Woolf D.K.
Environmental Research Institute
North Highland College
UHI Millennium Institute
Thurso, UK

Zavatarelli M.
Department of Physics
University of Bologna
Bologna, Italy

Zibordi G.
Joint Research Centre
European Commission
Ispra, Italy

Zielinski O.
Bremerhaven University of Applied
    Sciences
Bremerhaven, Germany

Ziemer F.
Institute of Coastal Research
GKSS Research Center
Geesthacht, Germany

Section 1:

Introduction
to Remote Sensing of the European Seas

The European Marginal and Enclosed Seas: An Overview

Vittorio Barale

Institute for Environment and Sustainability, Joint Research Centre, European Commission, Ispra, Italy

Abstract. The European continent is confined by the North Atlantic Ocean, to the west and north-west, and by a score of marginal and enclosed seas, both to the north and to the south. The North Sea and other near-coastal, open water bodies – *i.e.* the Norwegian Sea, the Barents Sea and the White Sea, to the north; the Irish Sea, the Celtic Sea and the English Channel, as well as the Bay of Biscay and the Gulf of Cadiz, to the west – are considered marginal basins of the Atlantic, where oceanic influences dominate. Among the major enclosed seas, the Mediterranean Sea behaves like a concentration basin, while the Baltic Sea, the Black Sea and the Caspian Sea are essentially dilution basins. These seas exhibit a wide spectrum of environmental traits, ranging from sub-polar to sub-tropical climatic zones, from shallow continental shelves to deep abyssal plains, from pristine marine reserves to regions impacted by countless economic and recreational activities. Understanding the inner workings of these seas – aiming to reconcile the conflicting needs of protecting their ecological balance and exploiting their natural resources – requires adequate observation systems, integrating both *in situ* and remote sensing techniques.

1. Introduction

The concept of an island world, encircled and bounded by a great body of waters, goes back to the Indo-European and Mesopotamian mythology and spreads across the millennia into early European culture. It can be traced with an ideal continuity from the Samudra – the Sanskrit ocean, literally a gathering together of waters, recurring in the Rigveda, the great collection of Vedic hymns in the Hindu tradition – to the great river Okeanos, flowing around the habitable hemisphere in the Greek and later Roman beliefs; and even to the mighty sea serpent Jörmungandr – which grasps its own tail in the great ocean that surrounds Midgard, the middle enclosure inhabited

3

V. Barale, M. Gade (eds.), *Remote Sensing of the European Seas.*
© Springer Science+Business Media B.V. 2008

by men – in the Scandinavian folklore. Indeed, the idea of a land surrounded by the sea, which suited so well the ancient Europeans, has deep roots in a vision of the world imagined to be enclosed not only by the marine vastness all around, but also by a celestial ocean above the heavens and an ocean of the underworld below ground.

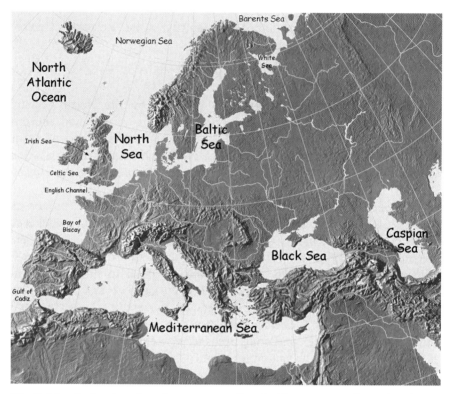

Fig. 1. Marginal and enclosed seas surrounding the European continental region.

Aside from its eastern connection to the Asian continent, the European land mass is really surrounded by the ocean, the North Atlantic Ocean, to the west and north-west, and bounded by a score of marginal and enclosed seas both to the north and to the south (Figure 1). Off the northern coasts, the Norwegian Sea and the Barents Sea edge onto the Artic Ocean. The semi-enclosed North Sea, the Baltic Sea and the White Sea occupy the northern shelf, while the Mediterranean Sea and the Black Sea wash the southern boundaries. At the western edge of the continent lies a series of marginal basins, again on large continental shelves, going from the Irish Sea, the Celtic Sea and the English Channel, to the Bay of Biscay and the

Gulf of Cadiz. At the eastern limit of the continent, separated from the Atlantic waters, is the totally enclosed Caspian Sea.

These marine basins, which in the following shall be referred to as the European Seas, correspond rather well to the classical definition of being "large expanses of (more or less) saline waters, connected to the world's ocean by a series of natural outlets". Some of them can be considered true marginal seas, portions of an outer ocean, partially enclosed by land forms such as islands or peninsulas, where water dynamics are mainly wind-driven. Others can be best classified as semi-enclosed or – as preferred in the following – enclosed seas, independent basins separated from the outer ocean, with which they have limited, if any, exchanges of water, and where water dynamics are dominated by thermohaline processes.

In broad terms, enclosed seas can be further classified into concentration or dilution basins. In a concentration basin, water output by evaporation exceeds input from precipitation and continental runoff. Thus, salinity increases and is generally higher than in the outer ocean. The basin's water exchange, if a connecting channel exists, consists of an inflow of fresher oceanic water in the upper layer and an outflow of saltier water in the lower layer. In a dilution basin, water input from precipitation and continental runoff exceeds output by evaporation. Thus, salinity decreases and is generally lower than in the outer ocean. The basin's water exchange through the connecting channel consists of an outflow of fresher water in the upper layer and an inflow of saltier oceanic water in the lower layer.

The ecological dynamics of enclosed seas are strongly impacted by their freshwater balance. When evaporation exceeds freshwater inflow, surface waters tend to become denser and sink, leading to convection and even complete overturning of the water column. Thus, nutrients accumulated in deep waters become available to production in the surface layer, while oxygen acquired from the atmosphere becomes available in deeper layers to foster life. When the inflow exceeds evaporation, the lighter freshwater tends to remain at the surface. In such a stratified system, convection is limited, or even prevented; so, mixing in the water column may not be sufficient to supply nutrients to the surface layer or oxygen to the deeper layers. River runoff may help in fertilizing surface waters, but, if the basin is isolated from the ocean, *e.g.* by a shallow sill, oxygen becomes depleted, or absent, in the bottom water and the variety of life forms much reduced.

An important feature of marginal and enclosed seas is the basin residence or turnover time, which can range from years to centuries. This is more relevant in enclosed seas, where exchanges with the outer ocean are limited, rather than in the more open marginal seas. In either case, anyway, residence time has a direct influence on bio-geo-chemical dynamics and on how contaminants are retained or accumulated in the marine ecosystem.

According to the above definitions, the North Sea and a score of other near-coastal marine regions – *i.e.* the Norwegian Sea, the Barents Sea and the White Sea, to the north; the Irish sea, the Celtic Sea, the English Channel and the Bay of Biscay, to the west – can be considered marginal basins of the Atlantic Ocean, where the oceanic influence prevails. Instead, among the main enclosed seas, the Mediterranean Sea behaves like a concentration basin, while the Baltic Sea, the Black Sea and the Caspian Sea are essentially dilution basins. These seas exhibit a wide spectrum of environmental traits, ranging from sub-polar to sub-tropical climatic zones, from shallow continental shelves to deep abyssal plains, from pristine marine reserves to regions impacted by countless economic and recreational activities. In the following, the main geographic attributes of the Atlantic Ocean will be recalled, together with some key oceanographic features of its northern section. Subsequently, the characteristics of each of the major European Seas shall be briefly reviewed, in order to place into perspective the various basins' environmental peculiarities and special challenges.

2. Atlantic Ocean

The Atlantic Ocean is the single element connecting all of the marginal and enclosed seas that surround the European continent. Named after Atlas, one of the Titans of Greek mythology, and first mentioned in The Histories of Herodotus, around 450 BC, the Atlantic is the second largest of the Earth's Oceans. It began to form about 150 million years ago, in the Jurassic period, when a rift opened up in the supercontinent of Gondwana, resulting in the separation of Africa and South America. The separation continues today along the Mid-Atlantic Ridge at a rate of about 2.5 cm/year. The basin width varies from 2,848 km, between South America and Africa, to about 4,830 km, between North America and Africa.

The Atlantic Ocean occupies about 20% of the planetary surface, representing an area of approximately 82,400,000 km^2 (106,400,000 km^2 if its adjacent seas are included), with an average depth of 3,926 m (3,338 m with adjacent seas), and a volume of 323,600,000 km^3 (354,700,000 km^3 with adjacent seas). At its deepest point, the Milwaukee Deep in the Puerto Rico Trench, the bottom is 8,605 m below the surface. Submarine rises extend between the Mid-Atlantic Ridge and the continental shelves, dividing the ocean floor into a series of abyssal plains.

Given its planetary size, the Atlantic Ocean displays a variety of climatic conditions. Due of its great capacity for retaining heat, maritime climates are moderate and free of extreme seasonal variations. Ocean currents contribute to the climatic control by transporting warm, or cold, waters from one region to another. Adjacent land areas are affected by the winds that are warmed, or cooled, when blowing over these currents.

Surface salinity remains relatively constant, ranging between 35 to 36 psu, with extremes from 33 to 37 psu – values that make the Atlantic the saltiest of the world's oceans – and varies with both latitude and season. These values are influenced by evaporation, precipitation, continental runoff, and melting of sea ice. Surface water temperatures, which vary with latitude, currents, and season, and reflect the latitudinal distribution of solar energy, vary over a wide range, from −2°C to 29°C.

In the Atlantic water column, four major water masses can be distinguished. The North Atlantic (NA) and South Atlantic (SA) central waters constitute the layer closer to the surface (origin NA and SA; depth 100–500 m and 100–300 m; salinity 35.10–36.70 and 34.65–36.00; temperature 8–19°C and 6–18°C). The sub-Antarctic intermediate water extends below this layer, from the south (origin SA, Antarctica; depth 500–1,000 m; salinity 33.80; temperature 2.2°C). Next come the North Atlantic Deep Water (NADW), formed at the highest northern latitudes (origin NA; depth 1,300–4,000 m; salinity 34.90–34.97; temperature 2.2–3.5°C) and the underlying Antarctic Bottom Water (ABW), which occupies the deepest basins (origin SA; depth >4,000 m; salinity 34.66; temperature 0.4°C).

The general circulation of the Atlantic Ocean, as part of the so-called global "Conveyor Belt" system, is composed by a prevalent northward flow in upper layers, and by a prevalent southward flow in lower layers. At the surface, the circulation pattern consists of two main basin-wide gyres, circulating in a anticyclonic direction in the North Atlantic, and in a cyclonic direction in the South Atlantic. The northern current system isolates a large, elongated body of water known as the Sargasso Sea, in which the salinity is noticeably higher than average. Further north, the convergence of relatively saline water and seasonal cooling contribute to the NADW formation. Specific interactions with the main marginal and enclosed seas of the North Atlantic region will be dealt with in the following sections.

The open oceans, due to their vast area, great depth and efficient water circulation, are still relatively unaffected by human activities, compared with marginal and enclosed seas. Hence, most of the North Atlantic ecosystem is relatively unpolluted, and degradation is mostly limited to near-coastal zones. A growing alarm concerns the deterioration of fish stocks (*e.g.* in productive near-coastal and tidally mixed waters), the reasons for which are not fully understood, but could include overfishing and climatic variability. The state of the North East Atlantic marginal and enclosed seas, which sometimes have limited exchange of water with the adjacent ocean, is extremely varied and strongly influenced by local processes.

3.   North Sea

The North Sea is a shelf sea, widely open towards the North Atlantic (Figure 2). The basin extends from 50°N to about 60°N and from 3°E to 10°E, with an area of 750,000 km^2 and a volume of 94,000 km^3. It has dimensions on the order of 800 km, in the north-south direction, by 500 km, in the east-west direction. It accounts for approximately 6,000 km of coastlines, including the south-western arm, leading into the English Channel, and the north-eastern Skagerrak (the Kattegat, further to the south, is usually regarded as part of the Baltic Sea).

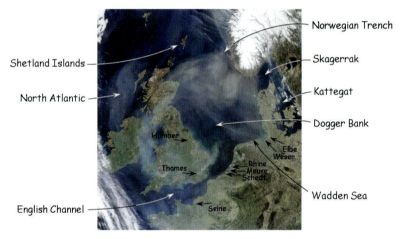

Fig. 2. North Sea. MODIS data, simulated true-colour, 19 April 2003.

The coastlines display a wide variety of landscapes, ranging from mountains, rocky islands and deep fjords, to cliffs, major estuaries, sand dunes and beaches – and the largest stretch of unbroken mudflats in the world, along the Wadden Sea. The name Wadden Sea identifies a body of water, and its associated coastal wetlands, lying between the coast of north-western continental Europe and the North Sea. It stretches from the Netherlands, in the south-west, past the river estuaries of Germany, to its northern boundary in Denmark, with a total length of nearly 500 km and a total area of about 10,000 km². It is a region typified by extensive tidal mud flats, deeper tidal trenches and the islands that bound them.

The European continental shelf gradually slopes from the north-eastern coast towards the Atlantic Ocean, forming a shelf sea with an average depth of only 95 m. A large part of the basin's bottom topography was shaped during the last glacial period, as evidenced by the presence of river valley systems and deep fjords. Although the shallowest regions are generally found off the southern coast, very shallow areas (less than 20 m), such as the central Dogger Bank, can be found also offshore. The deepest region is the Norwegian Trench, running along the coast of Norway. It reaches its maximum depth of 700 m in the Skagerrak.

The climate of the North Sea is strongly influenced by the inflow of Atlantic waters and the large-scale westerly air circulation pattern, which frequently carries low pressure systems. The air flow patterns produce large variations in wind direction and speed, high levels of cloud cover and rainfall. The annual precipitation is around 425 mm/year, with a peak of about 1,000 mm/year occurring along the Norwegian coast, due to the up-lift of moist air against the steep mountain ranges.

The North Sea has a catchment area of about 850,000 km^2. The freshwater input is on the order of 300 km^3/year, one-third of which comes from snowmelt in Scandinavia and two-thirds from major rivers such the Elbe, Weser, Rhine, Meuse, Scheldt, Seine, Thames and Humber. In addition, a large amount of freshwater – indeed the dominant freshwater source for the North sea – is supplied through the brackish inflow coming from the Baltic Sea, via the Kattegat and Skagerrak.

The hydrography of the North Sea is controlled by the combination of Atlantic water inflow, strong tidal action, freshwater input and climatic conditions. Most oceanic waters enter through the passage north of the Shetland Islands, and flow south along the western slope of the Norwegian Trench. Smaller amounts enter through the passage south of the Shetland Islands and through the English Channel. Most of these inflows, as well as the Baltic Sea waters entering the Skagerrak, flow back into the Atlantic Ocean via the Norwegian Trench, creating a general cyclonic circulation in the basin. The volume of Atlantic water entering the North Sea shows large annual variations, owing to the fluctuations of the North Atlantic Oscillation[1] (NAO). Sea surface temperature is quite stable in the northern part of the basin, with an annual mean around 9.5°C, but shows significant seasonal variability in the southern part, due to its shallow depth and large freshwater input. Salinity in open sea waters is around 35 psu and varies only between 32 and 34.5 psu in coastal waters. In the Skagerrak, and in the Norwegian Trench, salinity can be much lower, ranging from 10 to 34 psu, due to the brackish waters flowing episodically through the Kattegat from the Baltic Sea. The areas where lower salinity persists have a stable density stratification that is maintained throughout the year. In summer, solar heating causes thermal stratification over large portions of the basin, but this quickly disappears in winter through wind-driven vertical mixing.

---

[1] Air pressure oscillation between the Azores at ~30° N and Iceland at ~60° N, one of the major modes of variability of the Northern Hemisphere atmosphere.

No stratification develops in the shallower parts of the southern North Sea, due to intense tidal mixing.

Perennial macroalgae beds densely cover the littoral zones of the North Sea, although most seagrasses, once abundant along the coasts, now occur only in a few scattered areas. About 230 species of fish inhabit the North Sea, 13 of which are the main targets of major commercial fisheries. Sizeable populations of marine mammals are also present. Species diversity is lowest in the shallow southern part and in the English Channel.

Approximately 184 million people live within the North Sea catchment area. Population density is higher along the southern coast. A wide range of intense human activities, such as industrial production, as well as agriculture, oil and gas extraction, and fisheries, affect the North Sea environment. The two most important issues of environmental concern currently are eutrophication and contamination by trace organic compounds.

4.   Baltic Sea

The Baltic Sea is the largest brackish sea in the world (Figure 3). It comprises the Baltic Proper, the Gulf of Bothnia (further divided in Bothnian Bay, to the north, and Bothnian Sea, to the south), the Gulf of Finland and the Gulf of Riga. Other subdivisions, around of within the Baltic Proper, include the Archilpelago Sea and the Gotland Sea, in the northern and central sections, the Bornholm Sea and the Arkona Sea, in the southern section. The connection with the outer ocean is ensured via the Belt Sea and the Kattegat (the Skagerrak, further to the north, is usually regarded as part of the North Sea). The basin extends from 54°N to 66°N and from 10°E to 30°E. It is about 1,610 km long, in the north-south direction, and 193 km wide, on average, in the east-west direction, with an area of about 415,266 km^2 and a volume of 21,721 km^3. It accounts for 8,000 km of continental coastlines, but the Swedish and Finnish coastal archipelagos include a multitude of unaccounted islets, rocks and skerries[2] (30,000 in the Stockholm region alone), which may increase considerably the total figure.

A characteristic feature of the Baltic Sea is the presence of several interconnected basins, with depths ranging from 47 to 245 m, separated by shallow sills that restrict water exchanges. The overall average depth is 57 m, while the deepest point, the Lansort Deep in the western Baltic Proper, is 459 m. The fact that the main basin does not originate from the collision of tectonic plates, but is a glacially scoured river valley created after the last ice

---

[2] A reef or rocky island covered by the sea at high tide or in stormy weather.

age some 10,000 years ago, accounts for its relative shallowness. Its only links with the Kattegat and the open sea are the straits in the Belt Sea. The central Great Belt is the widest, has a sill depth of 18 m and accounts for about two-thirds of the water flowing in or out the Baltic Sea. The Little Belt, to the west, is very narrow, and most of the remaining flow occurs through the Sound, with a sill depth of only 8 m, to the east.

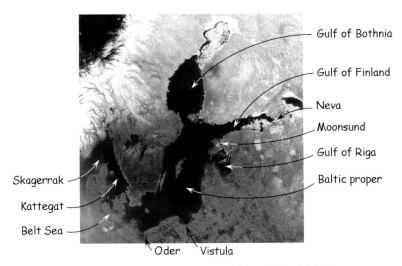

Fig. 3. Baltic Sea. MODIS data, simulated true-colour, 5 April 2004.

The Baltic Sea extends from a relatively mild and humid climate zone in the south to the Arctic region in the north. The average air temperature is 4.6°C, ranging from 7.2°C in the Baltic Proper to 0.3°C in the Gulf of Bothnia. The average surface water temperature in the Baltic Proper is about 1–2°C in February–March and about 16–17°C in July–August, while it ranges from 0 to 15°C in the northernmost part of the basin.

The restricted water exchange through the Belt Sea straits implies a residence time of about 25–35 years for Baltic Sea waters. In contrast to the slow rate of seawater input, large amounts of freshwater are supplied from a catchment area extending over 1,700,000 km^2, and accounting for more than 200 rivers, with an inflow of about 480 km^3/year. The northern areas contribute the greatest share of runoff, but 3 rivers alone, the Neva, Vistula and Oder, provide about 25% of the total inflow. Precipitation over the basin roughly equals evaporation, so that the total freshwater input can be equated to river runoff (although there are local and seasonal departures from this pattern, with maximum runoff occurring in spring, during the snow-melting period). As a result, salinity is very low (mean around 7.3 psu) and has a strong north-south gradient. The freshwater input drives an outgoing, lower-salinity (mean 8.7 psu), surface current into the North Sea, coupled to an incoming, higher-salinity (mean 17.4 psu), bottom current.

There are no significant tides in the Baltic Sea. However, seasonal variation in the water level can be more than 1.5 m due to the changes in atmospheric pressure and winds. Due to the large freshwater input and limited tidal excursion, the water column of the Baltic Sea has a stable density stratification (which would prevent oxygenated surface water from mixing downward in the water column), which is overcome only through wind-driven convection. Major inflows of saline waters are caused by large-scale pressure differences between the North Sea and the Baltic Sea areas, while persistent westerly winds can generate short-term inflows, so that the water in the deepest parts of the basin is periodically renewed. Between these (extreme) inflows, when vertical mixing is limited, the deep waters are stagnant and eventually become anoxic.

Due to its high latitude, relatively poor mixing and low salinity, ice cover is a characteristic feature of the Baltic Sea. The ice-covered area during a normal winter includes the Gulfs of Bothnia, Finland and Riga, as well as the Moonsund archipelago enclosure. In the long-term average, about 45% of the basin surface is frozen, for up to 6 months a year. The Baltic Proper does not freeze during a normal winter, with the exception of sheltered bays and shallow lagoons. The ice reaches its maximum extent in February or March. Typical ice thickness in the northernmost areas is about 70 cm for landfast sea ice. The thickness decreases southward.

The Baltic Sea is characterised by low biodiversity – even if its fauna includes species of fresh, brackish and marine waters – and by a simplified food web. This is because the sea is geologically very young, while physically it has very low oxygen levels, low water temperature and fluctuating salinity. Species distribution is largely determined by these physical factors. Salinity can be regarded as the single major factor constraining species distribution, since diversity tend to decrease with decreasing salinity.

The drainage basin of the Baltic Sea is more than four times larger than the entire sea surface area. It is densely populated (around 85 million people, 15 of which live within 10 km of the coast), heavily industrialised, and includes large areas devoted to intensive agriculture. The enclosed nature of the basin makes it very sensitive to pollution. Water quality is under pressure from the various anthropogenic activities above and by extensive shipping of cargo and petroleum products. Even potentially dangerous World War II chemical deposits, known to exist in relatively shallow areas, contribute to increase environmental concerns. Over-fishing is also an issue of major concern. The main environmental threats are eutrophication and pollution by persistent organic compounds, heavy metals and oil. Eutrophication, due to excessive imput of nutrients from rivers, coastal point sources and atmospheric deposition, as well as recurrent extensive (summer) blooms of cyanobacteria (*e.g. Nodularia spumigena)*, can be major factors in the formation of large oxygen-depleted areas, often resulting in large-scale death of benthic fauna. Petroleum products, which decompose rather slowly in the cold Baltic waters, also poses a significant threat to the Baltic ecosystem and wildlife.

5.   Mediterranean Sea

The Mediterranean Sea, cradle of western civilization, constitutes a unique scale model of the world's oceans, due to the variety of physical and bio-geo-chemical processes taking place in the basin (Figure 4). It is divided into a western basin, where the Alboran Sea, the Ligurian-Provençal-Balearic basin and the Tyrrhenian Sea present distinct traits, and an eastern basin, including the further enclosed Adriatic Sea and Aegean Sea, the Ionian Sea and the Levantine basin.

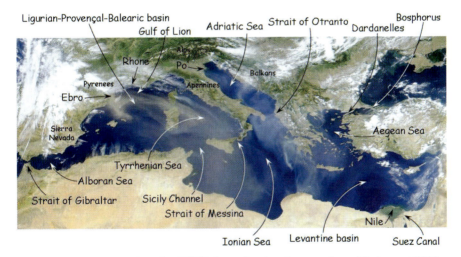

Fig. 4. Mediterranean Sea. SeaWiFS data, simulated true-colour, 25 August 2000.

The basin was commonly thought to be a remnant of the ancient Tethys Ocean. It is now known to be a structurally younger ocean basin, called Neotethys, which formed during the Late Triassic and Early Jurassic rifting of the African and Eurasian plates. The geologic history of the Mediterranean basin is complex, involving first the break-up and then the collision of the African and Eurasian plates, and the Messinian Salinity Crisis in the late Miocene, when the whole sea dried up almost completely.

The Mediterranean Sea extends from 30°N to 45°N and from 6°W to 36°E, with an area of 2,700,000 km^2 – two orders of magnitude below that of the major oceanic basins, corresponding roughly to 1% of the Earth's surface – and a volume of 3,700,000 km^3. It has dimensions of about 3,860 km in the east-west direction and a maximum of 1,600 km (average around 800 km) in the north-south direction. It accounts for 46,000 km of coast-lines, one-third of which are due to the mainland and islands of the Aegean Sea. Several mountain ranges (*e.g.* the Sierra Nevada, the Pyrenees, the Alps and Apennines, the Balkans) are distributed along the northern side of the basin. Since these mountains mostly slope steeply into the sea, the northern drainage basin is relatively small. And since the southern side is mainly covered by desert – and only a few large rivers (*e.g.* the Nile, Rhone, Po and Ebro) flow into the Mediterranean Sea – these combined factors tend to limit the input of continental freshwater.

The Mediterranean Sea consists of a series of deep depressions, connected to each other, with an average depth of 1,500 m. Actual depths vary between 2,500 and 3,500 m in western basin, and between 3,500 and 4,000 m in the eastern basin, where a maximum of 5,267 m is reached in the Calypso Deep of the Ionian Sea. The two main basins are separated by the Sicily Channel (with a 430 m sill depth) and the Strait of Messina (80 m), while the Strait of Otranto (800 m) separates the Adriatic Sea from the eastern basin. The Mediterranean Sea is connected to the Atlantic Ocean by the 22-km-wide Strait of Gibraltar (with a 290 m sill depth), in the west; to the Black Sea by the Bosphorus, and the Dardanelles (55 m), in the north-east; to the Red Sea via the Suez Canal (12 m), in the south-east.

The Mediterranean climate, famous for its abundant sunshine, is subject to both sub-tropical and mid-latitude weather systems, and is also influenced by the northern mountain ranges. The region's seasonal cycle is charac-terized by mild winters, when the region receives most of its rainfall, and by a relatively hot and dry summers. Strong local winds, such as the cold, dry, northerly Bora and Mistral, and the hot, dry southerly Sirocco, typify the region, particularly in winter. In general, air temperature differences between winter and summer are limited to about 15°C, although local geographic and meteorological factors can result in extreme conditions such as air temperatures up to 50°C on the African coast. The mean temperature of surface waters vary between minima of about 14–16°C (west to east), in winter, and maxima of about 20–26°C (again, west to east), in summer, but the total excursion can reach up to 20°C, as in the shallow parts of the Adriatic Sea, where the range is between the 8–10°C of winter and the 26–28°C of summer.

Evaporation greatly exceeds precipitation and river runoff (with an es-timated freshwater deficit of 2,500 km^3/year), so that the Mediterranean

Sea is characterized by very high salinity (mean around 38 psu), a fact that is central to water circulation within the basin. Evaporation is especially high in the eastern Mediterranean, causing the water level to decrease and salinity to increase (up to 39 psu) eastward. This pressure gradient pushes cooler, lower-salinity (about 36 psu) water from the Atlantic Ocean across the entire basin. The relatively less dense Atlantic water flows into the Mediterranean Sea through the Strait of Gibraltar, in the surface layer. The incoming water warms and becomes saltier as it travels east, and is eventually turned into denser Mediterranean waters through evaporation. It sinks in the Levantine basin (as well as in other areas where deep waters are formed), due to winter cooling, moves back westward and ultimately spills over the sill of the Strait of Gibraltar, in the bottom layer, and out into the Atlantic Ocean. This distinct Mediterranean Intermediate Water can be traced in the Atlantic Ocean to the northernmost latitudes, where it contributes to the NADW formation process. The complete cycle determines a residence time in the basin of about 80 to 100 years.

Tidal amplitudes are small, in the Mediterranean Sea, and the narrow continental shelves prevent tidal amplification along the coast. However, a substantial amount of vertical mixing is provided by strong regional (winter) wind regimes. Deep waters are formed in the Adriatic Sea and the Aegean Sea, although with different characteristics, and outspill into the eastern basin over the sills in the Otranto Strait and around Crete, respectively. Deep waters are also formed in the Ligurian-Provençal Sea, due to the effect of the Mistral wind, which increases the density of surface waters through intense evaporation and cooling, particularly in the Gulf of Lion region. This can lead to a complete overturning of the water column, with deep convection processes taking place over 2000 m of water. The process ventilates the deepest parts of the Mediterranean Sea and triggers the onset of large algal blooms, sometimes covering the entire northwestern basin. The deep waters of the eastern and western basins do not communicate, due to the shallow sill of the Sicily Channel.

On average, the Mediterranean Sea is poor in nutrients, with consequent low phytoplankton biomass and primary production. Oligotrophy increases from west to east. Primary production in the open sea is considered to be phosphorus-limited, rather than nitrogen-limited as in most of the world's oceans. By contrast, a rich biodiversity characterizes the Mediterranean ecosystem: the fauna and flora are among the richest in the world, with over 10,000 marine species recorded, highly diverse and with a large proportion (28%) of endemic species. No species disappearances have been reported, but changes in species composition and richness have occurred in some areas. The introduction of exotic species – such as that of tropical species from the Red Sea, which occurred after the Suez Canal opening[3] – is a growing concern. Large populations of marine mammals (*e.g.* the fin whale, *Balaenoptera physalus*) have been documented, while other species are in danger (*e.g.* the monk seal, *Monachus monachus*, is critically endangered, while the marine turtles *Caretta caretta* and *Chelonya midas* are listed respectively as threatened and endangered).

The state of the Mediterranean's open waters is generally good, but coastal areas are subject to various environmental problems, including eutrophication and heavy-metal, organic and microbial pollution. Land-based activities (urbanization, industry and agriculture, particularly in the northwest) are the main sources of pollution. About 30% of the ever-increasing population of Mediterranean countries, approximately 150 million people, lives on the coast. Adding to the resident population, each year the Mediterranean Sea is also host to 150 million tourists. Population density, almost 100 people *per* km^2, twice that of the region as a whole, and related economic activities place a high pressure on coastal zones. Since the Mediterranean Sea is essentially oligotrophic, eutrophication is limited to near-coastal sites that receive anthropogenically enhanced nutrient loads from rivers, or direct discharges of untreated domestic and industrial wastewaters, and adjacent open waters. However, this is expected to worsen, together with the occurrence of Harmful Algal Blooms (HAB). Oil transport and pollution are also a growing concern, since of the 2000 cargos that sail in the Mediterranean at any given time, about 250–300 are oil tankers, transporting 400 million tons/year of petroleum products (*i.e.* 30% of world traffic). This translates into about 1 million tons/year of oil discharged at sea illegally (*i.e.* 20% of the total oil pollution in the oceans).

6. Black Sea

The Black Sea is a landlocked basin, located between south-eastern Europe and Asia Minor (Figure 5). The convexity of both the southern and the northern coastline, the latter extending south from the Crimean peninsula, suggest a further geographical subdivision into two sub-basins. In the south-west, the Black Sea is connected to the Mediterranean Sea via the Bosphorus, which opens first into the Marmara Sea and then, via the Dardanelles, into the Aegean Sea. In the north-east, it is also connected to the shallow Azov Sea via the Kerch Strait. The Black Sea extends from 41°N to 47°N and from 28°E to 42°E, with a maximum distance around 700 km,

---

[3] A phenomenon known as the Lessepsian Migration, after Ferdinand de Lesseps, the engineer who oversaw the Canal's construction.

in the north-south direction, and 1,200 km, in the east-west direction. Its total area is about 423,000 km^2 and its volume 547,000 km^3. It accounts for about 4,340 km of coastlines.

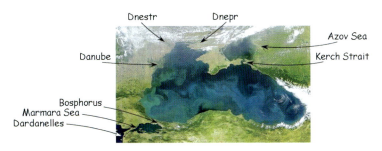

Fig. 5. Black Sea. SeaWiFS data, simulated true-colour, 13 June 2000.

The Black Sea basin was formed during the Miocene orogenesis, which uplifted the mountain ranges that divided the ancient Tethys Ocean into several brackish basins (the remnants of which include also the Azov Sea, Caspian Sea and Aral Sea). The Black Sea comprises a central deep basin, with an average depth of 1,240 m, rising to form a broad shelf in the north-west. In this region, the shelf is up to 200 km wide and has depth up to 160 m. To the south and east, the shelf is only 2 to 20 km wide and has a depth of less than 100 m. The continental slope is steep everywhere, descending with an average angle between 5° and 8° (but the gradient can reach 20–30° in some sections). The Euxine abyssal plain in the centre of the Black Sea has depths between 2,000 and 2,200 m, reaching a maximum (2,212 m) south of the Crimean peninsula. The Bosphorus, which provides the only connection with the world ocean, has a length of approximately 30 km, a maximum width of 3.7 km at the northern entrance, and a minimum of 0.7 km, and a depth varying from 36 m to 124 m in midstream.

A major part of the Black Sea experiences a Mediterranean climate, with warm, wet winters and hot, dry summers. The south-eastern area is surrounded by mountains and is characterized by a humid sub-tropical climate with abundant precipitations, warm winters and hot summers.

The water column of the Black Sea is characterized by a permanent density stratification, with a well pronounced pycnocline situated at a depth of 150 (to 300) m. This feature is created mainly by salinity, between 18.0 and 18.5 psu at the surface and near 22.5 psu in deep waters, regulated by the large input from rivers and rainfall, by the low evaporation rate and by the limitation of water exchanges with more saline basins – due to the narrow and shallow sills separating the Black Sea from the Mediterranean Sea. The lower salinity of the upper layer is preserved largely by the continuous inflow of freshwater (in excess of 300 km^3/year) from a vast catchment area, extending over 2,000,000 km^2 in central and eastern Europe. The higher salinity of the lower layer is maintained by the inflow of Mediterranean waters (also in excess of 300 km^3/year) through the Bosphorus. This deep influx of dense water from Mediterranean is balanced by a surface outflow of fresher surface water from the Black Sea into the Marmara Sea. Sea surface temperature varies seasonally from 8°C to 30°C, whereas the deep sea temperature is about 8.5°C. Intense winter cooling on the northwestern shelf can decrease surface temperature to 6°C, producing a cold intermediate layer at a depth of about 50–90 m, which constitutes one of the distinctive features of the Black Sea thermal stratification. The upper layer circulation is characterized by a basin-wide cyclonic gyre, *i.e.* the Rim Current that encircles the entire Black Sea (and that can divide into two man gyres in winter), and by mesoscale eddies capable of mixing coastal waters into the open sea.

The Black Sea is the world's largest meromictic[4] basin, with 90% of its volume occupied by anoxic waters. As seen above, the hydro-chemical configuration is primarily controlled by basin topography and fluvial inputs, which result in a positive water balance and a strongly stratified vertical structure. The permanent stratification prevents mixing between surface and deep waters and, thus, has created an anoxic environment below 150 m, with a high concentration of hydrogen sulphide. The surface oxygenated waters support a rich and diverse marine life.

In recent years, the diversity of species has changed substantially, with an increase in the number of Mediterranean species, especially in the south-west, while the abundance of many species has decreased rapidly. Due to the trend of various commercial fish species, fisheries have been declining too. The main causes behind this are thought to be overfishing, eutrophication, and the introduction of alien species (which has substantially disrupted the food web). In particular, the ctenophore *Mnemiopsis leidyi*, after its accidental introduction via ship ballast waters in the mid 1980's, flourished in the eutrophic Black Sea environment. By the 1990's, its biomass had increased to the point of damaging significantly the whole ecosystem. In the deep anoxic waters, some protozoa, bacteria and invertebrate species are known to exist, but little is known about them.

Various types of wetlands, providing productive spawning and feeding grounds for many species and refuge for migratory waterfowl, are found in the Black Sea, with limans[5], lagoons, estuaries and deltas being the most widespread. They are mainly located at the mouth of large rivers, such as

---

[4] A stratified basin, which has layers of water that do not intermix.
[5] A type of wetland formed by the flooding of seawater into river valleys.

the Danube, Dnster, Bug, Dnper, Don, Kuban, Rioni, Sakarya and Kizilirmak. The same rivers are the main source of excess nutrients discharged in the Black Sea, causing both a decline of water quality and extensive coastal eutrophication. The latter has drastically altered the ecosystem, by creating hypoxic conditions even in the upper layer and killing many bottom-dwelling organisms[6]. It has also led to the destruction, in the north-western basin, of massive meadows of red algae of the genus *Phyllophora*, heavily affecting the organisms dependent on them.

7. Caspian Sea

The totally landlocked Caspian Sea is the largest inland body of water on Earth (Figure 6). It is bordered by the Caucasus and Elburz Mountains, from west to south, by the Caspian depression and central Asian steppes and deserts, from north to east. Although it has no outlet, the Caspian Sea is connected to the Sea of Azov by the Manych Canal and the Volga-Don Canal, ultimately providing links with the Black Sea, the Baltic Sea and the White Sea, through an extensive network of inland waterways.

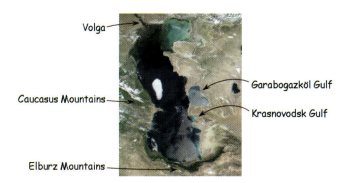

Fig. 6. Caspian Sea. MODIS data, simulated true-colour, 1 June 2002.

The Caspian has characteristics common to both seas and lakes. It is often listed as the world's largest endorheic[7] lake (by both area, 371,000 km², and volume, 78,200 km³), though it is not a freshwater lake. The basin is estimated to be about 30 million years old, but it became land-locked only about 5.5 million years ago. It extends from about 37°N to 47°N and from

---

[6] An estimated 60 million tons, including 5 million tons of fish, in the 70s and 80s.

[7] An endorheic basin – also called a terminal or closed basin – is a watershed from which there is no outflow of water, either at the surface as rivers, or underground by flow or diffusion through rock or permeable mate.

47°E to 55°E, with a maximum length around 1210 km, from north to south, and an average width of 320 km, from east to west. It has about 7000 km of irregular coastlines, with large gulfs in the east, including the Krasnovodsk Gulf and the very large, shallow embayment, known as the Garabogazköl Gulf, which acts as an evaporation basin. The Caspian Sea has an average depth of 184 m, with minima in the shallow northern part (averaging only about 5 m), intermediate depths (788 m) in the central part, and a maximum (1025 m) in the southern part.

The Caspian Sea crosses several climatic zones. In the north, it has a continental climate, with cold winters and hot, dry summers; in the west, the climate is temperate and warm; in the southwest, sub-tropical and wet; in the east, it has the harsh, dry traits of the steppe zone. The northern region further differentiates for sudden air temperature changes and low precipitation. The shallow northern part of the Caspian Sea freezes from November to March, when ice thickness can reach up to 2 m. Normally, the remainder of the basin is ice-free year-round, but in the coldest winters some ice can also be found in coastal areas of the deeper southern part.

In winter, the average water temperature ranges from –0,5°C in the north, to 3–7°C in the central basin, and to 8–10°C in the south. In summer, the temperature of the water surface is 24–26°C, with 29°C in the south. The average salinity is 12.7–12.8 psu, about one third the salinity of seawater, with maximum of 13.2 psu on the eastern coast and minimum around 0.1–0.2 psu near the northern river mouths.

The sea has numerous tributaries, notably the Volga (accounting for about 80% of the inflow), Ural, and Zhem, from the north, the Gorgan and Atrek, from the east, and the Kura, from the west. However, there is no natural outflow from the basin other than evaporation (except the Manych Canal). Thus the Caspian is a closed system, with its own water level history, independent of the eustatic level of the world's oceans. The level of the Caspian has fallen and risen, often rapidly, many times over the centuries. It is known that a prolonged minimum, about 2 to 4 m below current levels, occurred in the VII–XI centuries. The last short-term water level cycle started with a fall of 3 m from 1929 to 1977, followed by a rise of 3 m from 1977 until 1995. Since then, smaller oscillations, including a seasonal fluctuation of about 30 cm, have taken place. Currently, the water level is rather stable at about 28 m below the global sea level. Of course, these changes can cause major environmental problems.

In general, Caspian Sea levels have changed in synchronicity with the discharge of the Volga, which in turn depends on precipitation over its vast (1.400.000 km^2) catchment area (as well as on damming and diversions of river water for agricultural, industrial and residential use, to a lesser extent). Rainfall is related to variations in the amount of North Atlantic depressions that reach the continental interior, and these in turn are affected by cycles of the NAO. Thus, the water level in the Caspian Sea relates to atmospheric conditions in the North Atlantic thousands of km to the north-west. These factors make the Caspian Sea a valuable place to study the causes and effects of global climate change.

The Caspian Sea is home to a number of ecosystems. At the meeting point of Europe, Asia and the Middle East, it is bordered by deserts, steppe, highlands and mountain systems. Its coastal wetlands include many shallow, saline pools, which provide significant biodiversity, but in general flora and fauna of the Caspian Sea are comparatively poor. About 400 species, such as the endangered Caspian Seal (*Phoca caspica*), are endemic. The Caspian Sea has important fisheries. The native sturgeons – seven different species are present, some of which are found nowhere else in the world – are famous for the roe they produce[8], but the destruction of spawning areas and illegal fishing has greatly reduced their numbers.

Underlying the Caspian region are some of the world's largest oil and natural gas reserves, and access not only to mineral resources, but also to fishing grounds and to international waters is a matter of dispute between riparian countries. Decades of environmental mismanagement have led to severe pollution problems in the Caspian Sea. Discharges from offshore oil and gas drilling and inflows from the highly polluted Volga river have contributed to the degradation of water quality. Pollution poses a serious threat to animal life, while the stated overfishing has caused a dramatic depletion of fish stocks. Polluted coastlines mean that swimming in most areas of the sea is hazardous, and toxic waste threatens to contaminate drinking water supplies for people living in the region. The impact of environmental degradation on human health has been significant, with higher rates of cancer recorded in the area.

8. Outlook

An overview of the marginal and enclosed basins of the North Atlantic region can only highlight the great variety of environments that typify the European continent. Understanding the inner workings of these seas – aiming to reconcile the conflicting needs of protecting their ecological balance and exploiting their natural resources – requires adequate observation systems, integrating both *in situ* and remote sensing techniques.

---

[8] Sturgeon from the Caspian Sea accounts for 90% of the world's caviar industry.

Earth's observations offer a wide range of possibilities, complementing more conventional *in situ* data gathering, for the large-scale, long-term assessment of interacting physical and bio-geo-chemical processes at the regional – as well as the global – scale. In the case of small basins, the assessment of environmental indicators from the vantage point offered by satellites in Earth's orbit allows to observe at a glance the dynamical relationships of natural setting, water exchanges, basic ecological relations and their driving forces, as well as the problems faced by marine waters.

While several sites in the European Seas have been studied for a long time, others remain surprisingly unexplored. The systematic and synoptic appraisal of surface parameters by means of passive remote sensing techniques – which measure reflected visible and near-infrared sunlight, or surface emissions in the thermal infrared or microwave spectral regions – or active techniques – which use transmitted impulses of visible or microwave radiation, for a subsequent evaluation of the signal returned by the water surface – promises to help closing these gaps. An in-depth analysis of the specific merits and drawbacks of each technique and spectral region, can provide clues to help compose the unique mosaic of dynamical and bio-geo-chemical features of the European Seas.

Bibliography

Bhatt JJ (1978) Oceanography. Exploring the Planet Ocean. D Van Nostrand Company, New York Cincinnati Toronto London Melburne

Corso W, Joyce PS (1995) Oceanography. Applied Science Review. Springhouse Corporation, Springhouse

Grant Gross M (1995) Principles of Oceanography. 7th Edition. Prentice-Hall, Englewood Cliffs

International Hydrographic Organization (1953) Limits of Oceans and Seas. Special Publication 23, 3rd Edition. Imp Monégasque, Monte Carlo

Longhurst AR (1998) Ecological geography of the Sea. Academic Press, San Diego

Pinet PR (1992) Oceanography: an Introduction to the Planet Oceanus. West Publishing Company, St. Paul

Sverdrup KA, Duxbury AC, Duxbury AB (2005) An Introduction to the World's Oceans. 8th Edition. McGraw-Hill Publishers, Boston

Sverdrup HU, Johnson MW, Fleming RH (1942) The Oceans: Their Physics, Chemistry and General Biology. Prentice-Hall, Englewood Cliffs

Tomczak M, Godfrey JS (2003) Regional Oceanography: an Introduction. 2nd Improved Edition. Daya Publishing House, Delhi

Williams J, Higginson JJ, Rohrbough JD (1968) Air and Sea: The Naval Environment. Naval Institution Press, Annapolis

Remote Sensing of the European Seas:
A Historical Outlook

Tom D. Allan

Satellite Observing Systems, Godalming, Surrey, UK

Abstract. Satellite remote sensing of the sea surface started in 1978 with 3 NASA missions, which would demonstrate to a sceptical oceanographic community that useful measurements of colour, temperature, roughness and topography could be made from space. Today, there are over a dozen spacecrafts in orbit, routinely delivering a variety of such data products. Europe arrived comparatively late on the scene, launching its first satellite in 1991. Now, the European Envisat, launched in 2002, is considered one of the most sophisticated platforms in its kind. Significant contributions to our knowledge of ocean processes have been made using these, and other, tools. This work will continue, especially given the current emphasis on monitoring global climate change – in which, of course, the oceans play a vital role. Long-term time series of repeat measurements, for which satellites represent a powerful tool, are key elements of all future programmes. Shorter-term operational benefits will also be provided by dedicated European programmes. Over the years we have acquired a better understanding of the information content of signals reflected back to the satellite from the sea surface, and we know, more or less, how to translate these signals into a form that is useful to a wide range of end-users. Where we have been less successful is in delivering this information at a frequency to match the day-to-day requirements of marine operations. Coverage, speed, continuity, details – all these elements provide satellites with unique advantages. We have also come to accept that satellites on their own may not always provide the whole answer, and that a combination of satellite data, *in situ* observations, and forecasting models is often the best approach.

1. Introduction

It is almost 30 years since 3 pioneering spacecraft in polar orbit around the Earth demonstrated to a global community of largely sceptical oceanographers that variations at the sea surface could be observed from space to useful

V. Barale, M. Gade (eds.), *Remote Sensing of the European Seas.*

accuracies. The interest generated overnight, in 1978, by the observations transmitted to ground by the sensors on board Nimbus-7, Tiros-N, and SEASAT was in recognition of the remarkable detail that could be achieved from a height of 800 km, coupled to the coverage of a single spacecraft that completed over 14 orbits of the earth – that is, more than 500,000 km of track – in a single day. To a breed of scientist used to taking hydrographic casts from slow moving research vessels, and spending many months in their laboratories analysing the results, this indeed appeared like a revolution that might change their world. Colour, temperature and radar imagery that instantly revealed eddies, internal waves, ocean fronts, plankton blooms, surface swell, effluent discharges, river run-offs, and individual ships, were widely distributed and for a while no marine scientific paper was complete without an introductory satellite image.

SEASAT ceased to operate after 3 months due to a massive power failure and this new breed of enthusiastic satellite oceanographers were left to wait 13 years before the European Space Agency (ESA) launched its European Remote Sensing Satellite 1 (ERS-1) equipped with a suite of microwave sensors similar to that carried on SEASAT. Throughout the 1980's, temperature imagery of the sea surface continued to be widely distributed by a series of National Oceanic and Atmospheric Administration (NOAA) satellites carrying the Advanced Very High Resolution Radiometer (AVHRR), and for some 7 years, after the demise of SEASAT, the Coastal Zone Color Scanner (CZCS) on Nimbus-7 continued to provide colour imagery that proved particularly useful in pinpointing areas of coastal upwelling. The fact that optical and infra-red devices are unable to penetrate cloud did curtail their usefulness as an operational tool, especially over Europe's more temperate latitudes. This restriction was partly compensated, however, by their wide swaths and continuous data recording, which in one sweep could reveal a panoply of sea surface features stretching from the Arctic to the African shores of the Mediterranean.

With the launch of ERS-1, in 1991, followed in 1995 by ERS-2, and eventually in 2002 by the multi-sensor platform Envisat, scientists could benefit from access to contemporaneous colour, infra-red and radar data. The National Aeronautics and Space Administration (NASA) continued its sequence of colour imagery with the Sea-viewing Wide Field-of-view Sensor (SeaWiFS) and, in addition, made widely available the sea surface wind fields monitored by QuikSCAT. NOAA continues to this day the world-wide distribution of AVHRR imagery, for which there is a number of receiving stations spread across Europe. The French aerospace community and NASA joined forces to launch very precise altimeters carried on TOPEX/POSEIDON (in 1992) and on Jason (in 2002), both of which achieved almost an order of magnitude greater precision than the SEASAT instrument and provided, in fact continue to provide, invaluable information on sea level and the changing behaviour of eddies and surface circulation patterns across the globe. Further, Japan's short-lived Advanced Earth Observing Satellite (ADEOS) 1 (in 1996) and ADEOS 2 (in 2002) mission, and the recent launch of the Earth Observing Satellite (EOS) Terra (in 1999), Aqua (in 2002) and Aura (in 2004) orbital platforms, by NASA and other space agencies, have combined to provide continuity to this data stream, observing the oceans as well as land and atmosphere.

2.   Science and marine applications

Today there are around a dozen marine-oriented satellites in orbit providing oceanographers access according to their need. However, we are still some way off having an easily-accessible central pool of all processed satellite observations. Despite an apparent plethora of data, it would be fair to say that a more realistic appreciation of the strengths and limitations of these observations has replaced the euphoria of the late '70s. Marine satellite programmes are still largely driven for and by scientists. As we shall see in the sections which follow, substantial progress has been made in understanding and interpreting satellite signals reflected from or emitted by the sea surface. Over a period of several years, a wide variety of sea surface features have been laid bare, for example, in the fine-resolution imagery of Synthetic Aperture Radars (SAR) – *i.e.* internal wave trains, sea-ice concentration, areas of oil pollution, swell wave propagation, effluent discharges, the disposition of fishing vessels, and coastal bathymetry. SAR interferometry is a technique that has developed in the last few years to reveal very small displacements in land masses and is now being used to detect sea surface currents. Optical imagery can also reveal many more features, on the sea surface or even below it, such as coastal upwelling, oceanic fronts, harmful algal blooms and the like. However, there remains a gap to be bridged in many areas between the value of remote sensing to scientific research and its value to daily marine operations.

The difference is one of time scale, where for research purposes it may be sufficient to detect gradual changes in a surface feature over comparatively long timeframes, say several months or years. By contrast many marine situations – the approach of a harmful bloom or an oil spill, the intrusion of an unauthorised fishing vessel, or the onset of a sudden squall - may require a rapid response and hence the fast delivery of satellite observations. In those circumstances the record from a single satellite with a re-visit cycle of several days or weeks may render a routine, operational service difficult if not impossible even where the sensor swath covers a wide area. For narrow swath instruments such as the radar altimeter the situation can be more serious, especially since the presence of land falling within its footprint can render the observation of wave height and sea level unreliable across a coastal strip a few kilometres wide.

3.   The review of 1992

In 1992, some 14 years after the launch of those pioneering space missions, the International Journal of Remote Sensing (IJRS) published a Special Issue on European achievements, to which over 40 researchers contributed. It is worth recalling here the opening statement in the Preface to that review:

"Most previous attempts to predict the future development of remote sensing have proved to be over-optimistic. It is true that the achievements since 1972 have been spectacular but the real break-through to full commercial exploitation has still eluded us".

So, after another 14 years or more, where are we now? Europe has undoubtedly established its position as a major player in Earth Observation having proceeded to build, launch and support Envisat, one of the most sophisticated remote sensing platforms in the world. Its scientists have made good use of both European and non-European data and have made significant contributions to our understanding of marine processes. Over the last few years the world has become more aware of the possible consequences of global warming, so that the ability of scientists to guide and advise politicians on future Earth monitoring programmes has become more important than ever. Major advances have been made in using information coming from space to tackle the peculiarities and special problems posed by the European Seas.

4.   Commercial viability

That said, those contributing to the 1992 review – especially those in the European Commission (EC) – made little attempt to conceal their disappointment at what they perceived to be the slow progress made towards making remote sensing 'commercially viable'. Various reasons were advanced at that time to explain this situation: underestimating the time to introduce a new technological tool; inappropriate hardware and software for inexperienced users; inadequate training programmes; poor marketing directed at potential end-users; and the entrenched attitude of those who believe that existing sources of information-gathering systems are adequate. These were the most popular reasons put forward at that time.

How much have things changed? What reasons would we advance today? It is always easy to blame poor marketing but if the product is not suited to the task no amount of marketing will persuade clients to sign up. Attitudes would perhaps be less 'entrenched' if the remote sensing alternative was clearly seen to be superior and the cost reasonable. Perhaps the marketing effort needed to be directed the other way, not so much to reluctant clients but to the EC itself and/or the space agencies, to design spacecraft with clearer operational objectives but which could also generate benefits to ongoing scientific programmes.This approach has never been tried but more recently, in the context of detecting tsunami waves by employing a suite of fit-for-purpose microsatellites, attention has been drawn to the benefits that would also accrue to global climate research from such an approach. Although it is accepted by everyone that it is essential for many environmental research programmes to continue as far as possible the time series of satellite repeat observations, there are signs that ESA and the EC in jointly supporting the Global Monitoring for Environment and Security (GMES) programme are seeking to look at a wider range of benefits to Europe. Science is no longer the only occupant of the driver's seat.

In the sections that follow many examples are presented of the ability of the SAR to reveal a variety of surface signatures. The techniques for extracting, processing and interpreting these signals have steadily improved over the years especially since the launch of Envisat in 2002. Similar progress has been made in interpreting multi-spectral optical imagery in terms of, for example, blooms, discharges, tidal flats or coastal upwelling. We may have reached a cross-roads. Already we have alluded to the fact that for a system to become operationally useful two conditions must be met:

(*i*) can the signal from the satellite be interpreted and presented to 'clients' in the form most useful to them?

(*ii*) can this information be delivered regularly at a frequency matched to the development of the feature so that clients may take appropriate action?

5. European progress

In Europe, as this volume illustrates, we have made significant progress in the first of these, but there is some catching up to do with the second.

There are, however, some encouraging signs within a number of ongoing European programmes that seek to combine 'real-time' satellite observations with well-established *in situ* services. The rapid examination of the number of vessels occupying fishing grounds, as revealed in a SAR image, can now be cross-checked with the number notified to the national authority, so that discrepancies can be identified and interlopers intercepted. Similarly the agencies responsible for combating oil spills around Europe's coastline are working more closely with organisations that have developed algorithms to extract and classify oil spills on the SAR image.

In a programme supported by ESA, the surface wind and wave observations made by the altimeters carried on Envisat and Jason are extracted from the records in near real-time and forwarded to the Norwegian Met Office to compare with forecasts previously issued to ships following a proscribed route. Discrepancies between the observed and the forecast greater than a certain critical value can then be used to issue a correction. It has also been demonstrated by the European Centre for Medium-Range Weather Forecasts (ECMWF) that the introduction of wind and wave observations made today from orbiting spacecraft creates a measurable improvement in their global forecasts. These are just a few examples of how 'real-time' data from satellites can now be used to inform and update ongoing operations. There is, however, another important application of remote sensing data which continues to improve each year and which does not require instantaneous delivery of the observations. This refers to the conclusions that may be drawn from a statistical analysis of long-term satellite records. The seasonal migrations and their variations from year to year of a well-established ocean frontal systems is a prime example of a product important to naval operations. There are now over 30 years of archived infra-red imagery to draw on. Likewise the records of the Medium Resolution Imaging Spectrometer (MERIS), the Moderate Resolution Imaging Spectrometer (MODIS) and SeaWiFs , and before that of CZCS, allow an analysis of the annual and seasonal variations in coastal upwelling, and the causes may be linked to the monthly statistics on average and extreme wind conditions also extracted from many years of satellite observations.

The altimeter in particular has proved an old and reliable work-horse. Uninhibited by cloud, global coverage is guaranteed. The seas and oceans can be divided up into, say, 2 degree squares – a compromise between spatial resolution and gathering enough independent data points in a small area to produce reliable statistics. Around the European seas, which support a thriving offshore exploration and production industry, the statistics of monthly mean and extreme measurements observed within these cells have proved of great value to rig designers and national safety authorities in estimating the likely height of the highest wave encountered over a certain time period. Clients for this type of information have included Naval Hydrographic Departments, marine insurance companies, ship-routing organisations, and ship certification authorities. Before the advent of satellites, global atlases of this type of information had to fall back on the very uneven coverage of ship and buoy observations.

If serious remote sensing of the marine environment can be said to have started 30 years ago, it is fair to ask how well the early promise has been fulfilled. We have already drawn attention to the doubts then being expressed in the 1992 review.

6.   Developments now and in the future

I wrote many years ago that the 1980's would be viewed as a period of consolidation – a decade of debate, deliberation and decision that saw NASA, ESA, as well as the space age ncies of Canada, France, Japan and others, all resolve to launch dedicated marine satellite systems. In the event they did, though not without unforeseen delays (we have seen that the microwave community had to endure a wait of 13 years after the demise of SEASAT, before ERS-1 was launched to take its place). The Director-General of the EC's Joint Research Centre (JRC) wrote, in his Foreward to the 1992 IJRS review, that the 1990's would be a decade of learning and experimentation that would ensure widespread acceptance and application of remote sensing data (mostly optical and thermal data as it turned out). To many, these hopes, sincerely expressed at the time, seem over-optimistic. What is the state-of-the-art now and what lessons may we draw?

Sensor accuracy and resolution have improved, but the 4 surface parameters measured to useful accuracy in 1978 – roughness, slope, temperature and colour – remain the same 4 monitored today from Envisat and other spacecraft. SAR interferometry was hardly foreseen in the early days, however, and has recently emerged as probably the most promising new application. What is not always fully appreciated is that the early caveats from seasoned oceanographers about satellites viewing only the sea surface – although essentially justified – were offset by the wealth of detail revealed in the record of processes within the volume of the sea right down to the sea floor which produced a measurable surface signature.

One of the most remarkable applications of this feature around Europe is surely the detailed topography of the English Channel and North Sea where, depending on the state of the tide and the surface wind conditions, a SAR image can reveal the disposition of sandbanks and sand waves on the sea-floor. In a sweep of no more than a few seconds duration bathymetric detail is revealed, which, by the more conventional method of side-scan sonar from a research vessel, would have taken several days or weeks to complete. Other examples include the currents and tides revealed in great detail over all of the world's oceans by the radar altimeter. The fronts and internal waves already mentioned are further examples of processes taking place within the volume of the oceans that produce a measurable signal at the surface. And over the deep ocean a great number of previously uncharted sea mounts were discovered from the fact that their presence on the sea floor disturbs the gravity field to an extent that produces a measurable change of sea level over them.

Looking back over 30 years what do we now expect satellites to contribute to the world in general and to Europe in particular? What we have learned is that satellites alone are rarely the whole answer to any problem. We have drawn attention to the unique strength of satellite observations in some areas. But such has been the advance in other technologies - especially computer power - that we must now think in terms of integrated systems combining real-time satellite observations with in situ operations and computer models, backed up by reliable environmental statistics derived from a long-term sequence of repeat satellite passes.

The contribution of faster, more powerful computers has also had a direct effect on the rate with which satellite data can now be processed and analysed. Those who worked with the original SEASAT SAR imagery may recall that the processors available at that time could take several hours to process a single 100 km × 100 km scene transmitted at a density of 100 mb/s. Or that the atmospheric correction of a single CZCS image could take the greater part of a day's work.

7.   Satellites – a key source of environmental data

We are reminded every day of the predictions that modern computers can make on a changing global climate. But no matter how sophisticated the model, nor how fine the spatial resolution it can now produce, it remains conjecture frequently based on the assumptions made on the nature of non-linear interactions. It was Henry Stommel, one of the great ocean modellers of his day, who remarked:

"The chief source of ideas in oceanography comes, I think, from new observations……. On the whole, when it comes to the phenomenology of

the ocean, there are more discoveries than predictions. Most theories are about observations that have already been made".

The ability of satellites to fly year after year above the weather, maintaining a continuous record of those observations at the sea surface, represents in itself a remarkable technical achievement and one in which Europe can justifiably take pride. There remains the thorny issue of commercial return so devoutly wished by those who fund the satellite programmes. We have drawn attention to the value of long time series both to science and to those marine operations that require more reliable knowledge of what to expect. There is, however, limited commercial return from atlas type of information which changes only slowly with time. A more economic proposition is the establishment of services that deliver information in near-real time on a regular basis to ships, offshore platforms, forecasting centres, insurance companies, public authorities, and ministries of defence. But this can rarely come about from data generated from satellites designed for research. That perhaps is the biggest lesson of the last 30 years. It may be time to change the mould.

In conclusion, within Europe's coastlines and inland seas are to be found a rich diversity of marine conditions – from the freezing Arctic waters in the north to the warm Mediterranean currents that lap the shores of North Africa. The seas around Europe are more than likely to witness the effects of future global warming – rising sea levels, a reduction in the thickness of the Greenland ice cap, melting polar ice and, in the Atlantic, a possible change in major circulation patterns. Europe needs to maintain its strong scientific heritage in understanding and predicting the behaviour of the seas that lap its shores. The greater part of European goods are transported by sea. For several decades Europe has extracted the oil and gas that lie beneath its continental shelves. Fisheries also represent a major European resource – but now, by universal agreement, these have to be carefully managed. The sea also poses a constant threat from sudden storms, from pollution both natural and man-made and, more recently, from potential terrorists. For all of those reasons it has become so important to maintain a continuous watch it is little wonder that polar-orbiting satellites are now regarded both by ESA and the EC as an essential monitoring tool.

Selected Bibliography

Allan TD (1981) Oceanography from Space – a European contribution? In: Gower J (ed) Oceanography from Space. Plenum Press, New York, pp 19–27

Allan TD, ed (1983) Satellite Microwave Remote Sensing. Ellis Horwood, Chichester

Barale V, Schlittenhardt P, eds (1993) Ocean Colour: Theory and Applications in a Decade of CZCS Experience. Kluwer Academic Publishers, Dordrecht

Ewing GC, ed (1965) Oceanography from Space. Woods Hole Oceanographic Institute, Woods Hole

Gower JFR, ed (1978) Passive Radiometry of the Ocean. D Reidel Publidshing Company, Dordrecht Boston London

Miller RL, Del Castillo CE, McKee BA, eds (2005) Remote Sensing of Coastal Aquatic Environments: Technologies, Techniques and Applications. Remote sensing and Digital Image Processing series. Springer, Dordrecht

Robinson IS (2004) Measuring the Oceans from Space: the principles and methods of satellite oceanography. Springer Praxis Books series, Geophysical Sciences subseries. Springer/Praxis, Berlin

Savigear RAG, ed (1992) Special Issue "European Achievements in Remote Sensing". International Journal of Remote Sensing, vol 13 (6 & 7)

Section 2:

Visible & Thermal Infrared Passive/Active Remote Sensing

Ocean Colour Remote Sensing of the Optically Complex European Seas

Jean-François Berthon, Frédéric Mélin, and Giuseppe Zibordi

Institute for Environment and Sustainability, Joint Research Centre, European Commission, Ispra, Italy

Abstract. Within the framework of "ocean colour" remote sensing, the European Seas can be considered optically complex waters, as an important fraction of their surface is distributed over the coastal shelf or in enclosed basins. In such waters, the origins of particulate and dissolved materials, playing a significant optical role, are manifolds. They result in a high variability of the marine inherent optical properties, from the strongly absorbing Baltic waters to the scattering North Sea waters and the transparent Eastern Mediterranean Sea waters, furthermore complicating the inversion of the remotely sensed water-leaving signal and the derivation of marine data products.

1. Introduction

Since the late 1970's and the launch of the Coastal Zone Color Scanner (CZCS), the exploitation of ocean colour remote sensing data has considerably improved. The reduction in the uncertainty of the derived atmospheric and marine data products benefited from improvements in the sensors instrumental characteristics but also in the processing models and inversion algorithms, as the knowledge on the optical properties of the atmosphere and water increased. Thanks to these improvements, recent ocean colour sensors – *e.g.* the Sea-viewing Wide Field-of-view Sensor (SeaWiFS), the Moderate Resolution Imaging Spectrometer (MODIS), the Medium Resolution Imaging Spectrometer (MERIS) – in the last decade have allowed for a quasi-daily coverage of European waters.

However, important uncertainties still persist for the retrieval of marine products like the phytoplankton chlorophyll *a* (Chl-a) in areas where the absence of covariance among the main seawater components occurs. In these waters, which mostly include coastal regions, the sources of particulate and dissolved materials are numerous (biological production, rivers

V. Barale, M. Gade (eds.), *Remote Sensing of the European Seas.*

outflow, bottom re-suspension, atmospheric deposition, …) and contribute to an increase of the uncertainty associated with the inversion of the surface optical signal. The proximity of the continent also induces the presence of specific aerosols (*e.g.* absorbing aerosols, desert dust, etc.) making the atmospheric correction procedure more challenging.

A considerable fraction of the European seas is located in the "coastal zone" and, from the Baltic Sea to the Mediterranean Sea and to the Black Sea, the dynamic range of in-water materials concentration and of optical properties can reach 4 orders of magnitude. A SeaWiFS-derived climatology (April) of surface Chl-a in European waters is shown in Figure 1. Absolute values, derived through the application of the open ocean 'OC4v4' algorithm and thus to be considered with caution in coastal zones, vary between 0.03 and 10 mg m^{-3} for the ultra-oligotrophic waters of the eastern Mediterranean Sea and the turbid Baltic Sea and North Sea, respectively.

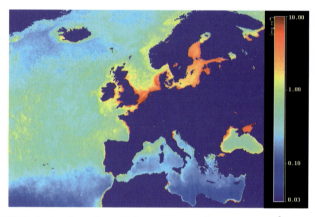

Fig. 1. SeaWiFS derived surface Chl-a concentration (in mg m^{-3}) in the European waters for the climatological month of April (1998–2004 climatology).

After a brief reminder of the general principles of ocean colour remote sensing, the main marine optical characteristics of the European basins will be here presented and compared. The emphasis is put on the variety and the resulting complexity of the optical properties of these open ocean and coastal waters by presenting original data and results from the literature.

2.   Ocean colour remote sensing

The recall of the ocean colour remote sensing principles is here restricted to a simple and essential description of the variables related to the interaction of light with the in-water optical components. A comprehensive analysis is given in Mobley (1994).

The remote sensing of ocean colour can be defined as the derivation from the top-of-atmosphere radiance at a certain wavelength λ, $L_T(\lambda)$, reaching a space sensor, of the spectral upward radiance, $L_w(\lambda)$, leaving the water surface in the direction of the sensor. $L_w(\lambda)$ results from the interaction of the downward irradiance $E_d(\lambda)$ having penetrated the sea surface, with the different optically significant components (particles, dissolved matter, sea water). The so-called atmospheric correction consists in subtracting from $L_T(\lambda)$ the radiance due to photons having only interacted with the atmosphere (possibly with the sea surface too but without having penetrated it) as well as accurately representing the atmospheric transmission of the photons really emerging from the sea surface. All these processes are wavelength and geometry (sun zenith and sensor viewing angle, sun-sensor azimuth difference) dependent. The principles of the atmospheric corrections are not evoked here and the reader is addressed to Gordon (1997), and to Ruddick *et al.* (2006) for atmospheric corrections over turbid coastal waters.

The retrieval of in water components involves optical quantities like the "normalized water-leaving radiance", $L_{WN}(\lambda)$ or the reflectance, $R(\lambda)$. $L_{WN}(\lambda)$ is a water-leaving radiance made independent from the illumination conditions. The "just below" surface reflectance $R(0^-, \lambda)$ is the ratio of the up- to down-welling irradiances just beneath the sea surface, $E_u(0^-, \lambda)/E_d(0^-, \lambda)$. E_u is related to L_u, the upwelling radiance (and thus to L_W) through the so-called "Q factor", which describes the anisotropy of the underwater light field. $R(0^-, \lambda)$ can be related to the inherent optical properties (IOPs) of the water, namely, the back-scattering coefficient, $b_b(\lambda)$, and the absorption coefficient, $a(\lambda)$, according to:

$$R(0^-,\lambda) = \left[Q(\lambda)\,L_u(0^-,\lambda)\right]/E_d(0^-,\lambda) = f'(\lambda)\,b_b(\lambda)/\left[a(\lambda) + b(\lambda)\right] \quad (1)$$

where $f'(\lambda)$ and $Q(\lambda)$ are function of geometry, atmospheric conditions, surface sea state, and optical properties of the in water optical components. $R(0^-, \lambda)$ is thus the surface optical expression of the water content and the quantity from which the different marine products are retrieved, after a proper inversion.

3.   The optical properties of the main European basins

The open ocean waters, for which the concentration of phytoplankton and its derivative products dominate the optical properties, have been simply classified as "Case 1" waters (Morel and Prieur 1977) by opposition to "Case 2" waters for which the concentration of organic/inorganic particles and/or dissolved organic matter from other origins (river outputs, bottom suspension, atmospheric deposition, ...) are the dominating components. Today, the Case 1 waters are commonly considered as those whose optical properties can be modelled solely on the basis of Chlorophyll a concentration (Chl-a). The more complex "Case 2" waters, for which there exists a low covariance among the optical constituents, are typical, although not exclusive, of coastal waters. Note that this simple binary classification is more and more questioned in favor of classifications allowing the representation of a *continuum* of optical properties. European Seas can be considered as pertaining to Case 2 waters in their majority as an important fraction of their surface is distributed over the coastal shelf or in closed basins.

3.1 The inherent optical properties (IOPs) of European waters

Regarding the absorption coefficient, the "optically significant constituents" are classically divided into three classes (Prieur and Sathyendranath 1981): i) the pigmented particulate material originating from the phytoplankton (PH); ii) the non-pigmented particulate material (NP), containing organic and non organic particles from different sources; iii) the coloured dissolved organic material (CDOM), also called "yellow substances" (YS), composed of humic substances originating from phytoplankton and/or terrestrial plants. The total absorption (excluding the almost invariant "pure" sea water component) is thus often written:

$$a(\lambda) = a_{PH}(\lambda) + a_{NP}(\lambda) + a_{YS}(\lambda) \qquad (2)$$

This ternary decomposition originates from the methodological concept used for the study of natural waters absorption properties. It has been proposed for the classification of Case 1 and Case 2 waters, with Case 1 waters supposedly located close to the phytoplankton apex. Regarding the back-scattering coefficient, a "total particulate" compartment ($PH + NP$) is often the only one considered, the contribution of YS to scattering being negligible. Note that the origin of back-scattering is still controversial and phytoplankton is probably a much less efficient back-scatter than its co-varying detrital material.

These simple classifications are probably to be revised, in particular for a better understanding of the absorption and (back-) scattering properties of natural waters (Stramski *et al.* 2004).

3.1.1 The absorption coefficient

Figure 2 presents spectra of the absorption coefficient of each of the three optical constituents (*PH, NP, YS*) and their budget, measured at surface during ship campaigns performed by the Joint Research Centre (JRC) in different European Seas between 2000 and 2006 as well as onboard an oceanographic tower in the northern Adriatic Sea between 1996 and 2005 (Zibordi *et al.* 2002, for methods).

The CDOM (YS) absorption

   The YS absorption spectra (Figure 2) show the classical monotonic and exponential decrease from 400 to 700 nm. High concentrations of CDOM are supposed to be found where land discharges are elevated in particular at the outputs of the large European rivers. The highest values (at $\lambda = 400$ nm) are found in the southern Baltic Sea (0.7 m^{-1} ± 0.5 m^{-1}). High values of absorption, up to 15 m^{-1} at 400 nm, have classically been measured in this low salinity basin (see Højerslev *et al.*, 1996 and references therein; and Table 1). Lower values can be found in the English Channel, in the North Sea and in the northern Adriatic Sea (Figure 2 and Table 1). The absorption is minimum in the Atlantic waters, in the western (Gulf of Lion) and eastern Mediterranean Sea (0.065 ± 0.04 m^{-1} at 400 nm for the latter, Figure 2). For the Black Sea, values at 400 observed here went from 0.2 in the open sea to 2 m^{-1} near the Danube river mouth (Table 1).

   The spectral dependency of the absorption by CDOM is classically represented by an exponential function of the type:

$$a_{YS}(\lambda) = a_{YS}(\lambda_0) \exp\left(- S_{YS}(\lambda - \lambda_0)\right) \qquad (3)$$

where the reference wavelength λ_0 is generally chosen in the blue or in the UV. Although an important part of the natural variability of S_{YS} reported in the literature has been attributed to the way it has been calculated (Twardowski *et al.*, 2004), it is nevertheless generally considered dependent on the CDOM composition (fractions of fulvic and humic acids).

   Babin *et al.* (2003a) reported little variations of $S_{YS}(\lambda_1 - \lambda_2)$ for different European seas (Table 1), whereas for the present work it varies between 0.011 (Eastern Mediterranean Sea) and 0.020 (Baltic Sea). Note that fulvic acid absorption would dominate at $\lambda < 500$ nm while humic acid absorption would do so for $\lambda \geq 500$ nm, suggesting the consideration of two different slopes for modeling $a_{YS}(\lambda)$ (Ershova and Kopelevich 1999). A low or a high slope on the whole interval would thus indicate a predominance of humic or fulvic acid, respectively.

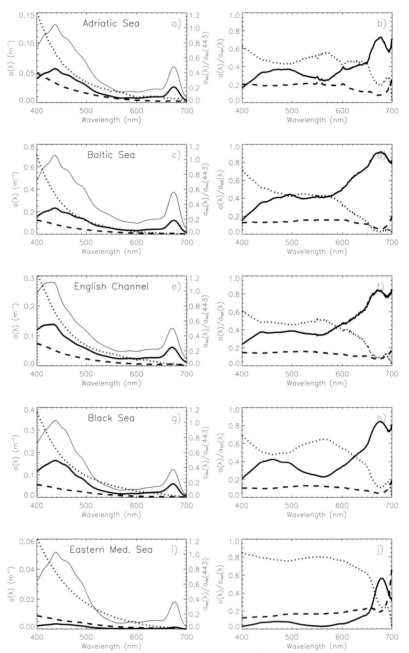

Fig. 2. (a,c,e,g,i) Average spectra a_{PH} (thick solid line), a_{NP} (dashed line) and a_{YS} (dotted line) and a_{PH} normalized to 443 nm (thin solid line, right ordinates scale); (b,d,f,h,j) Fraction of the total (minus water) absorption of *PH* (solid line), *NP* (dashed line) and *YS* (dotted line).

Table 1. Values of $a_{YS}(\lambda)$ (m^{-1}) at wavelength λ, and of the slope S_{YS} (nm^{-1}) computed for the wavelength range $\lambda_1 - \lambda_2$, for different European Seas. ± indicates a standard deviation.

$a_{YS}(\lambda)$	λ	$S_{YS}(\lambda_1-\lambda_2)$	$\lambda_1 - \lambda_2$	Area	Reference
0.7±0.5	400	0.020±0.001	350–500	Baltic Sea	Present work
0.2–3.3	400	0.018–0.020	–	Baltic Sea	Kowalczuk (1999)
0.3–0.7	443	0.019±0.001	350–500	Baltic Sea	Babin et al. (2003a)
0.3±0.1	400	0.016±0.001	350–500	Engl. Ch.	Present work
0.05–0.2	443	0.017±0.001	350–500	Engl. Ch.	Babin et al. (2003a)
0.2–1.1	440	0.013–0.018	350–500	Engl. Ch.	Vantrepotte et al. (2007)
0.4	443	0.017±0.0004	350–500	North Sea	Babin et al. (2003a)
0.01–0.1	443	0.017±0.003	350–500	Atlantic	Babin et al. (2003a)
0.01–0.35	400	0.006–0.018	350–500	Irish Sea	Tilstone et al. (2005)
0.01–0.3	443	0.017±0.003	350–500	G. of Lion	Babin et al. (2003a)
0.2±0.1	400	0.017±0.002	350–500	Adriatic	Present work
0.04–0.4	443	0.019±0.002	350–00	Adriatic	Babin et al. (2003a)
0.07±0.04	400	0.011±0.002	350–500	Eastern Med	Present work
0.2–2.0	400	0.016±0.001	350–500	Black Sea	Present work
0.04–0.4	400	0.018	300–600	Black Sea	Churilova and Berseneva (2004)

The non pigmented particulate (NP) absorption

This compartment includes both organic and mineral particles and the determination of their absorption properties generally results from the bleaching of the pigmented particulate material. Average absorption values at 443 nm presented here (Figure 2, Table 2) vary between 0.008 ± 0.003 m^{-1} in the Eastern Mediterranean Sea and 0.1 ± 0.1 m^{-1} in the Baltic Sea. A global correlation has been observed by Babin *et al.* (2003a) for different European waters between $a_{YS}(443)$ and the concentration of total suspended matter, well correlated itself with $a_{NP}(443)$. The spectral dependence of the absorption coefficient is modeled as in Eq. 3, substituting YS with NP. For S_{NP}, average values lower than S_{YS} have traditionally been reported (Table 2). As for S_{YS}, the calculation conditions have a great influence, and in particular the fact of forcing to zero the absorption in the range 745–750 nm. For the present data set, by adopting the last condition and fitting $a_{NP}(\lambda)$ between 400 and 750 nm, the resulting average S_{NP} also shows a striking constancy, with values around 0.012 (Table 2). Departure from a pure exponential shape or changes of slope can express the presence of mineral particles (more precisely, of iron oxide) among organic "detritus" as it is the case for saharian dust in the Mediterranean Sea, in particular (Babin and Stramski 2004).

The pigmented particulate (PH) absorption

The intensity and the shape of $a_{PH}(\lambda)$ are determined by the presence of chlorophylls and carotenoids pigments and thus depend on the distribution of the phytoplankton groups and their physiological state. The ratio of the peak at 440 nm to the one at 675 nm (see the normalized $a_{PH}(\lambda)$ spectra on Figure 2) varies from about 2.1 (Baltic, English Channel, Adriatic) to 2.6 (Black Sea, Eastern Med. Sea). Similar results were obtained by Babin *et al.* (2003a), with 2.8 for the western Mediterranean Sea and 3.2 for the Atlantic. For the Black Sea, Chami *et al.* (2005) reported values varying between 3.3 and 2.4 as the Chl-a concentration increased from 0.7 to 2.1 mg m^{-3}. High blue-to-red ratios are globally indicator of low Chl-a concentration and vice versa (Bricaud *et al.* 1995). Such a flattening of the spectra is due to increasing cell size and pigment packaging and is underlying in the relationship between the Chl-a specific absorption, $a^*_{PH}(\lambda)$ $(a_{PH}(\lambda)/Chl\text{-}a)$, and the Chl-a concentration itself (Bricaud *et al.* 1995):

$$a^*_{PH}(\lambda) = A(\lambda)\, Chl\text{-}a^{B(\lambda)} \tag{4}$$

A good agreement with the coefficients $A(\lambda)$ and $B(\lambda)$ in Bricaud *et al.* (1995) was reported for the English Channel (Vantrepotte *et al.* 2007), the Irish Sea (Tilstone *et al.* 2005), the Atlantic and Mediterranean waters (Babin *et al.*, 2003a). Significant differences have been observed for the Baltic Sea (Majchrowski *et al.*, 2000), the North Sea (Babin *et al.*, 2003a, Stæhr and Markager, 2004) and the Black Sea (Churilova and Berseneva 2004). A "flpeak of absorption between 400 and 440 (as in Figure 2e) has been observed by Astoreca *et al.* (2006) in the North Sea and is typical of diatoms dominated waters. Regarding the Black Sea, Churilova and Berseneva (2004) reported significantly higher absorption at 440 than Bricaud *et al.* (1995) for coastal waters off the Crimea (whereas the agreement was rather good at 678 nm) and explained such differences by a larger fraction of auxiliary pigments.

Beside the classical diatoms blooms, some of the European Seas are known for specific phytoplankton seasonal blooms: cyanobacteria (*e.g. Nodularia spumigena*) in the Baltic Sea (Kutser 2004), prymnesiophytes (in particular *Phaeocystis* sp.) in the English Channel and North Sea (Vantrepotte *et al.* 2007; Astoreca *et al.* 2006), coccolithophores in the Atlantic or in the Black Sea. As an example, the reinforced absorption around 620–630 nm for the Baltic Sea normalized spectrum on Figure 2c could be the indication of the phycocyanin pigments present in cyanobacteria (Kutser, 2004). Typical spectra of *Phaeocystis* dominated waters can be found in Astoreca *et al.* (2006) for the North Sea.

Table 2. Values of $a_{NP}(\lambda)$ (m^{-1}) at wavelength λ, and of the slope S_{NP} (nm^{-1}) computed for the wavelength range $\lambda_1 - \lambda_2$, for different European Seas. $\pm$ indicates a standard deviation.

$a_{NP}(\lambda)$	λ	$S_{NP}(\lambda_1 - \lambda_2)$	$\lambda_1 - \lambda_2$	Area	Reference
0.1±0.1	443	0.012±0.001	400–750	Baltic Sea	Present work
0.1–0.4	443	0.013±0.001	380–730	Baltic Sea	Babin et al. (2003a)
0.06±0.03	443	0.012±0.002	400–750	Engl. Ch.	Present work
0.007–0.1	443	0.012±0.001	380–730	Engl. Ch.	Babin et al. (2003a)
0.03–0.08	440	0.011–0.012	400–700	Engl. Ch.	Vantrepotte et al. (2007)
0.01–1.0	443	0.012±0.001	380–730	North Sea	Babin et al. (2003a)
0.001–0.04	443	0.013±0.002	380–730	Atlantic	Babin et al. (2003a)
0.02–0.14	442	0.008–0.013	350–550	Irish Sea	Tilstone et al. (2005)
0.05–0.3	443	0.013±0.002	380–730	G. of Lion	Babin et al. (2003a)
0.04±0.02	443	0.013±0.001	400–750	Adriatic	Present work
0.01–0.4	443	0.013±0.001	380–730	Adriatic	Babin et al. (2003a)
0.008±0.003	443	0.010±0.001	400–750	Eastern Med	Present work
0.05±0.08	443	0.011±0.001	400–750	Black Sea	Present work
0.03–0.1	443	–	–	Black Sea	Chami et al. (2005)

The budget of absorption

As expected, the contribution of a_{YS} is dominating at 400 nm (>60%) (Figures 2b,d,f,g,i). High contributions of *YS* (between 80 and 90%) have been reported for the Baltic Sea by Ferrari *et al.* (1996) and Babin *et al.* (2003a). The percentage in the Eastern Mediterranean Sea is here very high (85%) in spite of a probable intense photo-bleaching of CDOM. Note that submicron Saharan mineral dust, by being erroneously identified as CDOM, could partially explain such a high value (Oubelkheir *et al.* 2005).

At 443 nm, and excepting the Eastern Mediterranean Sea, the average pigments contribution (Chl-a mainly) is comprised between 35% (Adriatic, Baltic) and 40% (Black Sea). A relative constancy is also observable in Babin *et al.* (2003a) for different European waters (from 28% to 42%), excepting the Adriatic Sea (52%). An average 27 % has been reported for the Baltic Sea (Ferrari *et al.*, 1996), from 40% to 54% in the Irish Sea (Tilstone *et al.*, 2005), from 25% to 33% in the coastal Black Sea (Churilova and Berseneva 2004). At 675 nm, where a secondary peak of absorption by Chl-a is observed, the contribution of *PH* is almost systematically above 70% (<60% in the Eastern Med. Sea where a_{NP} contribution is reinforced) (Figure 2).

The contribution of the non pigmented particulate matter (NP) is here rather constant from 400 to 600 nm and varies between 10% (Black Sea) and 20% (Adriatic), in agreement with the 11–28% found in Babin *et al.* (2003a). The contribution of the three components appears more balanced (especially in the 500–550 nm interval) in the Adriatic Sea than in the other basins (Figure 2).

3.1.2 The back-scattering coefficient

The knowledge on variability and partition of $b_b(\lambda)$ in relevant components is much lower with respect to the absorption coefficient. The recent development of specific instrumentation, in particular *in situ* back-scattering (or scattering) profilers, partially filled that gap. In non-bloom conditions it is commonly assumed that small-sized non-living particles (organic and mineral) dominate the back-scattering (see Stramski *et al.* 2004 and references therein). However, specific conditions can perturb this general law: monospecific blooms, coccolithophore scales, micro-bubbles, deposed atmospheric dust, ... (Stramski *et al.* 2004).

Figure 3 presents back-scattering spectra of particles (measured at 442, 488, 510, 555, 620, and 665 nm) collected by the JRC using a "Hydroscat-6".(HOBI Labs Inc.). The spectra for the Mediterranean Sea were fitted according to a power law of the wavelength (see after). At 442 nm, there are more than three orders of magnitude between the Eastern Mediterranean Sea (0.0005 m^{-1}) and the Adriatic and Baltic Seas (0.15 m^{-1}). A strong variability is also observable within the Adriatic Sea, the Baltic Sea and the English Channel, and at a lower degree within the Black Sea. However, their average values are very similar (0.015-0.02 m^{-1}). Values of $b_{bp}(555)$ comprised between 0.003 and 0.01 m^{-1} were reported for the Black Sea (Chami *et al.*, 2005), between 0.001 and 0.12 m^{-1} for the English Channel and North Sea (Loisel *et al.*, 2007). Note that Babin *et al.* (2003b) observed values of the scattering coefficient $b_p(440)$ covering four orders of magnitude (from 0.05 to 40 m^{-1}) across European coastal waters.

The spectral dependence of $b_{bp}(\lambda)$, as well as the intensity of the back-scattering ratio (that is $b_{bp}(\lambda)/b_p(\lambda)$), depends on the composition of particles and their properties (size, index of refraction, absorption) (see Loisel *et al.*, 2007 for the English Channel and southern North Sea). The collected spectra have been fitted according to a power law of wavelength (with $\lambda_0 = 510$ nm):

$$b_{bp}(\lambda) = b_{bp}(\lambda_0)(\lambda/\lambda_0)^{-n} \qquad (5)$$

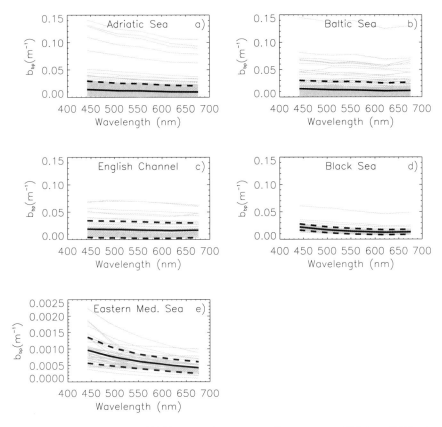

Fig. 3. Individual (thin solid lines) and corresponding average (thick solid line) ± standard deviation (dashed line) spectra of back-scattering by particles collected in different European basins.

Although the phytoplankton is not the principal backscatter, there exists an inverse relationship between the exponent n and the Chl-a concentration in the open ocean (Morel and Maritorena 2001), as the average size of phytoplankton cells increases with Chl-a. The average exponent computed here for each basin takes the values of 0.4 (English channel), 0.7 (Baltic Sea), 1.0 (Adriatic), 1.3 (Black Sea) and 1.9 (Eastern Mediterranean Sea). This increasing series of exponents is in rough agreement with the assumption that the exponent n varies from 0 in turbid coastal waters to 2 in oligotrophic waters. Departure of the $b_{bp}(\lambda)$ spectral shape from such a function are related to the absorption properties of the particles (Babin *et al.* 2003b; Chami *et al.* 2005).

3.2   The "colour" of the European waters

3.2.1   In-situ irradiance reflectance

Irradiance reflectance $R(0^-, \lambda)$ spectra in Figure 4 were collected by JRC using multi-spectral underwater radiometric free-fall or fixed profilers (Satlantic Inc.), at 412, 443, 490, 510, 555, 665 and 683 nm (Zibordi et al. 2002). The average spectrum measured in the northern Adriatic Sea shows a large maximum laying between 490 and 555 nm and absolute values at 412 nm ranging between 0.02 and 0.12. The relatively low absorption by CDOM (Figures 2a,b) combined to the high back-scattering leads to relatively high values of reflectance in the blue. The increase of the absorption by CDOM and non-pigmented matter induces a shift of the maximum reflectance from 490 to 510 and 555 nm. Differently, the spectra collected in the Baltic Sea show much lower absolute values at 412 nm (from 0.005 to 0.02), as well as maxima almost systematically located at 555 nm. The very large absorption by CDOM and its exponential decrease toward the green (Figures 2c,d) is somewhat "mirrored" in reflectance spectra (see Siegel et al.,1994; Darecki et al., 2003 and references therein). The English Channel shows intermediate situations between the Adriatic and Baltic waters, with spectral shape similar to the Baltic. The absorption by CDOM and non pigmented particulate matter in the blue is high enough to shift the reflectance maximum to 555 nm (see Astoreca et al. 2006). Higher values at 555 are observed as a consequence of a reinforced back-scattering, possibly by suspended minerals, with respect to the Baltic Sea (Figure 3).

In the Black Sea, there is a clear differentiation between spectra collected on the north-western shelf and showing a maximum at 555 nm (due to the CDOM absorption), and those collected in the open basin (bottom depth >1500 m) and presenting a maximum at 490 nm. A few spectra showed a broad maximum (with almost constant values between 490 and 555 nm), corresponding to high concentrations of coccolithophores, and most of all of detached coccolithes. These highly scattering calcite "disk-like" particles induce a flattening of the reflectance in the blue and blue-green regions resulting in an increased brightness of the waters. However, quantifying the respective contributions of coccolithophores and suspended matter brought by river discharges to the seasonal augmentation of back-scattering in the Black Sea is still difficult (Karabashev et al. 2006).

In the ultra-oligotrophic eastern Mediterranean Sea, the reflectance signal is almost zero at 665 nm due to the strong absorption by water and the low back-scattering by particles. In the blue part, the very low absorption and the back-scattering by the few particles and by water induce reflectance ratios $R(412)/R(555)$ nm up to 8.

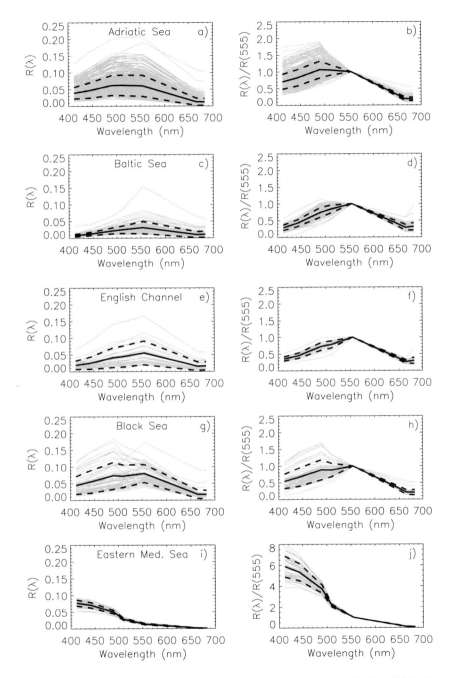

Fig. 4. Individual (thin solid lines) and corresponding average (thick solid line) ± standard deviation (dashed line) spectra of irradiance reflectance "just below" the surface (0⁻) collected in different European basins.

Note that the oligotrophic Mediterranean Sea is known for its deviation with respect to the typical Case 1 behaviour, possibly as a result of enhanced absorption in the blue and back-scattering in the green by desert dust (Claustre *et al.* 2002).

3.2.2 Remotely sensed "normalized water-leaving" radiance

Figure 5 presents average monthly climatological spectra of the normalized water-leaving radiance for different European basins as derived from 7 years of SeaWiFS data after the correction for the atmospheric perturbations (SeaDAS 4.9). They correspond to the averaging of satellite derived spectra within extended space and time domains. The resulting optical characteristics are not always strictly representing a unique ecosystem (an upwelling, a coastal area, etc …). The observations made in the previous section are also valid here although a shift of the reflectance maximum can be observed (English Channel, Black Sea). Figure 5 also shows the phytoplankton dominated (Case 1) situation of the Iberian upwelling and the Ligurian Sea in the spring bloom period (April) leading to relatively flat spectra in the range 412–490 nm. In addition, Figure 5 presents the seasonal variability observable from remote sensing in these basins. For example, a pronounced increase of the signal in July with respect to April is observed in the Southern Baltic Proper as the freshwater plume from rivers summer flood carries more scattering particles and as the spring phytoplankton bloom collapses (Kowalczuk *et al.* 1999). The northern Adriatic Sea shows a decrease of the signal and an increase of the ratio $L_{WN}(490)/L_{WN}(550)$ when passing from the particles rich waters in winter to the oligotrophic waters in summer (Berthon *et al.* 2002). The Ligurian Sea shows a shift of the reflectance maximum from 443 in July and October to 490 in April, together with a decrease of the absolute values, as the phytoplankton spring bloom develops.

4. Perspectives

This presentation of the optical properties of European waters does not aim at being exhaustive but at illustrating their potential variability. Such a variability calls for an improvement of the techniques presently allowing for the retrieval of the main optical components from ocean colour data in the European Seas. Although semi-analytical algorithms are more universal and physically sound, there is probably still a need for regional empirical algorithms. In fact, statistical techniques can be used for determining the domains of applicability of seasonal/regional empirical algorithms and for combining them (D'Alimonte *et al.* 2003). However, new models and applications will not be possible without an increased knowledge of the apparent and inherent optical properties of the waters under investigation.

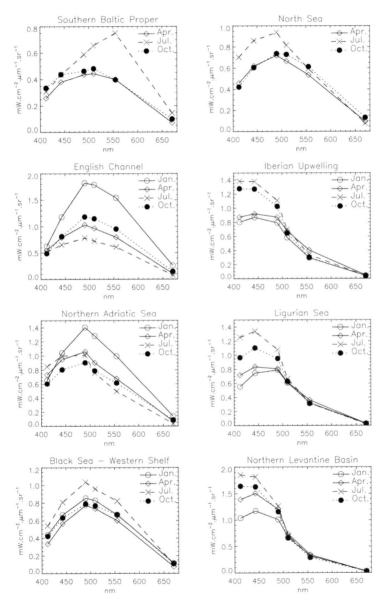

Fig. 5. Spectra of normalized water-leaving radiance obtained from a 7-year SeaWiFS climatology averaged over several domains of the European seas.

Future ecological and biogeochemical applications will require to go beyond the traditional products (Chl-a, CDOM, ...). Some recent works already put in evidence the possibility of deriving phytoplankton functional groups in the open ocean from multi-spectral ocean colour data (Alvain *et al.* 2005). Equivalent applications for the coastal waters may require the use of hyper-spectral remote sensing data as *in situ* measurements demonstrated the advantage of high spectral resolution for monitoring particular events like Harmful Algal Blooms. Another perspective improving our ability to assess the dynamics of marine coastal ecosystems could be provided by the operation of ocean colour geostationary platforms. Furthermore, studies integrating various sources of remote sensing data, including ocean colour, as well as dynamic, biogeochemical, and ecosystems models, will enhance our capacity to understand, monitor and predict the state and processes of marine ecosystems.

Acknowledgements

Acknowledgments are due to Dirk van der Linde and Elisabetta Canuti from JRC for their participation to field measurements and laboratory analyses.

References

Alvain S, Moulin C, Dandonneau Y, Bréon F-M (2005) Remote sensing of phytoplankton groups in case 1 waters from global SeaWiFS imagery. Deep-Sea Res 52: 1989–2004

Astoreca R, Rousseau V, Ruddick K, Parent J-Y, Van Mol B, Lancelot C (2006) Discriminating phytoplankton groups from space: Is it feasible in Case 2 belgian coastal waters ? Proceedings (CR-Rom) of the Ocean Optics Conference XVIII, Montreal, Canada, 09–13 October 2006

Babin M, Stramski D, Ferrari GM, Claustre H, Bricaud A, Obolensky G, Hoepffner N (2003a) Variations in light absorption coefficients of phytoplankton, non algal particles, and dissolved organic matter in coastal waters around Europe. J Geophys Res 108, C7, 3211, doi:10.1029/2001JC000882

Babin M, Morel A., Fournier-Sicre V, Fell F, Stramski D (2003b) Light scattering properties of marine particles in coastal and open ocean waters as related to the particle mass concentration. Limnol Oceanogr 48: 843–859

Babin M, Stramski D (2004) Variations in the mass-specific absorption coefficient of mineral particles suspended in water. Limnol Oceanogr 49: 756–767

Berthon J-F, Zibordi G, Doyle J-P, Grossi S, van der Linde D, Targa C (2002) Coastal Atmosphere and Sea Time Series (CoASTS), Part 2: Data analysis

NASA Tech. Memo. 2002-206892, vol 20, Hooker SB and Firestone ER (eds), NASA Goddard Space Flight Center, Greenbelt, Maryland

Bricaud A, Babin M, Morel A, Claustre H (1995) Variability in the chlorophyll-soecific absorption coefficients of natural phytoplankton: Analysis and Parameterization. J Geophys Res 100 (C7): 13,321–13,332

Chami M, Shybanov EB, Churilova TY, Khomenko GA, Lee MEG, Martynov OV, Berseneva GA, Korotaev GK (2005) Optical properties of the particles in the Crimea coastal waters (Black Sea). J Geophys Res 110, C11020, doi:10.1029/2005JC003008

Churilova TY, Berseneva GP (2004) Absorption of light by phytoplankton, detritus, and dissolved organic substances in the coastal region of the Black Sea (July–August 2002). Physical Oceanography 14: 221–233

Claustre H, Morel A, Hooker SB, Babin M, Antoine D, Oubelkheir K, Bricaud A, Leblanc K, Quéguiner B, Maritorena S (2002) Is desert dust making oligotrophic waters greener? Geophys Res Lett 29, 1469, 10.1029/2001GL014056

D'Alimonte D, Mélin F, Zibordi G, Berthon J-F (2003) Use of the novelty detection technique to identify the range of applicability of empirical ocean color algorithms. IEEE Trans Geosci Rem Sens 41: 2833–2843

Darecki M, Weeks A, Sagan S, Kowalczuk P, Kaczmarek S (2003) Optical characteristics of two contrasting Case 2 waters and their influence on remote sensing algorithms. Cont Shelf Res 23: 237–250

Ershova SV, Kopelevich OV (1999) A model of seawater optical properties in the UV spectral range in light of new data. Oceanology 39: 314–321

Ferrari GM, Dowell MD, Grossi S, Targa C (1996) Relationship between the optical properties of chromophoric dissolved organic matter and total concentration of dissolved organic carbon in the southern Baltic Sea region. Mar Chem 55: 299–316

Gordon HR (1997) Atmospheric correction of ocean color imagery in the Earth Observing System era. J Geophys Res 102: 17081–17106

Højerslev NK, Holt N, Aarup T (1996) Optical measurements in the North Sea-Baltic Sea transition zone, I. On the origin of the deep water in the Kattegat. Cont Shelf Res 16: 1329–1342

Karabashev GS, Sheberstov SV, Yakubenko VG (2006) The June maximum of normalized radiance and its relation to the hydrological conditions and coccolithophorid bloom in the Black Sea. Oceanology 46: 305–317

Kowalczuk P (1999) Seasonal variability of yellow substance absorption in the surface layer of the Baltic Sea. J Geophys Res 104 (C12): 30,047–30,058

Kowalczuk P, Sagan S, Olszewski J, Darecki M, Hapter R (1999) Seasonal changes in selected optical parameters in the Pomeranian Bay in 1996-1997. Oceanologia 41: 309–334

Kutser T (2004) Quantitative detection of chlorophyll in cyanobacterial blooms by satellite remote sensing. Limnol Oceanogr 49: 2179–2189

Loisel H, Mériaux X, Berthon J-F, Poteau A (2007) Investigation of the optical backscattering ratio of marine particles in relation to their biogeochemical composition in the eastern English Channel and southern North Sea. Limnol Oceanogr 52: 739–752

Majchrowski R, Woźniak B, Dera J, Ficek D, Kaczmarek S, Ostrowska M, Koblentz-Mishke OI (2000) Model of in vivo spectral absorption of algal pigments. Part 2. Practical applications of the model. Oceanologia 42: 191–202

Mobley CD (1994). Light and water. Radiative transfer in natural waters. Academic Press, San Diego

Morel A, Prieur L (1977) Analysis of variations in ocean color. Limnol Oceanogr 22: 709–722

Morel A, Maritorena S (2001) Bio-optical properties of oceanic waters: A reappraisal. J Geophys Res 106: 7,163–7,180

Oubelkheir K, Claustre H, Sciandra A, Babin M (2005) Bio-optical and biogeochemical properties of different trophic regimes. Limnol Oceanogr 50: 1795–1809

Prieur L, Sathyendranath S (1981) An optical classification of coastal and oceanic waters based on the specific spectral absorption curves of phytoplankton pigments, dissolved organic matter, and other particulate materials. Limnol Oceanogr 26: 671–689

Ruddick KG, De Cauwer V, Park Y-J, Moore G (2006) Seaborne measurements of near infrared water-leaving reflectance: The similarity spectrum for turbid waters. Limnol Oceanogr 51: 1167–1179

Sancak S, Besiktepe S, Yilmaz A, Lee M, Frouin R (2005) Evaluation of SeaWiFS chlorophyll-a in the Black and Mediterranean Seas. Int J Rem Sens 26: 2045–2060

Siegel H, Gerth M, Beckert M (1994) The variation of optical properties in the Baltic Sea and algorithms for the application of remote sensing data. Proceedings of the Ocean Optics XII Conference, SPIE Proceedings Series 2258: 894–905

Stæhr PA, Markager S (2004) Parameterization of the chlorophyll-a specific in vivo light absorption coefficient covering estuarine, coastal and oceanic waters. Int J Rem Sens 25: 5117–5130

Stramski D, Boss E, Bogucki D, Voss KJ (2004) The role of seawater constituents in light backscattering in the ocean. Progr Oceanogr 61: 27–56

Tilstone GH, Smyth TJ, Gowen RJ, Martinez-Vicente V, Groom S (2005) Inherent optical properties of the Irish Sea and their effect on satellite primary production algorithms. J Plank Res 27: 1127–1148

Twardowski MS, Boss E, Sullivan JS, Donaghay PL (2004) Modeling the spectral shape of absorption by chromophoric dissolved organic matter. Mar Chem 89: 69–88

Vantrepotte V, Brunet C, Mériaux X, Lécuyer E, Vellucci V, Santer R (2007) Bio-optical properties of coastal waters in the Eastern English Channel. Estuar Coast Shelf Sci 72: 201–212

Zibordi G, Berthon J-F, Doyle J-P, Grossi S, van der Linde D, Targa C, Alberotanza L (2002) Coastal Atmosphere and Sea Time Series (CoASTS), Part 1: A long-term measurement program. NASA Tech. Memo. 2002–206892, vol 19, Hooker SB and Firestone ER (eds), NASA Goddard Space Flight Center, Greenbelt, Maryland

Case Studies of Optical Remote Sensing in the Barents Sea, Black Sea and Caspian Sea

Oleg V. Kopelevich, Vladimir I. Burenkov, and Sergey V. Sheberstov

Shirshov Institute of Oceanology, Russian Academy of Sciences, Moscow, Russia

Abstract. Examples of variability of the bio-optical characteristics in the Barents, Black and Caspian Seas derived from SeaWiFS data in 1998–2005 are presented and analyzed. A common feature of these seas is high freshwater inflow, all of the seas are classified with Case 2 waters. Optical remote sensing of them requires modified processing algorithms regionally adapted. Some interesting phenomena were revealed from satellite data in the seas under study: coccolithophore blooms in the Black Sea and the Middle Barents, sharp increase of chlorophyll concentration and particle backscattering in the Caspian Sea in July–August 2001 attributed to a consequence of invasion of the ctenophore *Mnepiopsis leidyi*. The joint analysis of satellite and *in situ* measured data was carried out to explain the above-mentioned phenomena.

1. Introduction

The Barents, Black and Caspian Seas are classified as Case 2 waters in which, by definition, variability of seawater optical properties are influenced not just by phytoplankton and by the material associated with it, as in Case 1 waters, but also by other substances independent of phytoplankton (International Ocean-Colour Coordinating Group, IOCCG, 2000). In the seas of interest such substances are the particulate and dissolved matter brought in the most part with river runoff. The standard algorithms for processing of satellite ocean color data, derived mainly from Case 1 waters, break down in optically-complex Case 2 waters. As shown (Kopelevich 2004, 2005b), in the south-eastern part of the Barents Sea and in the northern part of the Caspian Sea, affected strongly by river runoff, the standard algorithm can overestimate chlorophyll-like pigment concentration (*Chl*) by a factor up to twenty, in the Black Sea the overestimation is more than twice. It can be explained to a great extent by that the standard algorithm is

53

V. Barale, M. Gade (eds.), *Remote Sensing of the European Seas.*
© Springer Science+Business Media B.V. 2008

based on a regression equation meaning definite relationship between the chlorophyll and yellow substance absorption coefficients. If seawater absorption increases due to enhanced content of yellow substance brought by rivers, the algorithm attributes the additional absorption to increasing chlorophyll concentration and overestimates the latter (Burenkov *et al.* 1999, 2000, 2001a).

The regional processing algorithms for the Barents, Black and Caspian Seas (hereinafter "the SIO RAS algorithms") were developed on the basis of our field data (Kopelevich *et al.* 2005a, b). For reasons of space we have no way of reporting these algorithms here, they were described in detail by Burenkov *et al.* (2001a, b), Kopelevich *et al.* (2004) and available at the site: http://manta.sio.rssi.ru. We should acknowledge that the algorithm validation was only performed in a limited number of areas and seasons, and further validation studies are needed. It is worth noting that the relative changes of the *Chl* values derived by the SIO RAS and the standard Sea-viewing Wide Field-of-view Sensor (SeaWiFS) OC4v4 algorithms are consistent with each other. For example, in all regions of the Barents Sea (see below) the correlation between monthly mean *Chl* values derived from SeaWiFS data by different algorithms was rather strong, with coefficients of determination (R^2) within 0.925–0.984 (Kopelevich *et al.* 2004).

The mean monthly distributions of chlorophyll "a" concentration, the particle backscattering and yellow substance absorption coefficients in the Barents, Black, Caspian Seas and Sea of Japan, as well as the monthly mean values of these characteristics in the sub-regions distinguished in the above-mentioned seas, were derived from SeaWiFS data in 1998–2004, and CD-ROM with 742 color maps, tables, the algorithm description, and brief analysis of the obtained results was issued (Kopelevich 2005b). The materials presented by the CD-ROM are also available at the above-mentioned site. In this paper we report some results obtained in the seas of interest from SeaWiFS data 1998–2005. For reasons of space we present here only examples of variability of chlorophyll "a" concentration and particle backscattering.

2.   Seasonal and inter-annual variabilities

Studying seasonal and inter-annual variability of the above-mentioned bio-optical characteristics, we examine variability of three important seawater components: photosynthetic phytoplankton, particulate matter and colored organic matter, and how much their changes are affected by the river run-off and by other factors. The effect of the river runoff on seawater optical properties depends primarily on total amount of particulate and dissolved substances brought by river as well as on processes of transformation of these substances in estuary and their further spreading into the sea. According to Lisitsyn (2004), a "marginal filter" on the river-sea boundary removes about 93% of particulates and about 40% of dissolved matter. But the matter deposited on the shelf bottom can be brought back to water body by storm and/or tide stirring up.

The processes of further spreading the river runoff into the sea, including currents, turbulence, and mesoscale eddies, determine impact of the river runoff on bio-optical characteristics of the open sea. Several sub-regions is distinguished in the seas of interest with respect to bathymetry, hydrological and hydrodynamic conditions (Kopelevich *et al.* 2004, 2005b).

2.1 The Barents Sea

The Northern, Middle and Southern Barents are distinguished reasoning from natural conditions determining formation of bio-optical characteristics. The Northern region is predominantly occupied by the flow of cold Arctic Basin Water whereas the Middle Barents is under influence of the warm Norwegian Current. We take the Polar Front as a boundary between these regions and define this boundary, for simplicity, along 75°N; its real position is, of course, not a straight line and changeable. The Southern region is a shallow basin comprising the Cheshskaya Bay and the Pechora Sea between Kolguev and Vaigach islands influenced strongly by the Pechora river runoff. For a lack of room we do not consider the White Sea although this sea is also studied by us with SeaWiFS data together with the Barents Sea and individually (Burenkov *et al.* 2004).

Variability of the monthly mean values of chlorophyll "a" concentration (*Chl*) and the particle backscattering coefficient (b_{bp}) derived by the SIO RAS algorithms in different sub-regions of the Barents Sea is demonstrated by Figure 1. As is seen from Figure 1a, variations of chlorophyll concentration in the Northern and Middle regions are in phase with each other, whereas they are out of phase with the ones in the Southern Barents. The correlation coefficient between the changes of chlorophyll concentration in the Northern and Middle regions is positive (*r*[*Chl*-N, *Chl*-M] = 0.719) and the one is negative between the Middle and Southern regions (*r*[*Chl*-M, *Chl*-S] = –0.525).

The maximum monthly mean *Chl* values in the Northern and Middle regions are observed in May and caused by spring phytoplankton bloom. In the Southern Barents the maxima are most observed in July that corresponds to the Pechora flood-time. The enhanced values of *Chl* are held to September (they can even be the maxima as in 1999 and 2001–2002) as the Pechora discharge is high throughout summer and autumn due to frequent rain freshets (Roshydromet 2001).

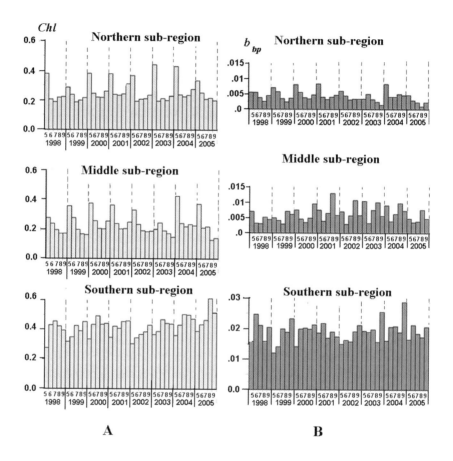

Fig. 1. Variability of the monthly mean values of chlorophyll concentration, *Chl,* mg m^{-3} (A) and the particle backscattering coefficient, b_{bp}, m^{-1} (B) derived by the SIO RAS algorithms in different sub-regions of the Barents Sea.

As is seen from Figure 1b, the sub-regions differ noticeably in variability of the b_{bp} values, and the coefficients of correlation $r[b_{bp}\text{-}N, b_{bp}\text{-}M]$ and $r[b_{bp}\text{-}M, b_{bp}\text{-}S]$ are very low (0.069 and –0.118). In the Southern region the changes of b_{bp} values, as well as of *Chl,* are caused by the Pechora runoff, in the Northern Barents by spring phytoplankton bloom. In the Middle

Barents two seasonal maxima of b_{bp} are observed: the first in May, the second in August (in September in 2000); in most years the August maxima in b_{bp} are higher than the May ones (Figure 1b).

The mean monthly distribution of b_{bp} in the Barents Sea derived from SeaWiFS data in August 2004 is shown on Figure 2. The b_{bp} values higher than 0.02 m^{-1} can be observed in the western part of the Middle Barents whereas they are lower than 0.005 m^{-1} and even 0.002 m^{-1} in the eastern part.

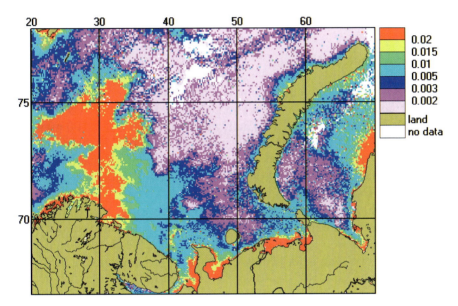

Fig. 2. The mean monthly distributions of b_{bp} derived from SeaWiFS data in the Barents Sea in August 2004.

The high b_{bp} values were attributable to the coccolithophore bloom, and the evidence of that was obtained in the *Professor Schtokman* cruise in August 2004. The water samples were taken in the area of 71–72°N, 28.5–32.5°E on 10–12 August, and the intense coccolithophore bloom was found from direct determinations by T. Rat'kova, SIO RAS. The coccolithophore *E. huxleyi* concentration exceeded 10^6 cells/L and even 10^7 cells/L. It is interesting to note that on 29 July 2004 there was no coccolithophore in the water sample taken in 74°27'N, 27°49'E in the interior of the red area (Figure 2), but were diatoms and dinoflagellates found in abundance: ~1.1 10^6 cells/L and 0.31 10^6 cells/L, respectively.

The analysis of the b_{bp} distributions in the Barents Sea derived from SeaWiFS data in 1998–2005 showed that position, contour, and area of the regions with high b_{bp} varied widely, and the connection between the distributions of b_{bp} and of the sea surface temperature (*SST*) was observed. The area with high b_{bp} was located within the area with *SST* higher than 7°C (mostly even 8°C). This is an area of the warm Norwegian Current propagation, and it explains why the coccolithophore bloom is not observed in the Northern Barents. The absence of the coccolithophore bloom in the eastern part of the Middle Barents supposedly connects with the salinity distribution: the climatological data (Climatic Atlas 1998) show that salinity in the eastern part is reduced due to propagation of the waters with low salinity from the White Sea and the Pechora Sea.

2.2 The Black Sea

The Black Sea as a whole basin was divided into eight sub-regions (Kopelevich *et al.* 2002). In the western shelf area we distinguish three inner (with depth of less than 50m) and two outer (depth of 50–200 m) shelf regions. Region #1 is under strong influence of water discharge of the Dnepr, Dnestr and Bug rivers, region #2 is influenced mainly by the Danube, region #3 is the south-western inner shelf. Regions #4 and 5 are the northern and southern outer shelf regions.

The general circulation of the Black Sea consists of the Rim Current following the narrow continental slope; the Rim Current limits the water and material transfer from coastal zone to the open Black Sea. The interior of the Rim Current zone is formed by two cyclonic cells occupying the western and eastern halves of the basin; for this reason the western (#6) and eastern (#7) open parts of the Black Sea were considered separately.

Region # 8 comprises the north-eastern, eastern, and southern shelf regions with depth less than 200m.

To investigate effects of river runoff on the open Black Sea, our present analysis is focused mainly on the variability of the particle backscattering coefficient, b_{bp}. Figure 3 shows its monthly mean values derived by the SIO RAS algorithms in the sub-regions #2, #6, #7, and #8. It is obvious that the changes of b_{bp} in all of the considered regions are in phase with each other, and the enhanced values of b_{bp} covering the whole basin are observed repeatedly in June.

A significant correlation (with a statistical probability more than 0.999) was found between changes of the b_{bp} values in the sub-regions #6, #7 and both #2, #8; the correlation coefficients $r[b_{bp}$-6, b_{bp}-2], $r[b_{bp}$-7, b_{bp}-2], $r[b_{bp}$-6, b_{bp}-8], and $r[b_{bp}$-7, b_{bp}-8] are equal to ~0.726, 0.730, 0.741, and 0.940, respectively. Variations of *Chl* in the sub-regions #6, 7 are weakly connected with the ones in the sub-region #2, and the corresponding correlation coefficients are low. A significant correlation is observed between the *Chl* changes in the sub-regions #6, 7 and #8; it can be explained by similarity of natural conditions in the central and eastern regions of the Black Sea.

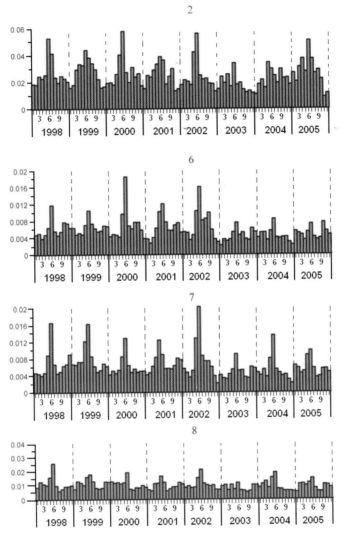

Fig. 3. Variability of the monthly mean values of the particle backscattering coefficient, m^{-1} derived by the SIO RAS algorithms in the sub-regions regions #2, #6, #7, and #8 of the Black Sea.

There are two hypotheses to explain the enhanced values of b_{bp} observed each June: (i) the particulate matter from the river runoff spreading to the whole basin across the boundary Rim Current by mesoscale eddies and via turbulent exchange; (ii) coccolithophore bloom covering the whole basin (Cokacar *et al.* 2001, 2004). To clarify that, the field studies were carried in the North-Eastern part of the Black Sea in 11–16 June 2004 (Figure 4).

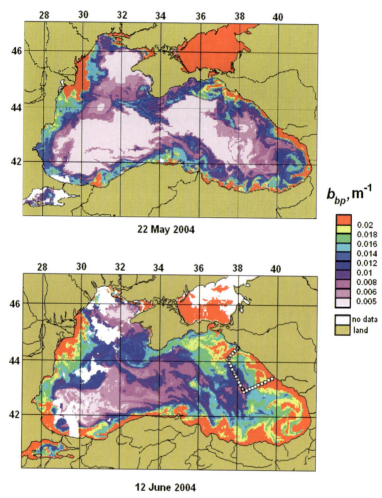

Fig. 4. The b_{bp} distributions in the Black Sea derived from MODIS-Aqua data on 22 May (left) and 12 June (right) 2004. On the latter white diamonds show the ship route and position of drift stations.

As is seen from Figure 4, the pronounced increasing of the particle backscattering coefficient b_{bp} started in the second half of May in the coastal zone; the particulate matter from there was transported to the open part by mesoscale eddies (Figure 4, top); three weeks later the water turbidity increased still further (Figure 4, bottom). The averaged values of b_{bp} in the eastern part of the Black Sea were highest just in the middle of June.

The joint analysis of satellite data and results of ship measurements (Burenkov *et al.* 2005) leads us to conclusion that enhancement of the particle content in the near-surface layer may be caused both by the coccolithophore bloom and the river runoff. The former is attested by a significant correlation between the coccolithophore concentration and the particle backscattering coefficient b_{bp}; the coefficient of determination (R^2) is equal to 0.65 (n = 21).

The river runoff is displayed best by low values of salinity, and Figure 5 shows that the changes of surface salinity S and of the particle backscattering coefficient b_{bp} along the ship track are in contrary directions. It is hard to separate out the considered effects because the river runoff has a pronounced impact on the coccolithophore bloom too.

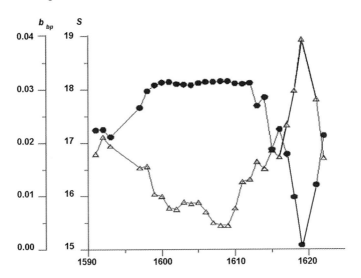

Fig. 5. Changes of surface salinity, S ‰ (circles) and the particle backscattering coefficient, b_{bp} m^{-1} (triangles) along the ship track (see Figure 4, bottom).

2.3 The Caspian Sea

By convention, the Caspian Sea is divided into the Northern (1), Middle (2) and Southern (3) regions (Fairbridge 1966). Kara-Bogaz-Gol with salinities of 300‰ is a very specific area and not considered here.

The Northern Caspian is a shelf area which depth is less than 10m. This region is under strong influence of river runoff; the largest European river Volga, the Ural and Terek rivers flow into the Northern Caspian. About 90% of the total river discharge in the Caspian Sea falls within the Northern region. The Middle Caspian is also influenced by the Volga river runoff. The Volga waters move mainly south along the western coast of the Caspian Sea and turn eastward near Apsheron forming cyclonic circulation in the Middle Caspian. The other part of the Volga waters comes into the Southern Caspian; this region is also influenced by the runoff of the Kura river and the rivers flowing down the Iranian coast.

Variability of the monthly mean values of chlorophyll concentration (*Chl*) in the Northern, Middle and Southern Caspian is shown in Figure 6.

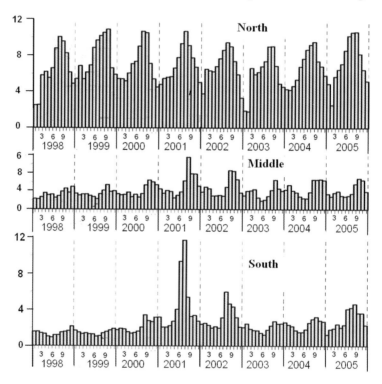

Fig. 6. Variability of the monthly means of chlorophyll-a concentration (mg m^{-3}) in the Northern, Middle and Southern Caspian Sea, derived using the SeaWiFS algorithms.

There is no strong interrelation between the *Chl* variations in the Northern and Middle sub-regions; the correlation coefficient $r[Chl$-N, Chl-M] is equal to 0.372. The connection between the *Chl* variations in the Middle and Southern sub-regions is closer: $r[Chl$-M, Chl-S$]$ is equal to 0.517.

The significant correlation was found between changes of the *Chl*, b_{bp} and a_g values within each of the sub-regions: the correlation coefficients $r[Chl, b_{bp}]$ ranged from 0.640 in the Northern Caspian to 0.985 in the Southern Caspian; $r[Chl, a_g]$ varied from 0.874 to 0.735, respectively.

The above results testify that the formation and variability of the bio-optical characteristics in each sub-region are mainly determined by processes within it, although the interrelationships between the sub-regions certainly exist. In particular, the above-mentioned coefficient of correlation $r[Chl$-M, Chl-S$]$ indicate a significant correlation (with a statistical probability more than 0.999) between the Middle and Southern Caspian.

The satellite data show sharp increase of *Chl* in the Middle and Southern Caspian in July-August 2001 (Figure 6) that was attributed to a consequence of invasion of the ctenophore *Mnemiopsis leidyi* (Kopelevich *et al.* 2004, 2005a,b). The level of chlorophyll concentration in the Southern Caspian in 2004 was similar to the one in 2000 and about 1.5 times higher than in 1998–1999. In 2005 anomalous algal bloom (AAB) occurred in the Southern Caspian in August-September; the algae responsible for the AAB was cyanobacteria *Nodularia*; the floating alga layer was tens centimeters thick (http://www.caspianenvironment.org/newsite/Caspian-AAB.htm).

Figure 7 shows the mean monthly distributions of *Chl* in the Caspian Sea in September 1999, 2000, 2001 and 2005. A similarity between the distributions of 2000 and 2005 can be observed; both of them and the distribution of 2001, especially, differ radically from the distribution of 1999 (before the *Mnemiopsis leidyi* invasion).

3. Conclusions

The results obtained demonstrate that satellite ocean color sensors provide us with important data to study variability of the bio-optical characteristics and the factors determining their changes.

In the Barents Sea the river runoff is localized in the shelf sub-region, and its impact on the open parts of the basin is not much pronounced. In the Southern Barents the seasonal changes of the bio-optical characteristics are mainly caused by the Pechora runoff, whereas in the Northern and Middle Barents they are determined by phytoplankton blooms. In the western part of the Middle Barents the coccolithophore blooms were revealed from satellite data and confirmed by direct determinations. The blooms were located within the area of the warm Norwegian Current where the sea surface temperature more 7°C.

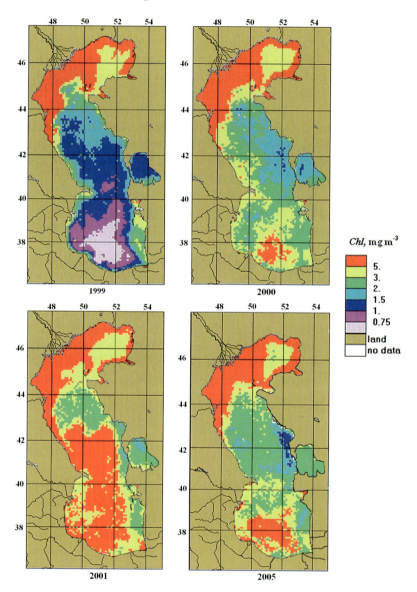

Fig. 7. The mean monthly distributions of chlorophyll concentration in the Caspian Sea in September 1999, 2000, 2001 and 2005.

In the Black Sea variations of the particle backscattering coefficient b_{bp} in the open parts and in the coastal regions are almost synchronous.

An interesting feature of the Black Sea revealed from satellite ocean color data is the enhanced b_{bp} values observed in June year after year. The joint analysis of satellite data and results of ship measurements leads us to conclusion that the enhancement of the particle content in the near-surface layer may be caused both by the coccolithophorid bloom and the river runoff. It is hard to separate out their effects because the river runoff has a pronounced impact also on the coccolithophore bloom.

In the Caspian Sea the Volga runoff determines the seasonal changes in the Northern sub-region and affects weakly the bio-optical characteristics in the Middle and Southern sub-regions. In the two last sub-regions intense phytoplankton blooms have been occurred in July-September since 2000 after invasion of the ctenophore *Mnemiopsis leidyi*; the increase in chlorophyll concentration was most pronounced in 2001.

Acknowledgements

The work was supported by the Russian Federal Agency on Science and Innovation. The authors are grateful to Vladimir Artemiev, Anatoly Grigoriev, Andrey Ivanov, Alexander Khrapko, Olga Prokhorenko, Tatiana Rat'kova, and Valery Taskaev for their contributions to this work.

References

Burenkov VI, Kopelevich OV, Sheberstov SV, Ershova SV, Evdoshenko MA (1999) Bio-optical characteristics of the Aegean Sea retrieved from satellite ocean color data. In: Malanotte-Rizzoli P., Eremeev VN (eds) The Eastern Mediterranean as a Laboratory Basin for the Assessment of Contrasting Ecosystems. Kluwer Academic Publishers, Netherlands, pp 313–326

Burenkov VI, Kopelevich OV, Sheberstov SV, Vedernikov VI (2000) Sea-truth measurements of ocean color: Validation of the SeaWiFS satellite scanner data. Oceanology 40: 329–334 (translated from Okeanologiya 40: 357–362)

Burenkov VI, Vedernikov VI, Ershova SV, Kopelevich OV, Sheberstov SV (2001a) Use of data from satellite ocean color scanner SeaWiFS for assessment of bio-optical characteristics of the Barents Sea. Oceanology 41: 461–468 (translated from Okeanologiya 41: 485–492)

Burenkov VI., Ershova SV., Kopelevich OV., Sheberstov SV., Shevchenko VP (2001b) An estimate of the distribution of suspended matter in the Barents Sea waters on the basis of the SeaWiFS satellite ocean color scanner. Oceanology 41: 622–628 (translated from Okeanologiya 41: 653–659)

Burenkov VI, Vazyulya SV, Kopelevich OV, Sheberstov SV (2004) Spatio-temporal Variability of the Suspended Matter Distribution in the Surface

Layer of the White Sea from the Data of the SeaWiFS Satellite Color Scanner. Oceanology 44: 461–468 (translated from Okeanologiya 44: 507–515)

Burenkov VI, Kopelevich OV, Pautova LV, Prokhorenko OV, Rusakov VY, Sheberstov SV (2005) Possible causes of the increased content of suspended particles in the northeastern part of the Black Sea in June. Oceanology 45, Suppl. I: S39–S50

Climatic Atlas (1998) Climatic Atlas of the Barents Sea 1998: Temperature, Salinity, Oxygen. Murmansk Marine Biological Institute, Russia. National Oceanographic Data Center Ocean Climate Laboratory, USA (CD-ROM)

Cokacar T, Kubilay N, Oguz T (2001) Structure of E. huxleyi blooms in the Black Sea surface waters as detected by SeaWiFS imagery. Geophys Res Lett 28: 4607–4610

Cokacar T, Oguz T, Kubilay N (2004) Interannual variability of the early summer coccolithophore blooms in the Black Sea: impacts of anthropogenic and climatic factors. Deep Sea Research Part 1: Oceanographic Research Papers 51: 1017–1031

Fairbridge RW, ed. (1966) Encyclopedia of Oceanography. Reinhold Publishing Corporation, New York

IOCCG (2000) Remote Sensing of Ocean Colour in Coastal, and Other Optically-Complex, Waters. Sathyendranath S. (ed.), Reports of the International Ocean-Colour Coordinating Group, No. 3, IOCCG, Dartmouth

Kopelevich OV, Sheberstov SV, Yunev O, Basturk O, Finenko ZZ, Nikonov S, Vedernikov VI (2002) Surface chlorophyll in the Black Sea over 1978–86 derived from satellite and in situ data. J Mar Systems 36: 145–160

Kopelevich OV, Burenkov VI, Ershova SV, Sheberstov SV, Evdoshenko MA (2004) Application of SeaWiFS data for studying variability of bio-optical characteristics in the Barents, Black and Caspian Seas. Deep-Sea Research II 51: 1063–1091

Kopelevich OV, Burenkov VI., Sheberstov SV., Lukianova EA, Prokhorenko OV (2005a) Construction of the long-term series of data in the bio-optical characteristics of the Russian seas from satellite ocean color data. In: Levin I, Gilbert G (eds) Current Problems in Optics of Natural Waters (ONW'2005). St.Peterburg. pp 293–298

Kopelevich OV, Burenkov VI., Sheberstov SV., Lukianova EA, Prokhorenko OV (2005b) Bio-optical characteristics of the seas of Russia from data of the SeaWiFS satellite ocean color scanner. CD-ROM. SIO RAS, Moscow

Lisitsyn AP (2004) World Ocean Geology in the third millennium – new approaches, achievements, and prospects. In: Vinogradov ME, Lappo SS (eds) New ideas in oceanology, vol. 2. Nauka, Moscow (in Russian)

Roshydromet (2001) River estuaries in the Barents Sea. CD-ROM. CD-5. Roshydromet, GOIN (in Russian), 2.8 MB

Variations in the Phytoplankton of the North-Eastern Atlantic Ocean: from the Irish Sea to the Bay of Biscay

Samantha J. Lavender[1,2], Dionysios E. Raitsos[1,3], and Yaswant Pradhan[1,2]

[1]SEOES & Marine Institute, University of Plymouth, Plymouth, UK
[2]Centre for Observation of Air-Sea Interaction & Fluxes (CASIX), Plymouth, UK
[3]SIR-Alister Hardy Foundation for Ocean Science, Plymouth, UK

Abstract. The North-Eastern Atlantic, including the western part of the European continental shelf, was studied by splitting the region up into 6 biogeographical areas. Biomass (chlorophyll-a concentration) and physical forcings (light, sea surface temperature and wind stress) were extracted from satellite data sources. The analysis included also Continuous Plankton Recorder information on phytoplankton functional groups. The results showed differences in biomass magnitude and phytoplankton group dominance, for the different areas, linked through the physical forcings.

1. Introduction

The majority of the Earth's surface (~71%) is covered by the aquatic environment, of which ~97% is oceanic and inhabited by microscopic plants called phytoplankton. Phytoplankton contribute to greater than 45% of the total primary production by plants on earth (Falkowski *et al.* 2004), influencing greenhouse gas (*e.g.* carbon dioxide) concentrations in the atmosphere and contributing to the biological pump. Also, being at the bottom of the food chain, these ecological drivers transfer energy to higher levels of the marine food web (Irigoien *et al.* 2002) and thus influence the biodiversity trends of other organisms such as zooplankton, fish, seabirds and marine mammals (Paerl *et al.* 2003).

Knowing distribution and abundance of phytoplankton biomass, both on spatial and temporal scales, is therefore of importance in understanding climate change and variability in the marine biological pump. Recent work has highlighted that there was a positive shift in the central North-Eastern (NE) Atlantic (Raitsos *et al.* 2005) chlorophyll levels during the 1980's. In this study, the NE Atlantic was divided in six areas so that the physical

V. Barale, M. Gade (eds.), *Remote Sensing of the European Seas.*

forcings that drive phytoplankton in these areas could be examined through a synergistic combination of *in situ* and satellite datasets.

The satellite data provided biological and physical information through a combination of remote sensing missions. However, as satellite datasets do not systematically provide functional type information (an active area of research; see *e.g.* Sathyendranath *et al.* 2004; Alvain *et al.* 2005), *in situ* data were used to understand the biomass fluctuations. The *in situ* ecological information was acquired from the Continuous Plankton Recorder (CPR), which is an upper layer plankton-monitoring programme running in the North Atlantic since 1931 (Reid *et al.* 2003).

2. Methodology

2.1 Geographical coverage

Figure 1 shows the NE Atlantic including the western part of the European continental shelf divided into six areas. As well as showing the areas in relation to the land masses, the 1000 and 100m bathymetric contours and CPR sampling points have been included to aid interpretation. The divisions were made subjectively, but with a knowledge of the biophysical regimes that occur within these locations; separation of the continental shelf water from open ocean water in a latitudinal context, within the limits of rectangular divisions and sufficient CPR data.

Area 5 is primarily Atlantic ocean with a small part of the continental shelf in the north eastern and south eastern corners (Figure 1). Area 3 is north of this and contains a greater percentage of the continental shelf; termed the Celtic Sea. The Celtic Sea is a transition zone (with distinct tidal-mixing fronts) between the Atlantic, shelf sea and coastal waters of the Bristol Channel and Irish Sea, and from May to November this region is dominated by thermal stratification (Pingree 1980). Area 1 is also oceanic and continental, including the western coast of Ireland and Porcupine bank (visible in Figure 1 as a westward extension of the 1000 m contour). Townsend *et al.* (1994) used biological simulation modeling to show that in offshore and open ocean waters the spring bloom is related to the weather, through controls on the light field and wind mixing. In shallow, coastal waters incident irradiation (cloudiness) was the dominant factor.

Area 2 is the Irish Sea, Bristol Channel and Celtic Sea (Figure 1). The Irish Sea is optical complex as a result of differences in tidal mixing, freshwater inflow and bathymetry that creates distinct hydrographic regions (Gowen *et al.* 1995). The offshore waters of the western Irish Sea become seasonally stratified each year (typically develops in early May), and to the south and north of the stratified region tidal mixing is sufficient to ensure that the water column is vertically mixed throughout most of the year (Tilstone *et al.* 2005). Seasonal stratification (with a strong semidiurnal variation and tidal mixing fronts) is also experienced by Liverpool Bay and Cardigan Bay (Simpson *et al.* 1990).

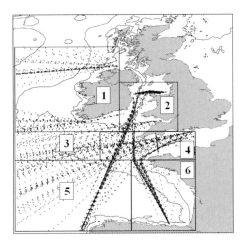

Fig. 1. Geographical locations for the bio-geographical areas (1 to 6) within the North-East Atlantic. Solid lines represent bathymetric contours (100 and 1000 m) and dots represent Continuous Plankton Recorder sampling points.

Area 4 is primarily the western English channel with a small part of the outer Celtic Sea. The western English channel undergoes an annual cycle of stratification in the spring through to autumn, and then undergoes wind driven mixing through the winter (Holligan and Harbour 1977; Pingree *et al.* 1977). Area 6 is primarily the Bay of Biscay, which is dominated by the continental shelf and has riverine inputs from estuaries such as the Gironde. Gohin *et al.* (2003) suggested that the freshwater extension of large rivers play an important part in physical forcing, which radically modifies the nutrient and light conditions of the oceanic environment. Longhurst (1998) suggested this area has 4 ecological seasons: mixed conditions and light limitation in winter; a nutrient-limited spring bloom; stratified conditions during summer, with a deep chlorophyll maximum; a second general bloom, when the autumn gales break down the summer stratification.

2.2 Satellite data

Sea-viewing Wide Field-of-view Sensor (SeaWiFS), OrbView-2. Version 5.1 level–3 monthly composites (9×9 km^2) for Chlorophyll *a* (Chl-a) and

Photosynetically Available Radiation (PAR) were acquired from the NASA Oceancolor website (http://oceancolor.gsfc.nasa.gov/). Ocean Colour 4 version 4 (OC4v4) is the standard Chl-a algorithm (mg m^{-3}), which has primarily been developed for Case 1 waters; previously defined as having Chl-a as the dominant optical constituent with additional optically important constituents, such as coloured dissolved organic matter (CDOM), co-varying. While shelf sea and coastal waters are often characterised as Case 2 because Chl-a is not the dominant optical constituent and so global chlorophyll algorithms (such as OC4v4) are less reliable (IOCCG, 2000). As an example, the chlorophyll concentration in the Bay of Biscay has been routinely retrieved from SeaWiFS data using a look-up table as described by Gohin *et al.,* (2002). However, for this paper the OC4v4 Chl-a algorithm was used and the optical properties were taken into account during the analysis. Finally, PAR (E m^{-2} day^{-1}) is used to indicate the incoming solar radiation or insolation that can generally be defined as the light intensity received at the surface of the Earth.

ERS-2 altimeter and SeaWinds, QuikSCAT (QS). Monthly mean ERS-2 and QS wind stress ($0.5° \times 0.5°$) were acquired from the IFREMER CERSAT website (http://www.ifremer.fr/cersat/en/index.htm). Preliminary comparisons of wind speed indicate that the sensors are compatible. Wind stress, which is responsible for vertical mixing in the water column, may have an indirect effect on phytoplankton biomass through nutrient availability as the spatial variation of wind stress over the ocean causes surface divergence of horizontal flow that in turn gives rise to vertical mass flux through Ekman pumping (Pond and Pickard 1983).

Advanced Very High Resolution Radiometer (AVHRR), NOAA satellites. The night-time AVHRR Pathfinder 5 monthly mean Sea Surface Temperature (SST) at 4×4 km^2 resolution were obtained from the NASA PO-DAAC website (http://poet.jpl.nasa.gov/). Only the night-time SST products were used so that the solar radiation bias (the diurnal fluctuation in SST) that can occur during the day-time could be avoided.

2.3 *In situ* data

Measurements of phytoplankton abundance (cell counts) were derived from the CPR survey. Samples are collected by a high-speed (~15–20 knots) plankton recorder towed behind 'ships of opportunity' in the surface layer of the ocean (~6–10 m depth), so one sample represents ~18 km of tow (Richardson *et al.* 2006). The plankton are filtered onto a constantly moving band of silk and the laboratory analysis involves the taxonomic identification of plankton and species cell counts for each sample.

Phytoplankton cells retained on the filtering silk, including those that disintegrate, produce a greenish colouration (Reid *et al.* 2003; Robinson *et al.* 1986). So, prior to species analysis, a visual estimation is undertaken with reference to a standard colour chart. This Phytoplankton Colour Index (PCI) is given four different 'greenness' values: 0 (no greenness, NG), 1 (very pale green, VPG), 2 (pale green, PG) or 6.5 (green, G). PCI was then converted to PCI Chl-a (mg m^{-3}) using NE Atlantic SeaWiFS matchups: NG = 1.03 ± 0.21 mg m^{-3}; VPG = 1.65 ± 0.16 mg m^3; PG = 2.61 ± 0.29 mg m^{-3}; G = 4.25 ± 0.98 mg m^{-3} (Raitsos *et al.* 2005).

2.4   Analysis methods

The satellite datasets were downloaded from September 1997, as this is the start of the SeaWiFS data, until December 2005. However, the CPR data was not available after December 2004 at the time of writing. The additional SeaWiFS year gives some additional information on the long term trends that (as with all datasets) are enhanced as further data is available.

The CPR species data was collated (total number per sample) into functional groups which are defined here as diatoms, dinoflagellates, coccolithophores and silicoflagellates as each group is composed of many species; detailed list in Richardson *et al.* (2006). Each individual product was then geographically split into the areas (Figure 1) and averaged spatially to create monthly values. The number of samples included within a satellite monthly average will be variable. For SeaWiFS, it is assumed that clouds do not statistically affect the results, but there will be a latitudinal effect from the solar zenith angle cut-off in the processing. Area 1 will loose some of its northernmost pixels in November and January, and in December only the southern half of areas 5 and 6 will be covered. In addition, the CPR data does not have a regular spatial distribution, see Figure 1, and so the spatial average will not be contributed to by the whole area.

3.   Results

Figure 2 shows the seasonal variation for the climatological (1997 to 2004) monthly CPR cell counts and Chl-a as described in Sections 2.3 and 2.4. All of the areas appeared to have a spring bloom with peaks for areas 1 and 6 in April, and the remaining areas in May. In general, the cell counts followed the biomass concentration, but the peak for area 3 is in April rather than May. It can also be seen that in all areas diatoms are the dominant (hence blooming) group in the spring. Distinct summer blooms, according to the Chl-a, are evident in areas 1 and 6 (Figure 2).

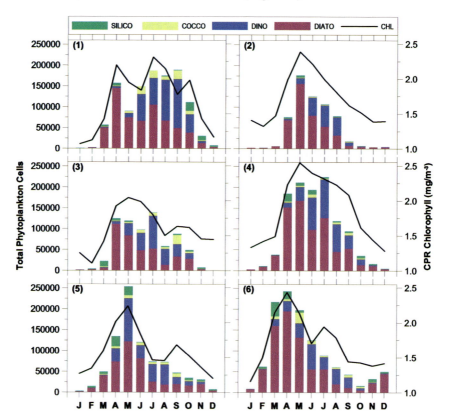

Fig. 2. Seasonal variation for the 6 areas (of Figure 1) as Continuous Plankton Recorder (CPR) total cell counts, split into functional groups (SILICO silicoflagellates; COCCO coccolithophores; DINO dinoflagellates; DIATO diatoms); PCI chlorophyll-a (CHL) concentrations (mg m^{-3}) as solid lines.

For area 1 there is an increase in the diatoms that can account for the summer peak, but for area 6 there is no obvious increase for any group. The lack of a cell count peak could be explained by an increase in the nanoplankton that would not be counted by the CPR analysis techniques. For area 1, the summer bloom also shows high cell counts from July to September that are dominated by dinoflagellates. This persistence of high numbers is also evident within area 4 (April to July) which shows a sustained diatom abundance, but also increase in dinoflagellates. Areas 1, 3 and 5 have a distinct autumn bloom according to the Chl-a (Figure 2); no obvious increase in any group and so nanoplankton could be contributing.

Figure 3a compares the SeaWiFS and CPR PCI (see Section 2.3) derived Chl-a concentration as time series plots showing the seasonal variation within the six areas. The plots were created using standard Kriging interpolation (uses linear least squares estimation algorithms) methods within the Golden Software Surfer (v8). Figure 3b is the SeaWiFS Chl-a Spring climatology produced by NASA overlaid with 100 and 1000m bathymetric contour lines. Similar patterns to Figure 2 can been seen in the seasonal variability, but Figure 3a also highlights inter-annual variations. Areas 2 and 4 have the highest overall concentrations and are dominated by continental shelf waters (Figure 1). The lowest concentrations are seen in areas 3 and 5 that are both dominated by oceanic waters. Areas 1 and 6 are a mixture of continental shelf water (that experience strong blooms) together with oceanic waters, *e.g.* the Porcupine Bank that can be seen as an area of high Chl-a concentrations west of Ireland (Figure 3b).

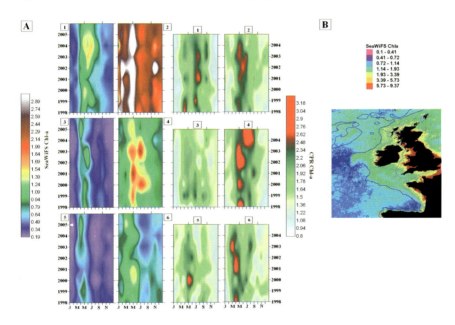

Fig. 3. [A] Time series plots of SeaWiFS (1997–2005) and Continuous Plankton Recorder (1997–2004) averaged Chl-a concentrations (mg m^{-3}) for the 6 areas (of Figure 1), and [B] SeaWiFS Chl-a (mg m^{-3}) Spring climatology (1997–2005), with bathymetric contour lines (100 and 1000m).

In terms of the seasonal variability similar patterns to Figure 2 can be seen, but Figure 3a also highlights inter-annual variations. Areas 2 and 4 have the highest overall concentrations and are dominated by continental shelf waters (Figure 1). The lowest concentrations are seen in areas 3 and 5 that are both dominated by oceanic waters. Areas 1 and 6 are a mixture, they have areas of continental shelf water that experience strong blooms together with oceanic waters, *e.g.* the Porcupine Bank in area 1 that can be seen as an area of high Chl-a concentrations west of Ireland (Figure 3b).

In the SeaWiFS Chl-a (Figure 3a), the spring bloom in area 1 has a consistent starting period, but the length of the blooming period varies with longer blooms in the later years (after 2001). However, area 6 appears to have an opposite pattern with a decrease in the intensity/length of the spring bloom in the latter years. For the CPR Chl-a the pattern is different, a strong spring bloom in the 1990s, with a continuous summer bloom. In area 6, there is a continuous spring bloom with a possible increase in the summer Chl-a levels. Areas 3 and 5 have a strong spring bloom in the SeaWiFS Chl-a with a much weaker increase in the autumn; this pattern is also evident in the CPR Chl-a. The SeaWiFS Chl-a for areas 2 and 4 has consistently, throughout the year, high concentrations. In area 2, the CPR Chl-a shows a spring bloom, but low concentrations during the winter. Area 4 shows high CPR Chl-a initially during the spring, but from 2001 onwards this seems to have shifted more towards the summer.

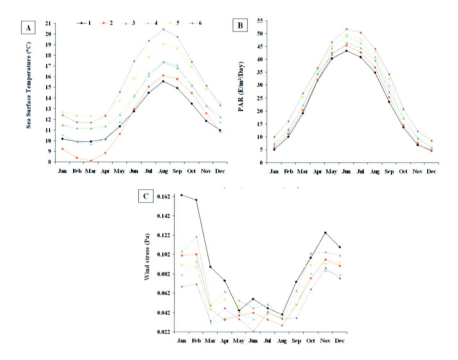

Fig. 4. Seasonal plots of [A] SST (°C) [B] PAR (E m^{-2} Day^{-1}) and [C] wind stress (Pa) for the 6 areas (of Figure 1), averaged for the area.

Plotting the seasonal cycles of the physical forcings offers the opportunity for a preliminary assessment of the environmental conditions occurring in relation to phytoplankton functional groups (Figure 2). Figure 4 shows the seasonal variation in SST, PAR and wind stress derived from satellite data sources as monthly mean values (1997–2005 average) for the areas shown in Figure 1. Figure 4a shows a strong seasonal cycle for the SST with all areas having their lowest value in February or March and their highest in August. From December to April there is a trough in the values, which is less well defined than the peak of the high temperatures in August. From April, the rate of change of SST increases dramatically and this steep increase continues until August before a sharp fall.

For PAR (Figure 4b), there is also a strong seasonal cycle with the lowest values in December and January and the highest values in June. The peak is in the spring and summer, April to September. Wind stress (Figure 4c) has a much more variable cycle and has an inverse pattern to PAR; wind stress is lowest during spring and summer and highest during autumn and winter. From March to December, areas 1, 3 and 5 tend to have the highest values while the lowest values are for area 6, then areas 4 and 2.

4. Discussion and conclusions

The succession of phytoplankton groups has been described as being typically characterised by an abundance of diatoms during the spring bloom, progressing to dominance by dinoflagellates in summer and back to diatoms in the autumn (Holligan and Harbour 1977; Horwood *et al.* 1982). The CPR seasonal plots (Figure 2) show this pattern of high diatoms in the spring and increasing dinoflagellates with the progression to summer, but a strong autumn diatom bloom is not so evident. Biological factors attributed to diatom blooming includes their high maximum growth rates (Furnas 1990; Egge and Aksnes 1992) and growth being unchecked because they are primarily grazed by mesozooplankton (Paffenhöfer 2002) that generally lag phytoplankton growth (Kelly-Gerreyn *et al.* 2004). In comparing SeaWiFS and CPR Chl-a (Figure 3) there are similarities, but differences can be attributed to the sampling methodologies and hence biases. SeaWiFS OC4v4 can perform less accurately in coastal waters and the CPR data is not being systematically spatially sampled (see Figure 1).

The consistently high SeaWiFS Chl-a for areas 2 and 4 may reflect poor algorithm performance (see Section 2.2). The magnitude of the August maximum SST for each area (Figure 4a) is strongly dependent on latitude (lowest latitudes have the highest temperatures), but the latitudinal pattern of the minimum SST is more variable. In winter, areas 2 and 4 are much colder than areas 1and 3 respectively, and to a lesser extent area 6 is colder than 5: probably a result of riverine input. A latitudinal dependence is also evident for PAR (Figure 4b), with the highest values in the southernmost areas. The highest wind stress values (Figure 4c), *i.e.* high vertical mixing and hence low stratification, correspond to oceanic areas (Figure 1).

In the NE Atlantic, Otto *et al.* (1990) suggested that stratification depth generally increases with latitude while solar radiation decreases. Thus, stratification, and therefore the spring bloom occurs later at higher latitudes. In Figure 4, stratification corresponds to low wind stress with a significant drop between February and March for all areas (Figure 4c). PAR also increased between February and March (Figure 4b), and an increase in the SST starts to occur between March and April (Figure 4a). This corresponds to a bloom initiation in March and April (Figure 2), but a latitudinal change was not evident. The succession goes from diatoms (opportunists), to dinoflagellates (stress-tolerators) and coccolithophores that are adapted to low nutrient and stable waters (Margaleff 1978; Drinkwater *et al.* 2003). An intermediate stage is flagellates, but these were not included. However, in coastal areas the water can remain well mixed and unstratified due to tidal turbulence (Pingree & Griffiths 1978). This would allow for a continued abundance of diatoms in summer that was clearly evident in area 4 (Figure 2). The phytoplankton will also retreat northwards and areas 1, 3 and 5 show that from September onwards diatoms/dinoflagellates are higher in the more northerly areas (Figure 2).

Continued work is needed if conclusions are not to be unduly biased by the intrinsic properties of a single dataset. This work has shown that by synergistically analysing satellite and *in situ* data an improved understanding of phytoplankton variation can be obtained. Further work includes advanced statistical techniques, such as neural networks, to predict the phytoplankton functional groups from ecological and physical parameters.

Acknowledgements

We thank staff of SAHFOS who have contributed to the maintenance of the CPR time series. Satellite data sources: SeaWiFS Project; NASA OBPG; NASA PO-DAAC; CERSAT at IFREMER. D.E. Raitsos was funded by a University of Plymouth scholarship. This research was supported by the UK Natural Environment Research Council through CASIX.

References

Alvain S, Moulin C, Dandonneau Y, Breon FM (2005). Remote sensing of phytoplankton groups in case 1 waters from global SeaWiFS imagery. Deep Sea Research Part I 52:1989–2004

Drinkwater KF, Belgrano A, Borja A, Conversi A, Edwards M, Greene CH, Ottersen G, Pershing AJ, Walker H (2003) The response of marine ecosystems to climate variability associated with the North Atlantic Oscillation. Geophysical Monograph 134: 211–234

Egge JK, and Aksnes DL (1992) Silicate as regulating nutrient in phytoplankton competition. Marine Ecology-Progress Series 83: 281–289

Falkowski PG, Katz ME, Knoll AH, Quigg A, Raven JA, Schofield O, Taylor FJR (2004) The evolution of modern eukaryotic phytoplankton. Science 305: 354–360

Furnas, MJ (1990) *In situ* growth-rates of marine-phytoplankton as approaches to measurement, community and species growth-rates. Journal of Plankton Research 12: 1117–1151

Gohin F, Druon JN, Lampert L (2002) A five channel chlorophyll concentration algorithm applied to SeaWiFS data processed by SeaDAS in coastal waters. International Journal of Remote Sensing 23: 1639–1661

Gohin F, Lampert L, Guillauda J-F, Herbland A, Nézand E (2003) Satellite and *in situ* observations of a late winter phytoplankton bloom, in the northern Bay of Biscay. Continental Shelf Research 23: 1117–1141

Gowen RJ, Stewart BM, Mills DK *et al.,* (1995) Regional differences in stratification and its effect on phytoplankton production and biomass in the northwestern Irish Sea. Journal of Plankton Research 17: 753–769

Holligan PM, Harbour DS (1997) The vertical distribution and succession of phytoplankton in the western English channel in 1975 and 1976. Journal of the Marine Biological Association of the UK 57: 1075–1093

Horwood JW, Nichols JH, Harrop R (1982) Seasonal changes in net phytoplankton of the west-central North Sea. Journal of the Marine Biological Association of the UK 62: 15–23

IOCCG (2000) Remote Sensing of Ocean Colour in Coastal, and Other Optically-Complex Waters, Sathyendranath, S. (Eds.), Reports of the International Ocean-Colour Coordinating Group, No. 3. IOCCG. Dartmouth, Canada

Irigoien X, Harris RP, Verheye HM, Joly P, Runge J, Starr M, Pond D, Campbell R, Shreeve R, Ward P, Smith AN, Dam HG, Peterson W, Tirelli V, Koski M, Smith T, Harbour D, Davidson R. (2002) Copepod hatching success in marine ecosystems with high diatom concentrations. Nature: 419 (6905): 387–389

Kelly-Gerreyn BA, Anderson TR, Holt JT, Gowen RJ, Proctor R (2004) Phytoplankton community structure at contrasting sites in the Irish Sea: A modelling investigation. Estuarine, Coastal and Shelf Science 59 (3): 363–383

Longhurst, A (1998) Ecological Geography of the Sea. Academic Press, San Diego, California

Margaleff R (1978) Life forms of phytoplankton as survival alternatives in an unstable environment. Oceanologica Acta. 1: 493–509

Nybakken JW (1997) Marine Biology: An ecological approach, 4th ed. Addison-Wesley Educational publishers Inc

Otto L, Zimmerman JTF, Furnes GK, Mork M, Saetre R, Becker G (1990) Review of the physical oceanography of the North Sea. Netherlands Journal of Sea Research 26:161–238

Paffenhöfer, GA (2002) An assessment of the effects of diatoms on planktonic copepods. Marine Ecology-Progress Series 227: 305–310

Pingree, RD (1980) Physical oceaonography of the Celtic Sea and English Channel. In: Banner, FT, Collins, WB and Massie, KS, Eds. The North-west European Shelf Seas: The Sea Bed and the Sea in Motion. II Physical and Chemical Oceanography and Physical Resources, Elsevier

Pingree RD, Griffiths DK (1978) Tidal fronts on the shelf seas around the British Isles. Journal of Geophysical Research 83: 4615–4622

Pingree RD, Maddock L, Butler EI (1977) The influence of biological activity and physical stability in determining the chemical distributions of inorganic phosphate, silicate and nitrate. Journal of the Marine Biological Association of the UK 57: 1065–1073

Pond S, Pickard GL (1983). Introductory dynamical oceanography. Fonte: Oxford; Butterworth-Heinemann

Raitsos DE, Reid PC, Lavender SJ, Edwards M, Richardson AJ (2005) Extending the SeaWiFS chlorophyll data set back 50 years in the northeast Atlantic. Geophysical Research Letters 32: L06603, doi:10.1029/2005GL022484

Reid PC, Matthews JBL, Smith MA (Eds.) (2003). Achievements of the Continuous Plankton Recorder survey and a vision for its future. Progress in Oceanography 58: 115–358

Richardson AJ, Walne AW, John AWG, Jonas TD, Lindley JA, Sims DW, Stevens D, Witt M (2006). Using continuous plankton recorder data. Progress in Oceanography 68 : 27–74

Robinson, GA, Aiken, J, Hunt HG (1986) Synoptic surveys of the western English Channel. The relationships between plankton and hydrography. Journal of Marine Biological Association of the United Kingdom 66: 201–218.

Sathyendranath S, Watts L, Devred E, Platt T, Caverhill C, Maass H. (2004). Discrimination of diatoms from other phytoplankton using ocean-colour data. Marine Ecology Progress Series 272: 59–68

Simpson JH, Brown J, Matthews J and Allen G (1990). Tidal Straining, Density Currents, and Stirring in the Control of Estuarine Stratification. Estuaries 13: 125–132

Tilstone GH, Smyth TJ, Gowen RJ, Martinez-Vicente V, Groom SB (2005) Inherent optical properties of the Irish Sea and their effect on satellite primary production algorithms. Journal of Plankton Research 27: 1127–1148

Townsend DW, Cammen LM, Holligan PM, Campbell DE, Pettigrew NR (1994). Causes and consequences of variability in the timing of spring phytoplankton blooms. Deep Sea Research Part I. 41: 747–765

Optical Remote Sensing of the North Sea

Kevin Ruddick[1], Geneviève Lacroix[1], Christiane Lancelot[2],
Bouchra Nechad[1], Youngje Park[1], Steef Peters[3], and Barbara Van Mol[1]

[1] Management Unit of the North Sea Mathematical Models, Royal Belgian
Institute for Natural Sciences, Brussels, Belgium
[2] Ecologie des Systèmes Aquatiques, Université Libre de Bruxelles, Brussels,
Belgium
[3] Instituut voor Milieuvraagstukken, Vrije Universiteit, Amsterdam, The
Netherlands

Abstract. Optical remote sensing in the North Sea is reviewed with a focus
on applications supporting environmental management. Optical remote
sensing provides estimates of Chlorophyll *a*, total suspended matter and
diffuse attenuation coefficient and related parameters. These are used for
harmful algal bloom detection, eutrophication assessment, ecosystem and
sediment transport modeling, and estimation of air-sea carbon fluxes.

1. Introduction

The objective of this section is to describe the applications currently being
addressed by optical remote sensing in the North Sea. The subject is
treated from the point of view of providing information that is needed by
users, such as environmental managers. Scientific and technological problems that affect data quality or availability are explained, but the detailed
description of the scientific data processing algorithms is left to other
works. The scope is limited to passive optical remote sensing, thus excluding LIDAR systems. The focus is on satellite systems.

1.1 Geophysical parameters

Optical remote sensing allows mapping of colour-related geophysical parameters. In deep oceanic waters phytoplankton is the main cause of variability of water colour, which goes from blue to green as Chlorophyll *a*
(Chl-a) and related pigments increase. In the North Sea, water colour is
also affected by: Coloured Dissolved Organic Matter (CDOM), originating

79

V. Barale, M. Gade (eds.), *Remote Sensing of the European Seas.*

primarily from the degradation of terrestrial vegetation and reaching the sea via rivers, and Non-Algae Particles (NAP), such as resuspended bottom sediments or particles discharged by rivers. For some sediment transport applications, the Total Suspended Matter (TSM) concentration is preferred to NAP. Further parameters that can be mapped from optical remote sensing include those related to water transparency (Doron *et al.* 2005) such as the diffuse attenuation coefficient (Kd). This can be considered spectrally or for broadband Photosynthetically Available Radiation (PAR) and is related to the commonly measured Secchi depth. Simple Red-Green-Blue (RGB) colour composite images are also often used in support of satellite-derived parameter maps. In addition to Chl-a, information on species composition and primary production may be derived for some applications.

Other phenomena such as floating layers of algae, oil slicks, whitecaps and waves may have optical signatures but are not considered here because rarely used in applications. For regions of clear water where the sea bottom is visible it may also be possible to determine to some extent properties of benthic fauna such as seagrass beds or coral reefs and/or water depth. However, for the North Sea the sea bottom is generally not visible from the surface and such applications are rare except for the extreme case of mapping intertidal vegetation (Thomson *et al.* 2003). At very high spatial resolution the detection of targets, such as ships or marine mammals becomes possible but is generally not cost-effective.

1.2   The north sea

The North Sea is a semi-enclosed temperate sea open to the Atlantic Ocean in the North and the South-West and to the Baltic Sea in the East. In terms of bathymetry the area can be broadly subdivided into the Channel (~50–100 m), the Southern North Sea (<50 m), the Central and Northern North Sea (~50–200 m) and the much deeper Norwegian trench (>200 m), which extends from the Norwegian coast to ~200 km offshore. This is a region of strong, mainly semi-diurnal tides. The shallow Southern North Sea is vertically well-mixed throughout the year except for limited regions such as river plumes. The Channel and the Central/Northern North Sea stratify thermally in summer with tidal mixing fronts running across the central North Sea separating well-mixed from stratified water. The physical oceanography of the region is reviewed by (Otto *et al.* 1990).

TSM concentrations are high in the shallow coastal waters of the Southern North Sea and lower offshore, in the Skagerrak and in the Central/Northern North Sea (Eisma 1981). Suspended sediments are supplied to the North Sea from a variety of sources: the Atlantic Ocean, and the Baltic, rivers, coastal erosion, seafloor erosion, dredging and mining operations, the atmosphere (dust) and primary production. Concentrations range from more than 100 mg/l in nearshore waters off Belgium, the Netherlands and South East England (Figure 1) to less than 0.2 mg/l in the northern North Sea and the Channel. These concentrations are highly variable in response to the seasonal and short-term variations in water movement and primary production.

The North Sea is a highly productive ecosystem. Nutrient concentrations have a strong seasonal cycle (Prandle *et al.* 1997), with a peak in December-January. Concentrations decline rapidly with the onset of the primary production during the spring bloom, until June-July when nutrients become limiting and cause a decline in phytoplankton production. Primary production varies considerably across the North Sea according to water depth and nutrient and light availability. The annual average Chl-a is shown in Figure 1 and temporal variability of the 2003 spring bloom is shown in Figure 2.

2. Applications

The North Sea coastline has a high population density, particularly along the continental European coast from France to Denmark. Intensive industrial and agricultural activities cause environmental impacts through the release of pollutants. Other activities with a major impact include fisheries and aquaculture, recreation and tourism, shipping, port construction and exploitation, sand mining, oil-drilling from platforms and offshore wind farming. The exchange of carbon between the sea and the atmosphere, and between the sea and land (via rivers), is an important part of the global carbon cycle. For most of these issues, the impacts of human activities must be assessed in a context of sustainable development. For some, as summarised in Table 1, optical remote sensing can provide information to assist in decision-making. The main market for optical remote sensing applications is composed by governmental users, particularly water quality managers. Another key user group is that of marine scientists, who benefit from the spatial perspective and relatively frequent data available from satellite remote sensing. These data are often combined with *in situ* measurements and/or mathematical models. Private industry may also use imagery, though generally in support of public authorities.

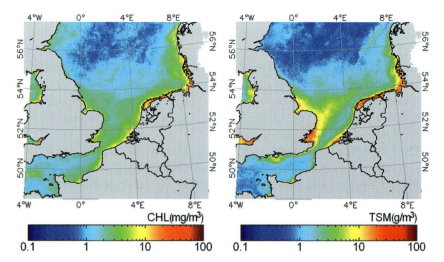

Fig. 1. Annual mean (left) Chl-a and (right) TSM distribution from 2003 MERIS data. Data provided by the European Space Agency (ESA) in the framework of AOID 3443. Processing includes the removal of low quality data as indicated by the standard product confidence flag.

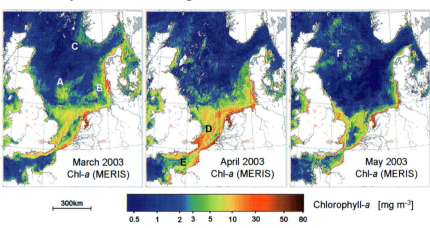

Fig. 2. Monthly median Chl-a for March-May 2003 from MERIS data, processed by the REVAMP project (Peters *et al.* 2005). Early blooms, mainly diatoms, occur on the Dogger Bank (A), in the German Bight/Danish coast (B) and the Norwegian Trench (C). In April, high biomass blooms, mainly *Phaeocystis globosa*, are found in the Southern Bight (D) and Eastern Channel (E). In the Northern North Sea (F) further blooms, probably coccolithophores, continue into May. Some regions may show persistently high artificial Chl-a because of the detection limit caused by high TSM.

Table 1. Summary of main applications of optical remote sensing in the North Sea with corresponding parameters and time and length scales.

Application	Geophysical Parameter	Length and Time Scales
Harmful Algal Bloom detection	Chl-a, RGB	100 m–1000 km 1–10 days
Eutrophication Assessment	Chl-a, Secchi	1–1000 km 1–10 years
Ecosystem Dynamics	Chl-a, Kd	1–1000 km 1 hour–10 years
Sediment Transport and Environmental Impacts	TSM, RGB, Kd	1 m–100 km 1 hour–1 year
Marine Research	Chl-a, RGB, TSM, Kd, CDOM	100 m–1000 km 1 hour–10 years
Carbon Cycle	Chl-a, PAR, Kd, Primary Production	1–1000 km 1–10 years

2.1 (Harmful) algae bloom detection

Certain algae produce or contain substances that are toxic to marine life or to the humans that consume fish and shellfish. Such toxic blooms can have a devastating economic impact on coastal fisheries and aquaculture e.g. *Karenia mikimotoii* in the Channel (Miller *et al.* 2006). Other algae occurring at high biomass may produce surface scums (Lancelot 1995) that are unpleasant and decrease the recreational value of a region. The biodegradation of high biomass algae blooms can also cause death of marine organisms by anoxia. For example in 2001 a bloom of *Phaeocystis globosa* that entered the Eastern Scheldt probably caused economic damage to the mussel industry estimated at millions of euros (Peperzak 2002). Such events have lead to the development of dedicated Harmful Algal Bloom (HAB) detection services in various North Sea countries (Brockmann Consulting; IFREMER; NERSC; Rutten *et al.* 2006). These services operate in near real-time giving web/email-based Chl-a maps together with *in situ* data, where available, and in some cases expert analysis of the event and/or model forecasts. Temporal chlorophyll anomalies (Stumpf 2001) are useful in emphasizing real bloom events and in removing systematic bias from satellite data (Park and Ruddick 2007).

A detailed review of satellite HAB detection can be found in (Stumpf and Tomlinson 2005). In general optical remote sensing provides excellent information on the horizontal distribution of high biomass surface blooms via Chl-a maps, but gives no direct information on harmfulness of algae. To know whether an algal bloom is a HAB (Ruddick *et al.* 2007) requires *in situ* data for species composition or expert knowledge of the system. The main limitations of these services are that clouds may prevent detection for many days and that low biomass or subsurface blooms may pass undetected.

For the accurate estimation of Chl-a in turbid coastal waters, data from the Medium Resolution Imaging Spectrometer (MERIS) has the advantage of better spectral resolution and especially a 709 nm band. The Moderate Resolution Imaging Spectrometer (MODIS) offers more frequent and more easily accessible data owing to the wide swath and open data policy.

2.2 Eutrophication assessment

The North Sea states have agreed within the Oslo and Paris Commission for the Prevention of Marine Pollution (OSPAR) to regularly assess the eutrophication status of their waters and the European Union's Water Framework Directive (WFD) imposes obligations to achieve "good" water quality where eutrophication does not occur. The monitoring requirements arising from OSPAR and the WFD can be very considerable especially for countries like the UK, France and Norway with extensive coastlines. While *in situ* data acquisition is still considered as the main monitoring tool there is a growing tendency (Sorensen *et al.* 2002) to use optical remote sensing as a supporting tool to achieve the monitoring requirements within severe resource constraints of available shiptime and manpower. Indeed the OSPAR Integrated Report 2003 on the Eutrophication Status of the OSPAR Maritime area notes that: (*in situ*) "data availability with respect to sampling frequency and spatial coverage was considered to be too low for some areas to make a proper assessment". To remedy this an atlas (Peters *et al.* 2005) of Chl-a for the North Sea was produced for 2003 as a demonstration product. However, the WFD focus on the first nautical mile from the coast is not well suited to satellite mapping because of nearshore data quality and spatial resolution problems.

Gohin *et al.* (2006) has used satellite data to support the eutrophication monitoring of French North Sea waters by calculation of the 90 percentile Chl-a over the period 1998–2004 from SeaWiFS. This screening shows regions where eutrophication may be most critical and provides a basis for optimisation of the *in situ* sampling strategy.

In a similar application for Belgian waters, satellite chlorophyll imagery has been used to test *in situ* sampling strategies for eutrophication assessment. Subsampling of the satellite data suggests (Figure 3) that the *in situ* monitoring of Chl-a could be made at a limited number of locations (9 instead of 20) without significant loss of spatial information.

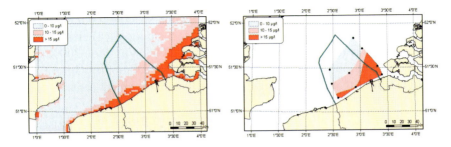

Fig. 3. (left) Mean Chl-a concentration derived from MERIS imagery for April using data from 2003–2005. (right) Same data subsampled to only 9 locations to simulate a possible cost-effective *in situ* sampling strategy. MERIS data was provided by ESA in the framework of Envisat AOID3443.

2.3 Ecosystem modelling: validation and light forcing

Optical remote sensing provides a valuable source of validation data for 3D ecosystem models which simulate phytoplankton distributions and dynamics in response to nutrient inputs from rivers (Druon *et al.* 2005; Edelvang *et al.* 2005; Lacroix *et al.* 2007) for e.g. eutrophication or aquaculture applications.

A second use of optical remote sensing in support of ecosystem modelling is through the model forcing for light attenuation, generally represented via the diffuse attenuation coefficient for scalar PAR irradiance. As an example, the ecosystem model of (Van Den Berg *et al.* 1996) uses a satellite TSM map as input for calculation of PAR attenuation. Another example is shown in Figure 4.

Theoretically, satellite Chl-a data could be integrated fully into ecosystem models via a suitable data assimilation scheme. At present however satellite-derived Chl-a has high uncertainty in most coastal waters and full data assimilation is an ambitious exercise.

2.4 Sediment transport

Similar to the use of satellite Chl-a with ecosystem models, satellite-derived TSM maps can be used to provide initial/boundary conditions or validation data for sediment transport models. Such models have been developed for many North Sea regions to explain the long-term transport of pollutants attached to particulate matter or to optimise dredging and dumping operations. In general, very few *in situ* measurements are available for validating sediment transport models and the good spatio-temporal coverage of satellite-derived TSM maps is valuable for supporting (Fettweis *et al.* 2007; Gemein *et al.* 2006; Pleskachevsky *et al.* 2005; Vos *et al.* 2000) simulations. Integration of satellite TSM maps with Geographical Information Systems (Eleveld *et al.* 2004) facilitates use in a management context.

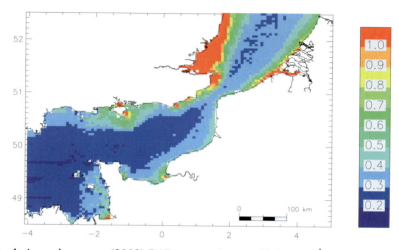

Fig. 4. Annual average (2003) PAR attenuation coefficient (m^{-1}) for the Southern North Sea and Eastern Channel calculated from and used by the 3D-MIRO&CO model (Lacroix *et al.* 2007) based on the following inputs: TSM maps compiled from SeaWiFS, CDOM deduced from the modelled salinity and Chl-a deduced from the modelled phytoplankton biomasses.

For most sediment transport applications MODIS seems the most attractive sensor because its wide swath gives more frequent data than MERIS and because of a more open data policy. For both MODIS and MERIS higher resolution (250 m and 300 m) data is becoming available. For smaller scale applications, e.g. in estuaries and ports, airborne sensors or high resolution satellite sensors such as SPOT or ASTER are suitable but cost considerations and less frequent temporal coverage may limit usage.

2.5   Other applications

In addition to the specific applications described above Chl-a is a standard oceanographic measurement relevant to many research activities in marine biology and chemistry as a basic indicator of the ecosystem. Chl-a maps can indicate fronts and water masses. Consequently most large (>30 m) North Sea Research Vessels now have the capability to receive imagery in near real time in order to guide measurements and provide a spatial context for *in situ* observations.

The flux of CO_2 across the air-sea interface and the related uptake of carbon by phytoplankton and possible export of carbon to bottom sediments are important elements of the global carbon cycle with relevance to climate change studies. Indeed the need for information on the carbon cycle at a global scale was the single justification for the SeaWiFS mission and future NASA support for satellite ocean colour missions is highly dependent on the need for Chl-a time series as a climate data record. The global air-sea CO_2 exchange takes place mainly in the oceans. However, there is growing interest in the possible contribution of the highly productive coastal waters (Borges and Frankignoulle, 2003) and in the potential of satellite remote sensing to deliver relevant information (Lohrenz and Cai 2006).

Professional and recreational diving and fishing operations may also benefit from information on water transparency.

3. Conclusions

During the last 10 years, the launch of SeaWiFS, MODIS and MERIS, the distribution by NASA of the SeaDAS processing software and the development of the Internet and low cost PCs have driven a revolution in the accessibility of high quality, near real-time optical remote sensing data. Algorithm improvements have also been impressive and a new focus on estimating the quality of retrievals is emerging. On the user side there is a growing realisation of the importance of putting very sparse *in situ* measurements into a wider spatial context and of the contribution that optical remote sensing can make. The applications of optical remote sensing reviewed here for the North Sea generally reflect this with a common theme of combining satellite estimates of Chl-a or TSM with *in situ* measurements and/or mathematical models, thus exploiting the complementarity of each information source. The main users are regional and national governmental authorities and marine scientists. The key advantages that optical remote sensing provides at the scale of the North Sea are the excellent spatial coverage and reasonable temporal coverage.

A number of limitations should also be recognized. Only a few water quality parameters can be detected optically and it will still be necessary to monitor many pollutants with *in situ* measurements. Clouds cause significant and unpredictable gaps in data, which are potentially very disruptive to near real time applications. Passive optical remote sensors can probe only the top few metres and subsurface processes such as deep Chl-a maxima or near-bottom sediment transport may be missed completely. There are still significant problems of data quality, particularly for CHL-A estimation in coastal waters with high absorption from CDOM or NAP and for situations with strong atmospheric scattering and absorption. Validation of satellite data products is very sparse in space and time, precisely because the satellites provide so much more data than can be efficiently collected at sea. While remote sensing may provide Chl-a estimates comparable to traditional *in situ* measurements (Duin *et al.* 2007; Peters *et al.* 2005), validation remains crucial to building user confidence for applications such as WFD reporting.

Many of the challenges for scientists in this field can be related to these limitations and future perspectives are outlined by (Ruddick *et al.* 2007). One major concern for the future is the possible gap in data continuity between the current generation of sensors and the future generation, probably including NPOESS and Sentinel-3.

Acknowledgements

This review was supported by funding from the Belgian Science Policy Office STEREO programme via the BELCOLOUR project (SR/00/03). The satellite imagery of Figure 3 was processed in the GMES-MARCOAST project funded by the European Space Agency. Figure 4 was prepared in the AMORE-II project (EV/36/19B) funded by the Belgian Science Policy Office. Colleagues in the MARCOAST and REVAMP projects are acknowledged for discussions regarding North Sea applications of optical remote sensing relating to water quality and harmful algal blooms.

References

Borges AV, Frankignoulle M (2003) Distribution of surface carbon dioxide and air-sea exchange in the English Channel and adjacent areas. J Geophys Res 108 (C5): 10.1029, doi:2000JC000571

Brockmann Consult. Water Quality Service System (www.waqss.de)

Doron M, Babin M, Mangin A, Fanton d'Andon O (2005) Retrieval of the penetration of light and horizontal visibility in coastal waters from ocean color remote sensing data, p 85–94. In Frouin RJ, Babin M, Sathyendranath S (eds), Remote Sensing of the Coastal Oceanic Environment. SPIE

Druon J-N, Loyer S, Gohin F (2005) Scaling of coastal phytoplankton features by optical remote sensors: comparison with a regional ecosystem model. Int J Rem Sens 26: 4421–4444

Duin R, Dury S, Roberti H (2007) Towards operational use of MERIS and SeaWiFS data for water quality monitoring: challenges for the end-user. In: Bochenek Z (ed), New developments and challenges in Remote Sensing. Millpress, pp 575–590

Edelvang K, Kaas H, Erichsen C, Alvarez-Berastegui D, Bundgaard K, Jorgensen PV (2005) Numerical modelling of phytoplankton biomass in coastal waters. J Mar Sys 57: 13–29

Eisma D (1981) Supply and deposition of suspended matter in the North Sea. Spec Publs Int Ass Sediment 5: 415–428

Eleveld MA, Pasterkamp R, Van Der Woerd HJ (2004) A survey of total suspended matter in the Southern North Sea based on 2001 SeaWiFS data. EARSeL eProc 3: 166–178

Fettweis M, Nechad B, Van Den Eynde D (2007) An estimate of the suspended particulate matter (SPM) transport in the southern North Sea using SeaWiFS images, in situ measurements and numerical model results. Cont Shelf Res 27: 1568–1583

Gemein N, Stanev E, Brink-Spalink G, Wolff J-O, Reuter R (2006) Patterns of suspended matter in the East Frisian Wadden Sea: comparison of numerical simulations with MERIS observations. EARSeL eProc 5: 180–198

Gohin F, Lozach L, Oger-Jeanneret H, Lampert L, Lefebvre A (2006) Assessing the evolution of the chlorophyll concentration at coastal stations using satellite and in situ data, p 1–10. OceanOpticsConference 2006. CDROM

IFREMER. NAUSICAA - Navigating through Satellite and In situ data over local areas (http://www.ifremer.fr/nausicaa/marcoast/index.htm)

Lacroix G, Ruddick K, Park Y, Gypens N, Lancelot C (2007) Validation of the 3D biogeochemical model MIRO&CO with field nutrient and phytoplankton data and MERIS-derived surface Chlorophyll a images. J Mar Sys 64 (1–4): 66–88

Lancelot C (1995) The mucilage phenomenon in the continental coastal waters of the North sea. Sci Total Env 165: 83–102

Lohrenz SE, Cai W-J (2006) Satellite ocean color assessment of air-sea fluxes of CO_2 in a river-dominated coastal margin. Geophys Res Lett 33, L01601, doi:10.1029/2005GL023942

Miller PI, Shutler JD, Moore GF, Groom SB (2006) SeaWiFS discrimination of harmful algal bloom evolution. Int J Rem Sens 27: 287–2301

NERSC. Near Real-Time Algal Bloom and Water Quality Monitoring Services for the North Sea and Skagerak Region (hab.nersc.no)

Otto L, Zimmerman JTF, Furnes GK, Mork M, Saetre R, Becker G (1990) Physical Oceanography of the North Sea. Neth J Sea Res 26: 161–238

Park Y, Ruddick K (2007) Detecting algae blooms in European waters. In: ENVISAT symposium. European Space Agency Special Publication SP-636

Peperzak L (2002) The wax and wane of Phaeocystis globosa blooms. PhD thesis. Rijksuniversiteit Groningen

Peters, SWM and 18 others (2005) Atlas of Chlorophyll-a concentration for the North Sea based on MERIS imagery of 2003. Vrije Universiteit, Amsterdam

Pleskachevsky A, Gayer G, Horstmann J, Rosenthal W (2005) Synergy of satellite remote sensing and numerical modelling for monitoring of suspended particulate matter. Ocean Dynamics 55: 2–9

Prandle D, Hydes DJ, Jarvis J, McManus J (1997) The seasonal cycles of temperature, salinity, nutrients and suspended sediment in the southern North Sea in 1988 and 1989. Est Coast Shelf Sci 45: 669–680

Ruddick K, Lacroix G, Park Y, Rousseau V, De Cauwer V, Sterckx S (2007) Overview of Ocean Colour: theoretical background, sensors and applicability for the detection and monitoring of harmful algae blooms (capabilities and limitations). In: Babin M, Roesler C, Cullen J (eds) Real-time coastal observing systems for ecosystem dynamics and harmful algal blooms. UNESCO Monographs on Oceanographic Methodology Series. UNESCO publishing, Paris

Rutten T, Roberti H, Ohm M (2006) Algenplagen beter voorspelbaar, Zoutkrant 2: 5

Sorensen K, Severinsen GG, Aertebjerg G, Barale V, Schiller C (2002) Remote sensing's contribution to evaluating eutrophication in marine and coastal waters. European Environment Agency

Stumpf RP (2001) Applications of satellite ocean color sensors for monitoring and predicting harmful algal blooms. Hum Ecol Risk Ass 7: 1363–1368

Stumpf RP, Tomlinson MC (2005) Remote sensing of harmful algae blooms. In: Miller R, Castillo CD, McKee B (eds), Remote sensing of coastal aquatic environments: technologies, techniques and applications. Kluwer.

Thomson AG, Fuller RM, Yates MG, Brown SL, Cox R, Wadsworth RA (2003) The use of airborne remote sensing for extensive mapping of intertidal sediments and saltmarshes in eastern England. Int J Rem Sens 24: 2717–2737

Van Den Berg AJ, Turner SM, Van Duyl FC, Ruardij P (1996) Model structure and analysis of dimethylsulphide (DMS) production in the southern North Sea, considering phytoplankton dimethylsulphoniopropionate (DMSP) lyase and eutrophication effects. Mar Ecol Prog Ser 145: 233–244

Vos RJ, Brummelhuis PGJT, Gerritsen H (2000) Integrated data-modelling approach for suspended sediment transport on a regional scale. Coast Eng 41: 177–200

Optical Remote Sensing Applications in the Baltic Sea

Herbert Siegel and Monika Gerth

Baltic Sea Research Institute (IOW), Rostock, Germany

Abstract. The main applications of ocean colour satellite data in the Baltic Sea (Case 2 water) are coastal discharge and phytoplankton blooms. These processes generate the variations of optically active water constituents. The phytoplankton development is characterised by a spring bloom of diatoms and dinoflagellates, and a summer bloom of nitrogen-fixating cyanobacteria. The blooms depend strongly on the meteorological conditions. Distribution of river discharge was intensely investigated in the Pomeranian Bight, the Oder River discharge area. Satellite data of different spectral and spatial resolution has been used. Information on oceanographic conditions was derived from Sea Surface Temperature. The implementation of satellite data improved the Baltic Sea research due to the synoptic character enables to transfer detailed ship-borne measurements to larger spatial and temporal scales.

1. Introduction to the Baltic Sea

The Baltic Sea is a semi-enclosed sea (Figure 1), with a limited water exchange to the ocean over the North Sea. The water exchange is limited by the morphometry of the transition area between North Sea and Baltic Sea. High freshwater supply by rivers and limited evaporation result in a positive water balance with a general outflow of low saline surface water. High saline and oxygen enriched surface water of the transition area, which sporadically enters the western Baltic Sea, sinks down. This generates permanent haline stratification. The dense bottom water reaches the deeper basins of the Baltic Proper by barotropic transport, which is the only way to transport oxygen into those regions (Matthäus 2006). The warming phase in early spring stabilises the top layer by thermal stratification. The stratified water delivers the starting condition for the spring bloom, because the phytoplankton remains in the euphotic zone. Upwelling processes and wind-induced mixing transport inorganic nutrients from the cold intermediate

V. Barale, M. Gade (eds.), *Remote Sensing of the European Seas.*

waters into the surface layer until the nutrients are consumed by biological uptake in both layers.

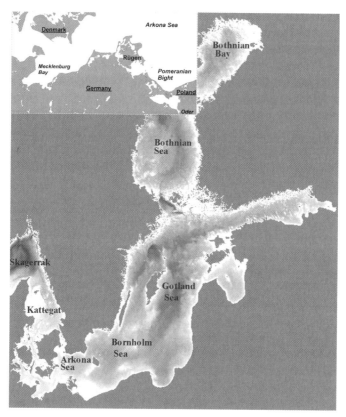

Fig. 1. Maps of the areas of investigation: Baltic Sea (larger map) and south-western Baltic Sea (smaller figure). Modified from Siegel and Gerth, 2000a.

After the stagnation period in late spring to early summer nitrogen-fixating cyanobacteria develop normally in July and August. The intensity of the cyanobacteria blooms depend strongly on the meteorological conditions in the summer. During calm conditions with high solar radiation, the cyanobacteria accumulate at the water surface where they are transported by wind or circulation processes. Strong wind-induced mixing during cloudy conditions may interrupt or terminate the bloom.

The application of satellite data, which started in the 1970s and intensified in the last 10 years, improved the Baltic Sea research because of the synoptic character. Coverage and repeating rate enables to transfer detailed ship-borne measurements to larger spatial and temporal scales, particularly, in studies of phytoplankton blooms and river discharge.

2.   Specificity of optical remote sensing in the Baltic Sea

The concentration and composition of suspended and dissolved optically active water constituents such as phytoplankton (chlorophyll *a*, Chl-a), total suspended matter (TSM) and coloured dissolved organic material (CDOM, or yellow substances) can be retrieved from ocean colour satellite date due to their specific optical properties. In the open ocean, variations in optical properties are dominated by phytoplankton (Case 1 waters). The Baltic Sea is essentially composed of Case 2 waters, where variations in optical properties are dominated by CDOM and other water constituents (Chl-a, TSM). The south–north increase of CDOM is due to the peat-land draining rivers in the northern Baltic Sea. Coastal discharge contains high concentrations of TSM and Chl-a due to the phytoplankton development starting in the lagoons. Satellite data contributed to biogeochemical river discharge studies because the water contains nutrients, dissolved and particle-bounded trace metals and organic pollutants. The main applications of such data are the investigation of river discharge and of phytoplankton blooms. Ocean colour data were often combined with satellite-derived Sea Surface Temperature (SST) to explore oceanographic conditions at large.

The application of ocean colour satellite data started in the 1970s with Landsat MSS before the first ocean colour sensors, the Coastal Zone Colour Scanner (CZCS), operated (1978–1986) in orbit. After a period of 10 years the ocean colour application continued with sensors such as the Multi-spectral Optical Sensor (MOS, 1996-), Sea-viewing Wide Field-of-view Sensor (SeaWiFS, 1997-), Moderate Resolution Imaging Spectroradiometer (MODIS, 1999-), and the Medium Resolution Imaging Spectrometer (MERIS, 2001-). Coverage, spatial resolution, and repeating rate enable studies on different spatial and temporal scales. Sensors of high spatial resolution such as the Wide Field-of-view Sensor (WiFS), LISS-III and PAN of the Indian Remote Sensing Satellites IRS-P3 and IRS-1C, 1D and Landsat Thematic Mapper (TM) and ETM+ are designed for land applications with broadband channels in the visible spectral range. They were implemented for local coastal studies to observe the detailed distribution of features. The Advanced Very High Resolution Radiometer (AVHRR), on board the meteorological satellites of the NOAA series (first launched in 1976) are widely used to derive SST maps.

The satellite sensor signal results from interaction between the incident sun radiation and the natural water strongly influenced by the atmosphere and water surface. The sensors are designed with narrow-band channels in the visible and near-infrared spectral range to derive optically active water constituents and to perform the atmospheric correction. Orbit and swath of most ocean colour sensors permit daily coverage of the Baltic Sea.

The atmospheric correction is difficult in this marginal sea because of changing influences of maritime and continental atmosphere. High absorption of CDOM limits the usable channels in algorithms. Therefore, atmospheric corrections implemented in the standard software distributed for different sensors do not work always properly, what validations for SeaWiFS and MERIS have shown (Siegel *et al.* 2001, Ohde *et al.* 2007).

Algorithms for derivation of biogeochemical products have to be adapted to the conditions of the Baltic Sea as well. Special algorithms for CDOM-dominated Case 2 waters are required. For example the absorptions band of chlorophyll at 443 nm is not applicable as in Case 1 algorithms. In Siegel (1991) regression analyses were performed between chlorophyll concentrations and colour ratios using all channels of selected ocean colour sensors, based on data from all seasons. The highest correlation was found for channels in the green and red spectral range. That means, the highest variations correlated to chlorophyll is related to scattering properties by phytoplankton particles in the open Baltic Sea. For SeaWiFS different combinations of ground-truth algorithms and atmospheric correction schemes have been tested. Best results were derived using MUMM atmospheric correction (Ruddick *et al.* 2000) and an IOW algorithms particular for chlorophyll using channels 4 and 5 of SeaWiFS (Siegel *et al.* 1999b, 2001). Darecki *et al.* (2005) and Vepsäläinen *et al.* (2005) provided SeaWiFS chlorophyll algorithms for the Baltic Sea and in HELCOM (2004) different SeaWiFS chlorophyll algorithms are compared. Newer methods on the basis of neuronal networks standard procedure for MERIS (Schiller and Doerffer, 1999, Doerffer *et al.* 1999) are in adaptation process to the conditions of the Baltic Sea (Ohde *et al.* 2007).

In shallow waters of high transparency sea bottom reflection may increase the spectral reflectance at the surface which results in an overestimation of geophysical products. Ohde and Siegel (2001) developed a correction algorithm to eliminate the bottom influence of the Oder Bank in the Pomeranian Bight of the south–western Baltic Sea.

3.   Coastal processes and river discharge

The investigation of water exchange between the coastal zone and the open sea started with application of AVHRR, CZCS, and Landsat data in the south–eastern Baltic Sea (Horstmann 1983, 1988; Brosin *et al.* 1988). Detailed studies on transport and turnover processes were performed in the early 1990s years in the Pomeranian Bight of the south–western Baltic Sea with the Oder River discharge. Because of the lack of ocean colour data (CZCS, 1979–1986; SeaWiFS 1997-) a systematization of the discharge patterns was based on SST and CZCS data were implemented for case studies. The river discharge has been related to wind-forcing (Siegel *et al.* 1996, 1999a) to identify the main transport directions and the accumulation areas. Prevailing easterly winds in spring induce upwelling along the Polish coast, guiding the Oder water along the German coast into the Arkona Sea. During dominating westerly winds in the other seasons, a transport band along the Polish coast established to the Bay of Gdansk. Examples are given in Figure 2. If the wind is changing back to east the Ekman transport spreads the band offshore and the water will be transported back into the Arkona Sea by large scale circulation. These results verified the geological findings stating that higher sedimentation rates occur in the Arkona Sea than in the Bornholm Sea.

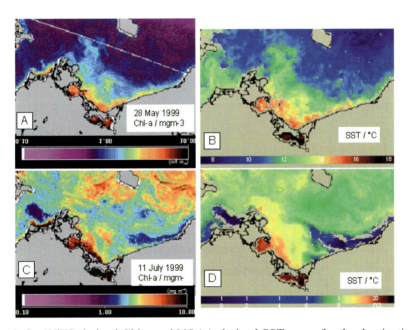

Fig. 2. SeaWiFS-derived Chl-a and NOAA-derived SST maps for the dominating westerly (A, B) and easterly winds (C, D). Adapted from Siegel *et al.* 2001.

During stable east-wind situations the Oder River discharge enters the central Arkona Sea and cause high Chla-concentration outside any bloom periods (Siegel *et al.* 2002). The systematisation performed for the Pomeranian Bight was extended to the entire coast of the German state, to study the dynamical features and processes there. The derived features were related not only to the dominating wind directions but also to transitions between these directions (Siegel *et al.* 2005b). The issued catalogue of processes supports the interpretation of monitoring data and the optimisation of the coastal monitoring programme of regional authorities. Furthermore, they allow the forecast of transport processes during special events, such as harmful algal blooms, floods, or accidents.

4.   Phytoplankton blooms

4.1   Spring bloom

The phytoplankton development in the Baltic Sea is characterised by two main blooms, spring bloom of diatoms and dinoflagellates, and summer bloom of cyanobacteria. Horstmann (1983), Dowell (1996), and Siegel *et al.* (1999c) applied CZCS data for the study of phytoplankton distribution. The spring bloom normally starts in the western Baltic Sea in March after the establishment of thermal stratification, develops in the Arkona Sea and Bornholm Sea end of March to April, and in the central Gotland Sea in May as shown at the example from 2000 in Figure 3.

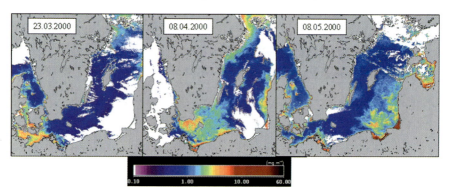

Fig. 3. SeaWiFS-derived Chl-a in the Baltic during the spring bloom on 23 March, 8 April and 8 May 2000, showing the spatial and temporal development.

4.2   Cyanobacteria bloom in summer

Cyanobacteria develop in warm summers in the Baltic Sea during stable thermal stratification and depletion of inorganic nitrogen in the top water layer. The dominant species, *Nodularia spumigena* and *Aphanizomenon sp.*

are able to use dissolved molecular nitrogen for the production of biomass. The initial temperature for the growth of cyanobacteria in 1980s and early 1990 was about 15–17°C (Kahru *et al.* 1994, Kahru 1997, Wasmund 1997). In the years 1998–2001 the initial temperature was approximately 13–15°C, starting day was earlier and rather different, between 8 June (2001) and beginning of July, and the starting areas were mostly the northern Gotland Sea and Gulf of Finland. The course of the bloom and duration of about 4–6 weeks depend on the meteorological situations. During low wind-mixing and high solar radiation, they form aggregates and float up to the surface due to their buoyancy-increasing intercellular gas vacuoles. There, they form subsurface clouds or surface accumulations. Surface circulation and wind transport the accumulations like passive tracers that form typical stripes which partly end in eddy-like structures (Kononen and Leppänen 1997). Some are toxic and in high concentrations they can be harmful for animals und humans. The patchy distribution of subsurface clouds and surface scum and their high spatial and temporal variability favoured ocean colour satellite data to investigate the development of blooms. Different reasons complicate the quantification by chlorophyll. Surface accumulations of a few centimetres thickness cannot be quantified by satellites sensor. Furthermore, the sampling of cyanobacteria is difficult due to the buoyancy of aggregates in the sampler. Kutser (2004) tried to derive chlorophyll concentrations from scum based on hyperspectral satellite data and optical modelling. Other absorbing phytoplankton species may develop in the aggregates and influence the chlorophyll content significantly. Marker pigment of cyanobacteria phycocyanin has an absorption maximum in the spectra range of 615–640 nm and would be a potential measure to quantify cyanobacteria, but biological laboratories are not able to determine them routinely. Seppälä *et al.* (2005) applied a phycocyanin index on MERIS data.

The first documentation of cyanobacteria using Landsat MSS data was published by Öström (1976) and by Ulbricht and Schmidt (1977) for the warm summer of 1975. Horstmann (1983) presented examples of CZCS scenes from the early 1980s. Strong inter-annual variations occurred in the 1980s with high accumulations in 1980 and without any in 1985 and 1986 (Siegel *et al.* 1999c). Mostly, the bloom started in the northern Gotland Sea and developed southward, reaching a maximum in August, and extending into the western Baltic Sea. Inter-annual differences were also documented by Kahru (1997), based on AVHRR data showing an increase in the occurrence of cyanobacteria in the early 1990s. In particular, at the end of the 1990s and at the beginning of this century, the accumulations increased with the increasing summer temperature in the Baltic Sea (Siegel *et al.* 2006). Strong surface accumulations with stable distribution pattern over several days were observed south of Gotland Island in August 1997 (Siegel, Gerth, 2000). From 1998 until 2004, the study was continued with SeaWiFS data (Schrimpf *et al.* 2005; Siegel *et al.* 1999b, 2001 and 2002; Mazur-Marzec 2006). The yearly cyanobacteria development is described in the biological assessments of the Baltic Sea (Wasmund *et al.* 2001). The monthly mean Chl-a maps from this period show strong seasonal and inter-annual variations in spatial distribution, intensity of the bloom, and influenced coastal regions. The bloom was very strong in the warm summer of 1999, 2001, and 2003 in the Baltic proper and particularly 2002 in the Gulf of Finland. The bloom mostly started in the northern and western Gotland Sea and extended, in 1999 and 2001, into the western Baltic Sea. In colder summers 1998 and 2000, the intensity of the bloom was much lower. In contrast to the early 1980's filaments were also observed in the Bothnian Sea, particularly in 2001. The monthly mean chlorophyll maps in Figure 4 show the low intense bloom in 1998 and high intense bloom in 2001.

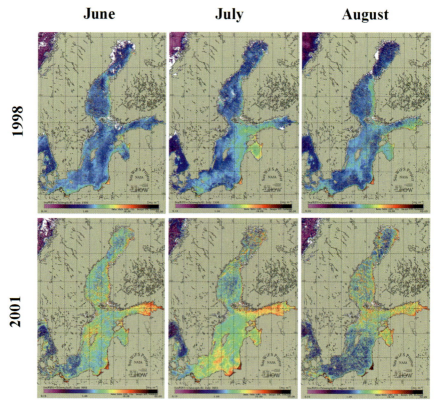

Fig. 4. Monthly mean chlorophyll concentration in June, July, and August 1998 and 2001 indicate the inter-annual variations in the cyanobacteria bloom.

The years 2005 and 2006 are characterised by rather different bloom developments. In 2005, the bloom maximum was already reached on 13 July in the central Baltic Sea and the western part was excluded (Figure 5). In 2006, the northern Baltic Sea was only shortly influenced and the western Baltic was affected from beginning of July until end of August. For the first time, satellite data indicated that the bloom was transported with the outflow of the Baltic Sea into the Kattegat.

Fig. 5. Maximum extent of the cyanobacteria bloom in the Baltic Proper in 2005, on 13 July, based on MODIS data.

5.   Summary and conclusions

Ocean colour remote sensing contributes significantly to the Baltic Sea research mainly to the investigation of river discharge and phytoplankton blooms. The synoptic coverage of the entire Baltic Sea as well as the spatial and temporal resolutions are the main advantages for the investigation of the course, extent and structures of blooms and for the spreading of river discharge. Limitations are the cloud coverage, the information content terminated to the upper few meters, and not yet sufficient quality of the derived products (atmospheric correction).

The course of the phytoplankton blooms in spring (diatoms and dinoflagellates) and in summer (cyanobacteria) is described in the yearly assessment of the state of the Baltic Sea (Wasmund *et al.* 2001). The combination with SST data supports to understand the physical forcing of phenomena. The implementation of satellite data of high spatial resolution provides detailed information on coastal processes and special features during phytoplankton blooms. For monitoring of special events like flooding and toxic algae blooms near real-time data access is essential for management questions. In regions of high cloud coverage, systematisations of coastal processes by satellite data support the interpretation of such events and forecast of the development for distinct wind forcing.

Currently, the number of ocean colour sensors in space guarantees the daily coverage, and the European activities in the Global Monitoring for Environment and Security (GMES) programme (*e.g.* MarCoast) for distributing MERIS products improve the applicability of satellite data for end-users like local authorities. Future hyperspectral data will improve the number of derived products like phytoplankton groups and geostationary satellites will improve the temporal resolution.

The procedures for deriving bio-geophysical data products from the different sensors need further improvements for the Baltic Sea particularly in the adaptation of atmospheric correction and ground-truth algorithms to the conditions of the Baltic Sea.

Acknowledgements

Ocean colour satellite data provided by the SeaWiFS Project at NASA GSFC, MODIS-Rapid Response System, and ESA. SST data provided by the German Federal Maritime & Hydrographic Agency, Hamburg.

References

Brosin HJ, Gohs L, Seifert T, Siegel H, Byckova IA, Viktorov SV, Demina MD, Lobanov VJ, Losinskij VN, Smoljanickij VM (1988) Mesoskale Strukturen in der südlichen Ostsee im Mai 1985. Beiträge zur Meereskunde 58: 8–18
Darecki M, Kaczmarek S, Olszewski J (2005) SeaWiFS oceancolour chlorophyll algorithms for the southern Baltic Sea. Int. J. Remote Sensing 26 (2): 247–260
Doerffer R, Sorensen K, Aiken J (1999) MERIS potential for coastal zone applications. Int. J. Remote Sensing 20: 1809–1818

Dowell MD (1996) Optically Active Components and Their Relationship with Meso-Scale Features in Baltic Coastal Zone. Marine Science Reports, IOW 19, pp 114–139

HELCOM (2004) Thematic Report on Validation of Algorithms for Chlorophyll a Retrieval from Satellite Data of the Baltic Sea Area. Baltic Sea Environment Proc. 94, pp 44

Horstmann U (1983) Distribution patterns of temperature and water colour in the Baltic Sea as recorded in satellite images: Indicators for phytoplankton growth. Berichte Institut für Meereskunde Kiel, 106, pp 145

Horstmann U (1988) Satellite remote sensing for estimating coastal offshore transports. In: Jansson BO (ed) Lecture Notes on Coastal and Esturine Studies, vol 22, Coastal-Offshore Ecosystem Interactions, pp 50–66.

Kahru M (1997) Using satellites to monitor large-scale environmental change: A case study of cyanobacteria blooms in the Baltic Sea. In: Kahru M, Brown CW (eds) Monitoring Algal Blooms. Springer-Verlag, Berlin Heidelberg New York, pp 43–57

Kahru M, Horstmann U, Rud O (1994) Satellite detection of increased Cyanobacteria blooms in the Baltic Sea: Natural fluctuations or ecosystem change? Ambio 23 (8): 469–472

Kononen K, Leppänen JM (1997) Patchiness, scales and controlling mechanisms of cyanobacterial blooms in the Baltic Sea: Application of a multiscale research strategy. In: Kahru M, Brown CW (eds) Monitoring Algal Blooms. Springer-Verlag, Berlin Heidelberg New York, pp 43–57

Kutser T (2004) Quantitative detection of chlorophyll in cyanobacterial blooms by satellite remote sensing. Limnol. Oceanogr., 49 (6): 2179–2189.

Matthäus W (2006) The history of investigation of salt water inflows into the Baltic Sea. Marine Science Report, IOW 65, pp 73

Mazur-Marzec H, Krężel A, Kobos J, Pliński M (2006) Toxic *Nodularia spumigena* blooms in the coastal waters of the Gulf of Gdańsk: a ten-year survey. Oceanologia, 48 (2): 255–273

Ohde T, Siegel H (2001) Correction of bottom influence in ocean colour satellite images of shallow water areas of the Baltic Sea. Int. J. Remote Sensing 22 (2/3): 297–313

Ohde T, Siegel H, Gerth M (2007) Results of MERIS Level-2 product validation in the Baltic Sea, the Namibian coastal area and the Atlantic Ocean. International Journal of Remote Sensing 28 (3–4): 609–624

Öström B (1976) Fertilization of the Baltic by nitrogen fixation in the blue-green algae Nodularia spumigena. Remote Sensing of the Environment 4: 305–310

Ruddick KG, Ovidio F, Rijkeboer M (2000) Atmospheric correction of SeaWiFS imagery for turbid coastal and inland waters. Applied Optics 39 (6): 897–912

Schiller H, Doerffer R (1999) Neural network for emulation of an inverse model – operational derivation of Case 2 water properties from MERIS data. Int. J. Remote Sensing 20: 1735–1746

Schrimpf W, Zibordi G, Mélin F, Djavidnia S (2005) Chlorophyll-a concentrations, temporal variations and regional differences from satellite remote sensing. HELCOM, http://www.helcom.fi/environment2/ifs/en_GB/cover/

Seppälä J, Ylöstalo P, Kuosa H (2005) Spectral absorption and fluorescence char-acteristics of phytoplankton in different size fractions across a salinity gradi-ent in the Baltic Sea. Int. J. Remote Sensing 26 (2): 387–414

Siegel H (1991) Empirical algorithms for the determination of chlorophyll by re-mote sensing methods. Beiträge zur Meereskunde, Berlin 62, pp 69–78

Siegel H, Gerth M (2000) The exceptional summer 1997 in the Baltic Sea – The warmest August, the Oder flood, and phytoplankton blooms. In: Halpern D (ed) Satellites, Oceanography and Society. Elsevier Ocean. Ser., pp 239–254

Siegel H, Gerth M, Mutzke A (1999a) Dynamics of the Oder River Plume in the Southern Baltic Sea - Satellite data and Numerical Modelling. Continental Shelf Research 18: 1143–1159

Siegel H, Gerth M, Schmidt T (1996) Water exchange in the Pomeranian Bight investigated by satellite data and ship-borne measurements. Continental Shelf Research 16 (14): 1793–1817

Siegel H, Gerth M, Ohde T (1999b) Remote sensing of phytoplankton blooming in the Baltic Sea using SeaWiFS data. In: IGARSS '99 Proc. IEEE Int. Geo-science and Rem. Sens. Symp., 28.6. – 2 .7.1999, Hamburg. Piscataway: IEEE 2: 837–839

Siegel H, Gerth M, Ohde T (2002) Remote Sensing Applications. In: Schernewski G, Schiewer U (eds) Baltic coastal ecosystems: structure, function and coastal zone management. CEEDES Series, Springer, pp 279–292

Siegel H, Ohde T, Gerth M (2001) SeaWiFS validation and application in the Bal-tic Sea. In: Proc. 4th Workshop on Ocean Remote Sensing, 30 May–1 June 2001, Berlin, pp 97–105

Siegel H, Gerth M, Neumann T, Doerffer R (1999c) Case studies on phytoplank-ton blooms in coastal and open waters of the Baltic Sea using Coastal Zone Colour Scanner data. Int. J. Remote Sensing 20 (7): 1249–1264

Siegel H, Gerth M, Ohde T, Heene T (2005) Ocean colour remote sensing relevant water constituents and optical properties of the Baltic Sea. Int. J. Remote Sensing 26 (2): 315–330

Siegel H, Seifert T, Schernewski G, Gerth M, Reißmann J, Ohde T, Podsetchine V (2005b) Discharge and transport processes along the German Baltic Sea Coast. Ocean Dynamics 55 (1): 47–66

Ulbricht KA, Schmidt D (1977) Massenauftreten mariner Blaualgen in der Ostsee auf Satellitenaufnahmen erkannt. DFVLR-Nachrichten, No. 22, pp 913–915

Vepsäläinen J, Pyhälahti T, Rantajärvi E, Kallio K, Pertola S, Stipa T, Kiirikki M, Pulliainen J, Seppälä J (2005) The combined use of optical remote sensing data and unattended fluorometer measurements in the Baltic Sea. Int. J. Remote Sensing 26 (2): 261– 282

Wasmund N, Pollehne F, Postel L, Siegel H, Zettler M (2001) Biologische Zustandseinschätzung der Ostsee im Jahre 2000. Marine Science Report, Bal-tic Sea Research Institute Warnemünde, 46, pp 74

Wasmund N (1997) Occurrence of cyanobacterial blooms in the Baltic Sea in rela-tion to environmental conditions. Int. Revue ges. Hydrobiol. 82: 169–184

Open Waters Optical Remote Sensing of the Mediterranean Sea

Rosalia Santoleri[1], Gianluca Volpe[1], Salvatore Marullo[2], and
Bruno Buongiorno Nardelli[1]

[1] Istituto di Scienze dell'Atmosfera e del Clima (ISAC)
[2] Ente per le Nuove tecnologie l'Energia e l'Ambiente (ENEA), Centro
Ricerche Frascati

Abstract. Recent technological and scientific advances related to the
Mediterranean Sea surface chlorophyll retrieval from space are discussed
in this chapter. In particular, a complete review on the definition and validation of specific regional ocean colour bio-optical algorithms for this area
is presented. The need for specific algorithms in the Mediterranean is explained mainly by the observed failure of standard procedures. Latest
results, mostly focusing on biological response to deep water formation
events, concerning the interactions between physical and biological processes in the Mediterranean Sea are discussed.

1. Introduction

The Mediterranean Sea (MED) has been often defined as "a laboratory basin".
This definition finds its rationale in the number and kind of processes that
occur in the basin as compared to the global ocean. These processes range
from the intermediate and deep water formation to a variety of mechanisms
such as mesoscale instabilities, vertical and horizontal mixing processes and
large scale water mass exchanges. From the biological point of view, the
MED is a mid-latitude, predominantly oligotrophic or ultra-oligotrophic
basin. However, higher biomass may locally and seasonally occur in regions
affected by river runoff or by deep convection events (see *e.g.* Antoine *et al.*
1995).

The MED physical and biological properties have been extensively in-
vestigated by means of satellite measurements from the Coastal Zone
Color Scanner (CZCS, 1978–1986) and the more recent Sea-viewing Wide
Field of view Sensor (SeaWiFS, 1997-). Several studies aimed at validating

V. Barale, M. Gade (eds.), *Remote Sensing of the European Seas.*

global and regional bio-optical algorithms for surface phytoplankton biomass retrieval by means of CZCS and/or SeaWiFS data.

The open ocean biological variability in the MED has been found to depend mainly on the seasonal cycle, air-sea interaction processes and oceanic internal dynamics. However, the MED area is also strongly affected by anthropogenic atmospheric emissions from continental Europe and by desert dust from northern Africa that could play a fundamental role in the MED biogeochemistry in terms of nutrients input (Bonnet *et al.* 2005).

The presence of these aerosols makes it difficult to apply standard remote sensing procedures for the atmospheric correction (Moulin *et al.* 1997). Moreover, the peculiar aerosol composition in the MED was believed one of the factors determining its different optical properties with respect to the global ocean (Claustre *et al.* 2002). This constitutes one of the reasons why an increasing number of papers have been concerned with the definition of regional ocean colour bio-optical algorithms for this area.

This chapter will first review the most recent approaches used to evaluate the chlorophyll concentration from satellite ocean colour data in the MED. Successively, the analysis of the main features characterizing the bio-physical dynamics in the MED is presented.

2.   Mediterranean empirical bio-optical algorithms

An extensive calibration and validation activity has been performed on the SeaWiFS data by the SeaWiFS[1] and SIMBIOS[2] Projects. The result of this CAL/VAL activity was the development of empirical bio-optical algorithms (OC2v4 and OC4v4, O'Reilly *et al.* 2000) for the retrieval of the phytoplankton pigment concentration. These empirical bio-optical algorithms are based on the inverse dependence of Chlorophyll *a* (Chl-a) on the ratio, $R_{Rrs(Green)}^{Rrs(Blue)}$ [3], between Remote Sensing Reflectance (Rrs) measured in the blue and green portion of the light spectrum.

At global scale, these algorithms were demonstrated to perform adequately (O'Reilly *et al.* 2000; Gregg and Casey 2004), presenting uncertainties in the range expected by space agencies. However, at regional scales these standard algorithms have been shown to perform generally and sensibly worse. More specifically, in the MED, the standard NASA algorithms (OC2v4 and OC4v4) lead to a significant overestimation of the

---

[1] http://oceancolor.gsfc.nasa.gov/cgi/postlaunch_tech_memo.pl/
[2] http://neptune.gsfc.nasa.gov/publications/
[3] A simpler way of writing can be $R_{Rrs(Green)}^{Rrs(Blue)} = R_{Green}^{Blue}$

SeaWiFS derived Chl-a (above 70% for Chl-a below 0.2 mg m^{-3}) (among others Bricaud *et al.* 2002; Claustre *et al.* 2002; D'Ortenzio *et al.* 2002).

Within the last decades, there have been many attempts to characterize the MED bio-optical properties for application to satellite data (mostly to SeaWiFS but also to CZCS and MERIS) either through the elaboration of bio-optical models or by fitting in situ Chl-a and Rrs ratios. As referred to the latter, all the MED-adapted empirical algorithms used a two-band Rrs ratio (blue-to-green, one channel to define the blue band and one for the green). Nonetheless, it is known that using multiple Rrs ratios decreases the noise-to-signal ratio, therefore enhancing the algorithms' performance (O'Reilly *et al.* 1998). Actually, Volpe *et al.* (2007) demonstrated that the performances of two MED algorithms, BRIC (Bricaud *et al.* 2002) and DORMA (D'Ortenzio *et al.* 2002), where highly correlated to the Chl-a and more directly to which band ratio constituted the MBR[4]. In particular, when MBR coincides with R_{555}^{443} BRIC algorithm performs better. The opposite is true when MBR coincides with R_{555}^{490} (band ratio used by D'Ortenzio *et al.* 2002). Volpe *et al.* (2007) quantified the uncertainties of some regional and global algorithms for the MED area (DORMA, BRIC and OC4v4) and identified and developed an optimal algorithm (MedOC4) for the production of high quality ocean colour datasets.

This new bio-optical algorithm is based on a fourth power polynomial regression between log-transformed Chl-a and log-transformed MBR. The MedOC4 algorithm – calibrated with the most representative bio-optical dataset for open waters ever used for algorithm refinement in the MED – matched the requirements of unbiased satellite chlorophyll estimates and significantly improved its accuracy. Moreover, the authors gave some explanations about the observed difference between global and regional algorithms. They discussed and dismissed some hypothesis such as the impact of atmospheric correction algorithms, the difference in instrument calibration, or the impact of different types of water stratification. They pointed out that one of the possible hypothesis to explain the different bio-optical regime of the MED is the difference in bio-optical characteristics at the regional scale, due to ecological reasons (such as the presence of specific phytoplankton groups). Moreover, they argued that such remarks are likely to be valid for other ocean colour sensors such as MERIS or MODIS.

Following Volpe *et al.* (2007), MED bio-optical algorithms for MERIS (MedOC4ME) and MODIS (MedOC3) sensors have been evaluated here and are reported in Table 1. For means of comparison the two SeaWiFS algorithms are also reported. Figure 1 shows the in situ bio-optical dataset used to retrieve the three MED-adapted algorithms. It shows that for all the

---

[4] MBR is the Maximum Band Ratio: $MBR = MAX\left(R_{555}^{443}, R_{555}^{490}, R_{555}^{510}\right)$.

three sensors the standard algorithms (dashed lines) are above the in situ dataset (crosses) for Chl-a below approximately 1 mg m^{-3}. This means that once the MBR is estimated from the remote Rrs measurements, the chlorophyll retrieval via standard algorithms is higher than that obtained via MED-adapted algorithms. Moreover, standard algorithms overestimate low Chl-a and underestimate high concentrations. Therefore the use of regional adapted algorithms is strongly recommended.

Table 1. Coefficients and functional forms for three regional MED algorithms adapted for MODIS (MedOC3), MERIS (MedOC4ME) and SeaWiFS (MedOC4). The standard algorithms are also shown. R is the MBR between R_{551}^{443} and R_{551}^{488} for MODIS; among R_{560}^{442}, R_{560}^{490} and R_{560}^{510} for MERIS and as defined in the footnote 4 on previous page for SeaWiFS.

	Algorithm	Functional Form and Coefficients
MODIS	MedOC3	$Chl = 10^{\left(0.380 - 3.688\,R + 1.036\,R^2 + 1.616\,R^3 - 1.328\,R^4\right)}$
MODIS	OC3 *	$Chl = 10^{\left(0.283 - 2.753\,R + 1.457\,R^2 + 0.659\,R^3 - 1.403\,R^4\right)}$
MERIS	MedOC4ME	$Chl = 10^{\left(0.46030 - 4.26343\,R + 4.98364\,R^2 - 4.38507\,R^3 + 1.20416\,R^4\right)}$
MERIS	OC4ME **	$Chl = 10^{\left(0.40657 - 3.63030\,R + 5.44357\,R^2 - 5.48061\,R^3 + 1.75312\,R^4\right)}$
SeaWiFS	MedOC4	$Chl = 10^{\left(0.4424 - 3.686\,R + 1.076\,R^2 + 1.684\,R^3 - 1.437\,R^4\right)}$
SeaWiFS	OC4v4 *	$Chl = 10^{\left(0.366 - 3.067\,R + 1.930\,R^2 + 0.649\,R^3 - 1.532\,R^4\right)}$

* O'Reilly *et al.* (2000)
** Morel and Antoine (2000)

3. The seasonal cycle of the chlorophyll field

A new MED chlorophyll dataset has been produced by re-processing all the available Local Area Coverage Level-1A SeaWiFS data (more then 6000 passes, covering the 1998–2005 period) with the MedOC4 algorithm (Volpe *et al.* 2007). This dataset has been used to compute the monthly average chlorophyll maps for the 1998–2005 period and corresponding climatological mean, from which the general oligotrophic character of the MED is evidenced (Figure 2). The average yearly value of 0.19 mg m^{-3} results from the contribution of extreme conditions, such as the very oligotrophic regime of the Levantine basin and the periodically blooming areas of the northwestern MED (~0.05 and ~0.26 mg m^{-3}, respectively).

Generally, chlorophyll increases in the coastal areas due to runoff from the continental margins, while offshore (e.g. in the Ligurian-Provençal basin or north Tyrrhenian Sea), the increased Chl-a is caused by spring bloom events. However, these values can be affected by uncertainties in near-coastal waters due to the presence of other optically active materials.

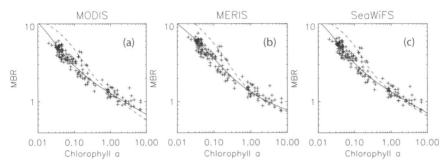

Fig. 1. Bio-optical dataset for MODIS (a), MERIS (b) and SeaWiFS (c). Continuous lines represent the regional algorithms while dashed lines represent standard algorithms (see Table 1).

The spatial distribution of the monthly "climatological" chlorophyll field is shown in Figure 2. Higher chlorophyll values are evident from December to April in the Ligurian-Provençal-Catalan region and in the Alboran-Algerian basin. In the former region, the increased springtime nutrient availability and Chl-a may be explained to a large extent by local meteorological forcing, as discussed in details in section 4.1. In the latter area, the enhanced productivity can be related to other mixing processes. In particular, it may depend on the dynamics of the quasi-permanent gyres in the Alboran Sea or be linked to the Algerian current system progressing eastward down to the Sicily Channel (Barale *et al.* 2005). Similar effects can be seen east of the southern tip of Sicily, where the so-called Capo Passero filament marks the eastward transport of the upwelled Sicilian coastal waters by the Atlantic-Ionian Stream (Buongiorno Nardelli *et al.* 2001).

In the Eastern MED averaged monthly values rarely exceed 0.1 mg m^{-3} with exceptions along the Egyptian coasts, the northern Ionian Sea off-shore the Calabrian coasts (the "Calabrian Gyre", D'Ortenzio *et al.* 2003), the north Aegean Sea and the well known Rhodes gyre. In the Calabrian Gyre, a significant offshore enhancement of surface biomass is observed in March and April, with values up to 0.4–0.5 and 0.2–0.3 mg m^{-3} (D'Ortenzio *et al.* 2003). In the Rhodes Gyre region, where a permanent cyclonic structure is present (Ozsoy *et al.* 1993), a moderately eutrophic regime is observed (average values 0.2–0.3 mg m^{-3}), as detected in the climatological March. In the other months, the gyre is either not visible or very weakly identified by minimal chlorophyll gradient. The north Aegean Sea always displays high Chl-a with peak values from roughly December to May and is plausibly related to exchanges between the MED and the adjacent eutrophic Black Sea.

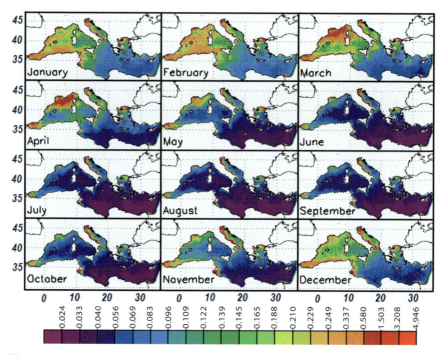

Fig. 2. SeaWiFS-derived (1998–2005) Chl-a climatological monthly means (MedOC4 algorithm). Units are [mg m^{-3}].

The Adriatic Sea colour field is largely dominated by the runoff from the north Adriatic rivers (mainly the Po River) along the Italian coast. The extension of these turbid waters varies with the seasons according to river discharge and wind. In particular, northeasterly Bora events seem to be the major responsible for offshore (eastward) export of such coastal waters (Bignami *et al.* 2006). The south Adriatic pelagic area is dominated by the presence of the south Adriatic gyre which is a well known site of dense water formation (see section 4.2). The monthly climatic means reveal that maximum Chl-a occur in March and April when values reach about 0.3–0.4 mg m^{-3} (see also Barale *et al.* 1986; Antoine *et al.* 1995; Santoleri *et al.* 2003).

4.   Open ocean deep convection and spring blooms

In the MED, atmospheric forcing is not strong enough to drive winter basin-wide open-ocean upward transport of nutrient-rich waters from intermediate and deep layers. Thus, except for coastal areas, high productivity occurs only at local sites of deep water formation (Antoine et al. 1995). However, as inter-annual variability in the extent and duration of deep-water formation is mainly driven by local buoyancy fluxes (Buongiorno Nardelli and Salusti 2000, and references therein), springtime nutrient availability and Chl-a are determined to a large extent by variations in meteorological forcing (Backhaus et al. 1999).

   Deep-water formation events are detectable in the satellite ocean colour imagery: the vertical mixing, consequence of deep water formation, pushes plankton away from the euphotic layer, and leads to a local decrease of the surface Chl-a . The subsequent depletion of phytoplankton cells results in a "deep blue hole" in the satellite images.

   In the Western Mediterranean basin, the spring bloom is more pronounced in the northern part, involving the Northern Tyrrhenian/Ligurian Sea and the Gulf of Lion, that are well-known sites of dense water formation (Morel and Andre 1991). On the other hand, the Eastern Mediterranean basin is also characterized by low annual productivity and a weak seasonal chlorophyll signal, with local exceptions such as the Rhodes Gyre and the Southern Adriatic Gyre (Antoine et al. 1995). High pigment concentrations and enhanced nutrient availability in the surface layer have been observed in these gyres following deep convection events (Zavatarelli et al. 1998; Crispi et al. 1999; Napolitano et al. 2000; Civitarese and Gacic 2001).

   A synthesis of the latest findings on the Gulf of Lion and Southern Adriatic Gyre bio-physical processes, based on SeaWiFS imagery, is presented in the following sections.

4.1   The Gulf of Lion

The NW MED has been the subject of numerous studies concerning physical and biological aspects (Levy et al. 2000, and references therein). Levy et al. (2000) used a physical-biological model with realistic forcing in order to understand the role of atmospheric input and of the mesoscale on the bloom onset. Vidussi et al. (2000) used data from a long-term station in the Ligurian Sea (DYFAMED) to characterize the phytoplankton pigment variations at the boundary of the bloom. However, a precise knowledge of the spatial and temporal scale of the phenomena is missing.

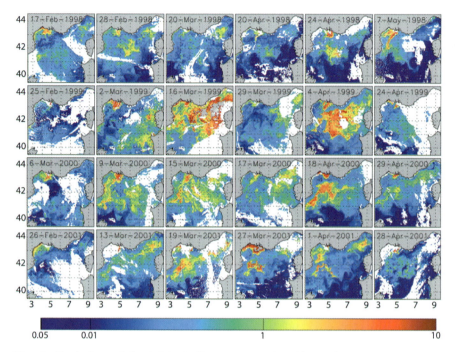

Fig. 3. Evolution of the NW Mediterranean spring bloom in the years 1998 to 2001 from SeaWiFS chlorophyll imagery. Units are mg m^{-3}.

Four years of SeaWiFS data (1998–2001) show that the spatial and temporal extension of the algal bloom is very variable (Figure 3). A "mixed-patch" is observed starting from the second half of January to the end of February in the years 1999, 2000 and 2001, always in the same location (42°N 5°E) and with rather constant shape and dimension (a circle of ~100 Km radius), in agreement with previous knowledge based on in situ data. In 1998, this feature is not clearly evident, even if the number of cloud free images is similar to that of the other years. The onset of the bloom generally occurs in the first days of March (except for 1998).

There is a direct correspondence between the end of deep water formation and the onset of the bloom, in agreement with numerous modelling studies (Tusseau-Vuillemin *et al.* 1998; Levy *et al.* 2000 and references therein). In contrast, there are some discrepancies with the analysis performed by Morel and Andre (1991), mainly in the timing of the events. They indicate mid-march as the end of the convection period and mid-April as the beginning of the bloom with one month lag between the two phenomena. On the contrary, from SeaWiFS images a much shorter time interval is seen between the end of convection and the increase of biomass, and in 1999 the two phenomena seem to happen at the same time. These discrepancies could be either due to the excessive smoothing achieved by Morel and Andre (1991) by using averaged images, or in a substantial change in the physical-biogeochemical conditions of the area in the 90's.

4.2 The Southern Adriatic Gyre

Earlier than the last decade, deep water formed in the Southern Adriatic permanent cyclonic Gyre (SAG) was considered the main source of Eastern Mediterranean Deep Water. During the abrupt climatic change known as Eastern Mediterranean Transient (EMT; see Malanotte-Rizzoli 2003), the Adriatic deep water formation was greatly reduced relative to the Aegean, with important consequences on the Mediterranean deep thermohaline circulation (Roether *et al.* 1996) and ecosystem.

Civitarese and Gacic (2001) observed increased nutrients below 200 m in the period from 1987 to 1995, i.e. during the EMT, and a decrease after 1997, suggesting the system was restored to the pre-transient conditions. They also found that this increase of the SAG internal nutrient pool was not effective in determining a significant increase in the new production because of the concomitant relaxation of the winter vertical convection.

Conversely, Santoleri *et al.* (2003) found that SeaWiFS Chl-a variability could be reproduced by their coupled physical-biological model simulations only taking into account the particular year's nutrient pool and not only the maximum convective depth. Thus, their explanation for the low Chl-a observed in 2000 (Figure 4) is the reduced nutrient pool associated to the restoring from the EMT.

Dutkiewicz *et al.* (2001) suggested that the phytoplankton production is a function of the mixing rates and identified regimes in which enhanced mixing can either increase or decrease the bloom intensity. These regimes were defined through the ratio of spring critical layer depth (hc) and winter mixed layer depth (hm). Dutkiewicz *et al.* (2001) found that in subtropical regimes ($0.6 <$ hc/hm $< {\sim}1$) the vertical mixing can increase surface productivity through enhanced nutrient supply, while in subpolar regimes (hc/hm < 0.4) the vertical mixing can delay the bloom due to transport of phytoplankton below Sverdrup's critical depth.

The hc/hm estimated by Santoleri *et al.* (2003) indicates that sub-polar regime is generally present in the SAG. However, in 1997 the hc/hm ratio is well within the sub-tropical regime. Gacic *et al.* (2002) report that in this year there was almost no winter convection with low phytoplankton production, as inferred from sediment trap data. The shifting from subpolar to subtropical regimes implies that SAG ecosystem's sensitivity to mixing can change significantly. During the EMT, generally characterized by milder winters, the mixed layer was typically shallower and tended towards a subtropical regime. Thus, a strong convective mixing rate led to enhanced phytoplankton production. In contrast, during pre- and post-Transient conditions, the system fell within the sub-polar regime with deepening of the mixed layer (hc/hm << 1), weaker sensitivity to convective mixing, and lower phytoplankton abundance as related to mixing increasing (Dutkiewicz *et al.* 2001).

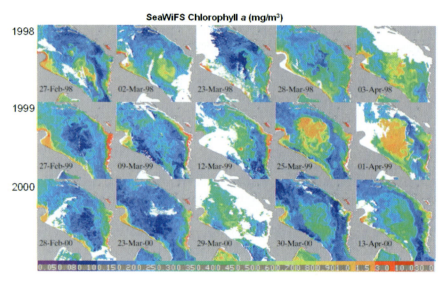

Fig. 4. Evolution of the South Adriatic spring bloom in the years 1998 to 2000 from SeaWiFS chlorophyll imagery (modified from Santoleri *et al.* 2003).

All these considerations suggest that the intense blooms seen by SeaWiFS in 1998 and 1999 are episodic signatures of the returning phase from EMT, when deep mixing events occurred, typical of the sub-polar pre-EMT regime, and nutrient concentration at intermediate levels were still higher then climatological levels. This hypothesis was further tested by Santoleri *et al.* (2003) examining the whole CZCS time series from 1979 to 1985. During these years only very moderate spring blooms (Chl-a between 0.3 and 0.5 mg m^{-3}) were observed. Even if the frequency of the CZCS observations was not as high as that of SeaWiFS, the number of available scenes and the typical duration of the event strongly support the hypothesis that the 1998 and 1999 blooms were "special events" while the weaker 2000 bloom can be considered as the normal south Adriatic situation. Santoleri *et al.* (2003) conclusions are also supported by more recent observations that indicate low values of biomass associated to the spring bloom (see Figure 4).

5.   Conclusions

Past analyses have demonstrated that the SeaWiFS OC4v4 ocean colour algorithm, when applied to the MED, performs worse than the two existing regional algorithms (BRIC, DORMA). Nonetheless, the latter algorithms present errors dependent on chlorophyll values. Thanks to an extensive integrated dataset, Volpe *et al.* (2007) recently introduced a new algorithm, MedOC4, which performs significantly better than all previous ones, matching the requirement of unbiased satellite chlorophyll estimates and reducing the satellite estimated APD and RPD errors from 117% and 103% (OC4v4) to 40% and 3% (MedOC4) over the whole range of chlorophyll, and from 134% and 133% (OC4v4) to 35% and 6% (MedOC4) in oligotrophic condition (Chl-a <0.4). It must be noted that the impact of MedOC4 on primary production estimates in the MED has been evaluated to be quite significant, on the order of 40% (Colella, 2006).

According to Volpe *et al.* (2007), the observed discrepancy between the global and the regional bio-optical algorithms in the Mediterranean, initially supposed to depend on a wrong atmospheric correction, due to the presence of peculiar aerosol in the region (D'Ortenzio *et al.* 2002; Claustre *et al.* 2002), could more likely depend on the inherent bio-optical properties of the basin. In particular, it should be investigated to what extent a different phytoplankton community structure and distribution could alter the spectral signature of the water column. This could be assessed only through more refined bio-optical measurements. Moreover, this target is expected to be at hand once the new generation of bio-optical sensors will be launched, and would certainly lead to a much deeper understanding of the Mediterranean ecosystem functioning, especially for what concerns the timing and evolution of the open ocean blooms described in the previous sections and for the possibility to retrieve reliable chlorophyll values in coastal areas. Multi-sensor approaches and proper modelling and assimilation of ocean colour data should also allow a deeper understanding of observed interannual variations.

Acknowledgements

This work was supported by the European Commission, within the framework of the MERSEA Integrated Project (Contract number SIP3-CT-2003-502885), and of the SESAME Integrated Project (Contract number 036949).

References

Antoine D, Morel A, Andre JM (1995) Algal pigment distribution and primary production in the Eastern Mediterranean as derived from Coastal Zone Color Scanner observations. Journal of Geophysical Research-Oceans 100 (C8): 16193–16209

Backhaus JO, Wehde H, Hegseth EN, Kampf J (1999) 'Phyto-convection': the role of oceanic convection in primary production. Marine Ecology-Progress Series 189: 77–92

Barale V, McClain CR, Malanotte-Rizzoli P (1986) Space and time variability of the surface color field in the Northen Adriatic Sea. Journal of Geophysical Research 91 (C11): 2957

Barale V, Schiller C, Tacchi R, Marechal C (2005) Trends and interactions of physical and bio-geo-chemical features in the Adriatic Sea as derived from satellite observations. Science of the Total Environment 353 (1–3): 68–81

Bignami F, Sciarra R, Carniel S, Santoleri R (2007) Variability of Adriatic Sea coastal turbid waters from SeaWiFS imagery. Journal of Geophysical Research-Oceans 112 C3: doi:10.1029/2006JC003518

Bonnet S, Guieu U, Chiaverini J, Ras J, Stock A (2005) Effect of atmospheric nutrients on the autotrophic communities in a low nutrient, low chlorophyll system. Limnology and Oceanography 50 (6): 1810–1819

Bricaud A, Bosc E, Antoine D (2002) Algal biomass and sea surface temperature in the Mediterranean Basin - Intercomparison of data from various satellite sensors, and implications for primary production estimates. Remote Sensing of Environment 81 (2–3): 163–178

Buongiorno Nardelli B, Salusti E (2000) On dense water formation criteria and their application to the Mediterranean Sea. Deep-Sea Research Part I-Oceanographic Research Papers 47 (2): 193–221

Civitarese G, Gacic M (2001) Had the Eastern Mediterranean transient an impact ore the new production in the Southern Adriatic? Geophysical Research Letters 28 (8): 1627–1630

Claustre H, Morel A, Hooker SB, Babin M, Antoine D, Oubelkheir K, Bricaud A, Leblanc K, Queguiner B, Maritorena S (2002) Is desert dust making oligotrophic waters greener? Geophysical Research Letters 29 (10): doi:10.1029/2001GL014056

Colella S (2006) La produzione primaria nel Mar Mediterraneo da satellite: sviluppo di un modello regionale e sua applicazione ai dati SeaWiFS, MODIS e MERIS. PhD Thesis, Univerisità degli studi di Napoli - Federico II: 162

Crispi G, Crise A, Mauri E (1999) A seasonal three-dimensional study of the nitrogen cycle in the Mediterranean Sea: Part II. Verification of the energy constrained trophic model. Journal of Marine Systems 20 (1–4): 357–379

D'Ortenzio F, Marullo S, Ragni M, Ribera d'Alcala M, Santoleri R (2002) Validation of empirical SeaWiFS algorithms for chlorophyll- alpha retrieval in the Mediterranean Sea - A case study for oligotrophic seas. Remote Sensing of Environment 82 (1): 79–94

D'Ortenzio F, Ragni M, Marullo S, Ribera d'Alcala M (2003) Did biological activity in the Ionian Sea change after the Eastern Mediterranean Transient? Results from the analysis of remote sensing observations. Journal of Geophysical Research-Oceans 108 C9: doi:10.1029/2002JC001556

Dutkiewicz S, Follows M, Marshall J, Gregg WW (2001) Interannual variability of phytoplankton abundances in the North Atlantic. Deep-Sea Research Part II-Topical Studies in Oceanography 48 (10): 2323–2344

Gacic M, Civitarese G, Miserocchi S, Cardin V, Crise A, Mauri E (2002) The open-ocean convection in the Southern Adriatic: a controlling mechanism of the spring phytoplankton bloom. Continental Shelf Research 22 (14): 1897–1908

Gregg WW, Casey NW (2004) Global and regional evaluation of the SeaWiFS chlorophyll data set. Remote Sensing of Environment 93 (4): 463–479

Levy M, Memery L, Madec G (2000) Combined effects of mesoscale processes and atmospheric high-frequency variability on the spring bloom in the MEDOC area. Deep-Sea Research Part I-Oceanographic Research Papers 47 (1): 27–53

Malanotte-Rizzoli P (2003) Introduction to special section: Physical and biochemical evolution of the Eastern Mediterranean in the 1990s (PBE). Journal of Geophysical Research-Oceans 108 C9: doi:10.1029/2003JC002063

Morel A, Andre JM (1991) Pigment distribution and primary production in the Western Mediterranean as derived and modeled from Coastal Zone Color Scanner observations. Journal of Geophysical Research-Oceans 96 (C7): 12685–12698

Morel A, Antoine D (2000) Pigment index retrieval in case 1 waters, MERIS Algorithm Theoretical Basis Document. ATBD2 9–25

Moulin C, Dulac F, Lambert CE, Chazette P, Jankowiak I, Chatenet B, Lavenu F (1997) Long-term daily monitoring of Saharan dust load over ocean using Meteosat ISCCP-B2 data.2. Accuracy of the method and validation using Sun photometer measurements. Journal of Geophysical Research-Atmospheres 102 (D14): 16959–16969

Napolitano E, Oguz T, Malanotte-Rizzoli P, Yilmaz A, Sansone E (2000) Simulations of biological production in the Rhodes and Ionian basins of the Eastern Mediterranean. Journal of Marine Systems 24 (3–4): 277–298

O'Reilly JE, Maritorena S, Mitchell BG, Siegel DA, Carder KL, Garver SA, Kahru M, McClain C (1998) Ocean color chlorophyll algorithms for SeaWiFS. Journal of Geophysical Research-Oceans 103 (C11): 24937–24953

O'Reilly JE, Maritorena S, Siegel DA, O'Brien MC, Toole D, Mitchell BG, Kahru M, Chavez FP, Strutton P, Cota GF, Hooker SB, McClain CR, Carder KL, Muller-Karger F, Harding L, Magnuson A, Phinney D, Moore GF, Aiken J, Arrigo KR, Letelier R, Culver M (2000) Ocean color chlorophyll a algorithms for SeaWiFS, OC2, and OC4: version 4. SeaWiFS Postlaunch Technical Report Series - NASA Technical Memorandum 2000-206892, Volume 11, NASA GSFC, Greenbelt, pp 9–23

Ozsoy E, Hecht A, Unluata U, Brenner S, Sur HI, Bishop J, Latif MA, Rozentraub Z, Oguz T (1993) A synthesis of the Levantine Basin circulation and hydrography, 1985–1990. Deep-Sea Research Part II-Topical Studies in Oceanography 40 (6): 1075–1119

Roether W, Manca BB, Klein B, Bregant D, Georgopoulos D, Beitzel V, Kovacevic V, Luchetta A (1996) Recent changes in Eastern Mediterranean deep waters. Science 271 (5247): 333–335

Santoleri R, Banzon V, Marullo S, Napolitano E, D'Ortenzio F, Evans R (2003) Year-to-year variability of the phytoplankton bloom in the southern Adriatic Sea (1998–2000): Sea-viewing Wide Field-of-view Sensor observations and modeling study. Journal of Geophysical Research-Oceans 108 C9: doi:10.1029/2002JC001636

Tusseau-Vuillemin MH, Mortier L, Herbaut C (1998) Modeling nitrate fluxes in an open coastal environment (Gulf of Lions): Transport versus biogeochemical processes. Journal of Geophysical Research-Oceans 103 (C4): 7693–7708

Vidussi F, Marty JC, Chiaveini J (2000) Phytoplankton pigment variations during the transition from spring bloom to oligotrophy in the northwestern Mediterranean Sea. Deep-Sea Research Part I-Oceanographic Research Papers 47 (3): 423–445

Volpe G, Santoleri R, Vellucci V, Ribera d'Alcala M, Marullo S, D'Ortenzio F (2007) The colour of the Mediterranean Sea: Global versus regional bio-optical algorithms evaluation and implication for satellite chlorophyll estimates. Remote Sensing of Environment 107 (4): 625–638

Zavatarelli M, Raicich F, Bregant D, Russo A, Artegiani A (1998) Climatological biogeochemical characteristics of the Adriatic Sea. Journal of Marine Systems 18 (1–3): 227–263

Optical Remote Sensing of Intertidal Flats

Carsten Brockmann and Kerstin Stelzer

Brockmann Consult, Geesthacht, Germany

Abstract. Intertidal areas are characterised by very high biodiversity and high productivity. Many of the intertidal flats worldwide are protected by the Ramsar Convention and other international monitoring programmes including also the Trilateral Monitoring and Assessment Programme for the Wadden Sea along the coasts of Denmark, Germany and The Netherlands. Remote sensing is a valuable tool for providing the cost efficient monitoring required by these directives. This includes mapping from airplanes, interpretation of aerial colour and infrared images, as well as classification of multispectral and hyperspectral data from airborne or spaceborne instruments. Information concerning different components of intertidal flats, namely the sediment type, macro- and microphytes and mussel beds, can be obtained by applying classification methods to these optical remotely sensed data. Comparison of the remote sensing data with in-situ measurements is an important step for improving the tuning parameters of the classification method and for validating the results. Examples are shown for a classification of the sediment type, macrophytes and mussel beds in the German Wadden Sea area. The methods presented here are currently being transferred into operational monitoring programs.

1. Introduction

Intertidal flats are defined as intertidal areas of bare mud and sand, drained and flooded through branching channels (Reise and de Jong 1999). These important wetlands have also a high ecological and economic value. Deppe (1999) has compiled a list of 350 intertidal flats worldwide, including 150 in Europe, 39 in Africa, 75 in America, 54 in Asia and 32 in Oceania. The largest connected intertidal flat is the Wadden Sea found between The Netherlands, Germany and Denmark. Most of these intertidal flats are protected by the Ramsar Convention, the European Fauna-Flora-Habitat Directive the European Water Framework Directive and the Trilateral Monitoring and Assessment Programme, all of which require regular

117

V. Barale, M. Gade (eds.), *Remote Sensing of the European Seas.*

monitoring. The intertidal areas are often remotely located and difficult to access. Field work in muddy areas is also very difficult and time consuming. Such difficulties favour the use of remote sensing techniques. For example, Bartholdy and Folving (1986) have investigated the use of Landsat Thematic Mapper (TM) for surface type mapping of the Danish Wadden Sea, Ruy *et al.* (2002) have undertaken a similar investigation for Korean intertidal areas and Kleeberg (1990) and Stelzer (1998) have studied remote sensing of the German Wadden Sea. In a systematic investigation with respect to the monitoring requirements of the Water Framework Directive (WFD), Stelzer (2004) found that sediment type and vegetation parameters (area, delineation of macrophytes and mussels) can be mapped operationally using optical remotely sensed data.

One very important advantage of satellite data is that it provides a synoptic overview of large areas. For example, the complete North Frisian and Lower Saxony Wadden Sea areas are covered by just one single Landsat TM overpass. In-situ measurements of the same area last for several months at least. One drawback, however, is that cloud and water coverage often limit the number of useful satellite data. An operational procedure which provides both reliable and temporally consistent results is currently the focus of research work in several European countries (Stelzer 2005, Deronde *et al.* 2006). The article presented here provides an overview of the currently applied methods and the results which can be achieved using automated procedures.

2. Instruments and data

Different kind of measurement systems are available today for optical remote sensing from airplanes or satellites. The data gathered with these instruments may differ in many respects:

The spatial resolution spans from a few centimetres up to more than a kilometre. The spectral characteristic of the instruments including the central wavelengths, spectral bandwidths and number of spectral bands also vary greatly. Typical multi-spectral instruments have 4–10 spectral bands, centred at wavelengths where specific absorption or scattering features exist and generally have a spectral bandwidth of ~60 nm. Hyperspectral instruments sample the spectrum in a large number of continuous spectral bands. By contrast, panchromatic instruments have only one spectral band which covers the whole visible part of the spectrum. There is often a trade-off between the spectral and spatial resolution (*e.g.* a high spatial resolution with panchromatic or a medium spatial resolution with hyperspectral band setting) due to the energy requirements of the measurements and/or due to data volume constraints. The repetition rate and spatial coverage (swath width) also have an impact on the availability of data over a given location. Satellites operate on fixed orbits. For example, Landsat has a repeat cycle of 16 days and a swath width of 185 km. In combination with the requirement of having low tide during the period of measurement together with cloud free skies, one typically obtains only 1–3 useful scenes of the Wadden Sea area during a given year. Airplanes are more flexible to operate and the flight height can be chosen to retrieve a pre-determined pixel size and spatial coverage. Finally, the costs vary from no cost up to 30 €/km² in the case of satellite data. Costs for airborne data are determined individually depending on the detailed measurement requirements.

For many applications today, aerial photography providing information with very high spatial resolution is the most commonly used remote sensing technique. Expert visual interpretation is the most common method of analysis applied to these data, *e.g.* operationally for the monitoring of saltmarshes or mussel beds. However, this method is not well suited for a large scale overview mapping and the results strongly depend on the interpreter and his or her knowledge of the area. The results cannot be easily transferred and this makes it difficult for different organisations responsible for the monitoring of a larger area to compare results. Digital multispectral data in combination with an automated classification method are more suitable in such cases. Satellite data, on the other hand, are consistent within single and across different scenes, provide a synoptic image and in most cases are cheaper than aerial photographs. The spectral resolution and the calibration of multi- or hyperspectral data is much better than those of RGB or CIR aerial photographs, thus enabling automated data analysis. It should be pointed out that present-day digital camera systems onboard aircraft also provide high quality digital imagery. However, the methodology for analysing these data is still being investigated. In summary, the decision to use a particular instrument depends on the question to be answered, namely on the spatial and spectral resolutions required, on the temporal requirements (point in time and frequency) as well as on the costs.

3. Methods

3.1 Optical properties of intertidal flat surfaces

Interpretation of optical remote sensing data is a process of inverting the radiative transfer from the sun to the surface and then to the sensor. The

measured radiance at the sensor is analysed with the aim of retrieving t
surface reflectance and relating this to the physical or biological paramc
ters of the intertidal flats. There are four main issues which need to be
taken into account in this process: (1) the spatial heterogeneity of intertidal
flats, (2) the relation between biological and morphological features and
the surface reflectance spectrum, (3) the water coverage of the surface aʳ
(4) the disturbing effect of the atmosphere.

The different surface types are characterized by their specific spec.....
reflectance signal. The spectral reflectance of sediments is determined by
its compositions of silica sand, clay, organic matter and water content. The
general shape of the spectrum is determined by the silica sand. Muddy
sediment has a higher fraction of clay and generally a higher content of organic matter, leading to higher absorption and consequently lower reflectances as compared to those of coarse sand. A weak absorption trough at
~670 nm is frequently observed in the sediment spectra. This is often
stronger in muddy sediments compared to sandy sediments and is likely to
be caused by the presence of microalgae (Hakvoort 1989, Thomson 1998).

The water content of the sediment causes a strong decrease in the reflectance with increasing wavelength, and is similar to the reflectance spectrum of pure water. The reflectance spectra of green macrophytes have a
typical vegetation spectrum dominated by chlorophyll-a absorption at
440 nm and 670 nm, causing a reflectance peak at 550 nm (green), and a
high reflectance in the NIR part of the spectrum (Mackinney 1941).

Some surface types such as mussel beds have a very inhomogeneous
appearance and are therefore difficult to detect spectrally. Other surface
types have similar spectral reflectances although belonging to different
types, for example green algae and seagrass or wet sand and dry mud.
When working with remote sensing data in intertidal flats, it is also important to consider that only the upper millimetre of the surface is detected.
During calm weather periods, sandy sediments can be covered by a thin
muddy skin, which may lead to miss-classifications. Due to the tides the
sediment surface is only part of the time under water and even when the
tide goes out water can remain on the surface as either a shallow layer or
even as a thin film. Both change the reflectance signal significantly due to
the strong absorption of water in the NIR/SWIR part of the spectrum. As
an illustration, Figure 1 shows typical reflectance spectra and the effect of
water coverage.

The elementary surface types described above alternate on all scales in
intertidal flats. Regardless of the pixel size the sub-pixel heterogeneity
needs to be taken into account in the classification. The sediment type
changes from muddy near to the shore, to coarse sand banks in the outer
regions of the flats. In between, all kinds of mixtures of sandy and muddy

sediments can be present. The sediment areas are also crossed by smaller or larger creeks and, during low tide, water remains on the sediment. This occurs on very small scales *e.g.* in ripple troughs, as well as on large scales, *e.g.* large-scale depressions. The sediments are also often covered by films of diatoms, small brownish microalgae, or assemblages of macrophytes and mussel beds. Intertidal flats are further characterized by gradual transitions; different surface types are seldom separated by sharp boundaries.

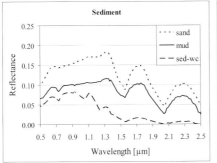 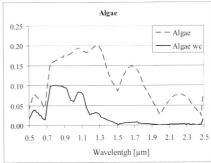

Fig. 1. Left: Spectral reflectance signals of different intertidal sediment types (sand, mud and water covered sediment [sed-wc]); right: Reflectance Spectra of dry and water covered green algae [Algae wc] (image spectra from HyMap sensor).

3.2 Visual interpretation method

Remote sensing methods currently in use for operational monitoring of the intertidal flats include mainly visual interpretation and mapping from airplanes. Objects are identified in analogue or digital airborne images based on their colour and/or texture and manual delineation of objects is performed. Due to the manual nature of this method the main disadvantage is a lack of objectivity and, as a consequence, the lack of repeatability (Schmidt 2003). Such visual interpretation of images is performed operationally primarily for mussel beds and macroalgae occurrence. For mapping mussel beds, stereographic images are used in some areas in order to use the height information of mussel beds, which are usually higher than the surrounding sediment. Besides the visual interpretation of images, a direct mapping of surface types such as seagrass beds (Reise 2004) from an airplane is also deployed operationally. However, this depends even more on the experience of the operator. Visual interpretation as well as airborne mapping is nearly always used in combination with ground truth campaigns.

3.3   Automated classification of digital image data

Even though advances in sensor technology continually improve the information content of airborne and spaceborne images, visual interpretation will always be part of the image analysis and should also be the first step to be undertaken before performing any automated classification. Automated analysis of multispectral or hyperspectral data involve the use of classification algorithms which are based on a statistical analysis of the spectral and/or textural characteristics of the data. Different classification methods can be used to classify intertidal flats. Standard classification methods are divided into either unsupervised or supervised classification techniques. Spectral modeling, such as linear spectral un-mixing, can also be used to retrieve the proportions of different surface types within each pixel. Several methodological investigations concerning the remote sensing of intertidal flats have been performed, including Kleeberg (1990), Stelzer (1998), Thomson *et al.* (1998), Ruy *et al.* (2002), Rainey *et al.* (2003), Smith *et al.* (2004), Deronde *et al.* (2006). These investigations were performed using a wide range of data sources obtained from both airborne and spaceborne sensors, *e.g.* Landsat (E)TM, CASI, DAIS and HyMap. A proper preprocessing is always a necessary first step and most importantly must include an appropriate atmospheric correction. A band selection may also be necessary in some cases, especially when using hyperspectral data. The following short descriptions summarize the procedures which are currently been used in intertidal flat applications.

- Preprocessing – Atmospheric Correction. This step includes the correction for scattering and absorption in the atmosphere and transforms the radiance measured at the sensor into a surface reflectance. Methods commonly used are, for example, 6S atmospheric correction (Vermote *et al.* 1995) or the Empirical Line Method for relative adjustment of images (Smith and Milton 1999).

- Preprocessing – Reduction of data space dimensions. This step is particularly important for hyperspectral data. The large number of spectral measurements contains a lot of redundant information and this step aims at preserving the independent information while at the same time separating this information from noise and reducing the data volume. Methods include the Sequential Floating Forward Search (SFFS) feature selection (applied in intertidal flat studies by Deronde *et al.* 2006) or the Principle Component Analysis PCA (Richard 1999).

- Classification – The classification step analyses the surface reflectance spectra and the texture information in order to retrieve classes of objects in the image. In the case of remote sensing of intertidal flats, these classes are typically the sediment type (mud – mixed – sand), vegetation type (macro algae, seagrass, Salicornia, Spartina, Diatoms, etc.) and mussel beds. Methods which have been shown to work well in intertidal areas are supervised classification or unsupervised classification with subsequent combination of classes and labelling (HIMOM 2005, Smith *et al.* 2004), Linear Spectral Un-mixing (Rainey *et al.* 2003) with subsequent decision tree labelling (Stelzer and Brockmann 2006), multiple regression between single bands and surface characteristics (Rainey *et al.* 2000, Smith *et al.* 2004) and Linear Discriminant Analysis (Deronde *et al.* 2006).

4. Validation

The validation of the classification results is a very important step. It helps improve all methods which have tuning parameters and particularly helps in the labeling step. Furthermore, and most importantly, it ensures the acceptance of the results by the user community. Validation of classification results is mainly done by comparison with in-situ measurements. It is important to understand the differences between the in-situ and remotely sensed parameters for this step. For example, in situ determination of the sediment type is usually performed by taking cores several centimeters in depth whereas the remotely sensed sediment type accounts for only the upper layer of less than 1 mm. In order to take such methodological differences properly into account it is necessary for local experts to work together with remote sensing experts during the validation step.

5. Applications

Remote sensing techniques are presently being used in several countries which have intertidal areas such as, for example, the trilateral Wadden Sea, estuaries in The Netherlands (*e.g.* the Westerschelde), the Wash in Great Britain or along the coast of Wales (Donoghue and Mironnet 2002).

In the example shown here the Linear Spectral Un-mixing (LSU) method together with a subsequent decision tree classification technique has been applied to the intertidal flats of the German Wadden Sea. The LSU divides the spectral reflectance measured in each pixel into the spectral reflectances of pure surface types contributing to this pixel. The number of so called endmember spectra (reflectance spectra of pure surface types) is mathematically limited by the number of bands. Different endmember sets can be used in order to consider the most important surface types of the scene. The derived band fractions are further classified using a decision tree classification based on pre-defined thresholds.

The surface characteristics of the intertidal flats can be modeled very well using this method primarily because both the small scale mixture of surface types and the fact that there are almost no sharp boundaries between different surface types which enables a good retrieval. Furthermore, the influence of water covering the sediment and the density of vegetation cover can also be modeled. An accurate atmospheric correction is, however, mandatory for the transferability of this method to other images. The resulting surface types retrieved by this method are different sediment type classes (ranging from sand to mud), macrophytes and mussel beds. A classification of the entire German Wadden Sea as shown in Figure 2 and in magnification in Figure 3 and was obtained using Landsat images from 2002, which cover the whole area in one overpass. It is important to note that the tidal level differs in different parts of this image. The applied method considers the influence of water coverage, and where possible, a thin water layer is corrected for by including a water covered sediment spectrum in the un-mixing procedure. If the water layer is too deep, the respective pixels are classified as shallow water. Validation of the result was performed by comparing this classification with in-situ measurements as well as an assessment of the results by intertidal experts.

5.1 Sediment type

A subset of Figure 2 showing the North Frisian Wadden Sea illustrates the classification results for sediment types (Figure 3). The sediment types are defined relative to each other. Dry sandy areas can be very well separated from mixed and muddy sediments, whereas a differentiation between mud and mixed sediments becomes more difficult.

If an accurate atmospheric correction has been performed, the classes are always comparable between images, irrespective of when the image was obtained. Figure 4 shows an example of such a time series of images to which this classification method has been applied. One image was taken in 2002, and the other two images in 2006 with only 2 weeks difference. The two classification results of 2006 are very similar and prove the transferability of the method, and the differences observed between the 2006 and 2002 images can be related to natural changes.

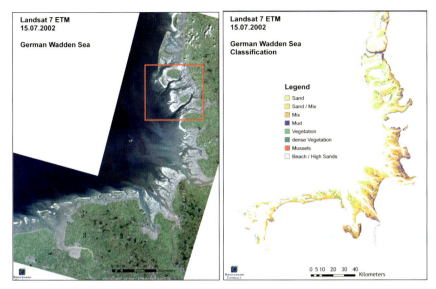

Fig. 2. Landsat images from 15.07.2002 classified by applying linear spectral Unmixing and subsequent decision tree classification (original data: Landsat 7 ETM © Eurimage 2002). The area of the red square is shown in Figure 3.

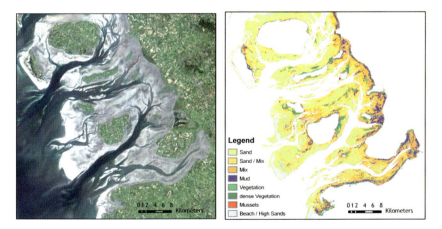

Fig. 3. Landsat image from 15.07.2002 of parts of the North Frisian Wadden Sea and respective sediment classification (magnification of Figure 2).

5.2 Macrophytes

The above described classification method also classifies different densities of vegetation coverage. The vegetation covering the sediment consists mainly of green and brown macroalgae (*Enteromorpha* and *Fucus*) and seagrass (*Zostera*). In general, vegetated areas can be spectrally very well differentiated from sediment, however the discrimination between green macroalgae and seagrass is less clear, especially when using coarse spectrally and spatially resolved images. The classification of macrophytes (macroalgae and seagrass) has been compared with the mapping results of seagrass from an airplane (Reise 2004) as part of the operational monitoring. Both methods agree in most areas, however, the classification of satellite images is more detailed in some cases and shows macrophytes classes not mapped from the airplane. These are probably macroalgae, which were not considered during the airborne campaign.

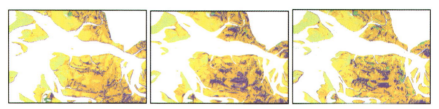

Fig. 4. Classification of images from different acquisition dates (left: 15.07.02, middle: 02.07.06, right: 18.07.06), original data: Landsat 7 and 5 © Eurimage.

5.3 Mussels

A map of the classification of mussel beds was compared with the polygons derived by stereoscopic image interpretation which is also used in the operational monitoring. Generally, brighter mussel beds can be detected quite well, whereas dark blue mussel beds in muddy sediment are difficult to detect. The main mussel beds could be detected very well if the water coverage was not too high.

Macroalgae, *i.e.* the brown algae, often settle on mussel beds which may lead to a misclassification as they obscure the underlying mussels. However, the knowledge about the location of mussel beds is in itself useful information. Mussel beds are further characterized by a very high variability in appearance. Brighter heart mussels or blue mussels have very different spectral signatures, although both have very high reflection in the near Infrared.

The results shown above demonstrate that a classification based solely on optical characteristics of surface types delivers important information but an agreement with in-situ measurements could be improved in some areas. This could be achieved by including complementary information concerning the structure of the surfaces and additional information derived from the existing long term monitoring knowledge.

6.   Conclusions and outlook

Monitoring of intertidal flats is essential for assessing the state of this ecological and economical important ecosystem. It is required by international regulations such as the Water Framework Directive. However, field work on intertidal flats is sometimes very difficult and often very time consuming and measurements are mainly performed at single points or on transects. It has been shown that mapping of sediment type, area and delineation of different macrophytes and of mussel beds using remote sensing data is possible. The method presented is transferable to different instruments and provides a consistent time series of classification results. However, many other parameters, required for an assessment of the ecological state of intertidal areas cannot, to date, be measured remotely. Therefore, remote sensing offers a very good and complementary data source to in-situ measurements in a rapidly changing ecosystem, providing in particular a synoptic and large scale overview.

However, the classification methods are based only on the spectral information of the object or surface type under consideration is, therefore, limited. Other information sources and knowledge based information layers should also be considered simultaneously. Such information could include the roughness of surfaces (derived from radar remote sensing), the height of the intertidal flats or potential areas of seagrass occurrence. The next step will involve integrating remote sensing techniques and automated classification procedures into the operational monitoring programmes.

References

Bartholdy J, Folving S (1986) Sediment classification and surface type mapping in the Danish Wadden Sea by remote sensing. Neth. J. Sea Res. 20(4): 337–345

Deppe F (1999) Intertidal Mudflats Worldwide. Practical course at the Common Wadden Sea Secretariat (CWSS) in Wilhelmshaven 1st June – 30th September 1999. cwss.www.de/news/documents/others/Mudflats-Worldwide-2000.pdf

Deronde B, Kempeneers P, Forster RM (2006) Imaging spectroscopy as a tool to study sediment characteristics on a tidal sandbank in the Westerschelde. Estuarine Coastal and Shelf Science. Vol. 69, Iss. 3–4, pp 580–590

Doerffer R, Murphy D (1989) Factor analysis and classification of remotely sensed data for monitoring tidal flats. Helgoland Marine Research Vol. 43 No. 3–4, 275–293

Donoghue DN, Mironnet N (2002) Development of an integrated geographical information system prototype for coastal habitat monitoring. Comput. Geosci. 28, 1 (Feb. 2002), 129–141

Hakvoort JHM, Heineke M, Heymann K, Kuhl H, Riethmuller R, Witte G (1998) A basis for mapping the erodibility of tidal flats by optical remote sensing. Mar Fresh Res 49:867–873

HIMOM (2005) Final Report EU Contract EVK3-CT-2001-00052: http://cordis.europa.eu/search/index.cfm?fuseaction=proj.simpledocument& PJ_RCN=5472389&CFID=10598594&CFTOKEN=73965844

Kleeberg U (1990) Kartierung der Sedimentverteilung im Wattenmeer durch integrierte Auswertung von Satellitendaten und Daten aus der Wattenmeerdatenbank der GKSS Diplomarbeit Universität Trier

Mackinney G (1941) Absorption of Light by Chlorophyll Solutions. Journal of Biological Chemistry, 1941

Rainey MP, Tyler AN, Bryant RG, Gilvear DJ, McDonald P (2000) The influence of surface and interstitial moisture on the spectral characteristics of intertidal sediments. Int. J. Remote Sensing, Vol. 21, No. 16, pp 3025–3038

Rainey MP, Tyler AN, Gilvear DJ, Bryant RG, McDonald P (2003) Mapping intertidal estuarine sediment grain size distributions through airborne remote sensing. Remote Sensing of Environment, 86, 480–490

Reise K, de Jong F (1999) Biology - The Tidal Area. In: CWSS (Eds.). Wadden Sea QSR 1999, Wilhelmshaven, Germany, pp 69–70

Reise K (2004) Vorkommen von Grünalgen und Seegras im Nationalpark Schleswig Holsteinisches Wattenmeer 2003,- Forschungsbericht

Richards JA (1999) Remote Sensing Digital Image Analysis: An Introduction. Springer, Berlin. 3rd Edition.

Ryu JH, Na YH, Won JS, Doerffer R (2002) A critical grain size for Landsat ETM+ investigations into intertidal sediments: A case study of the Gomso intertidal flats, Korea

Schmidt KS (2003) Hyperspectral Remote Sensing of Vegetation Species Distribution in a Saltmarsh. Dissertation. International Institute for Geo-Information Science and Earth Observation Enschede

Smith GM, Milton EJ (1999) The use of the empirical line method to calibrate remotely sensed data to reflectances. IJRS Vol. 20, No. 13, pp 2653–2662

Smith GM, Thomson AG, Möller I, Kromkamp JC (2004) Using hyperspectral imaging for the assessment of mudflat surface stability. Journal of Coastal Res., Vol. 20, No. 4, pp 1165–1175

Stelzer K (1998) Erfassung der Sedimentverteilung des Schleswig-Holsteinischen Wattenmeeres mit Hilfe multispektraler Fernerkundungsdaten. Diplomarbeit Universität Trier

Stelzer K (2004) Potenzial der Fernerkundung im Küstenraum für die Umsetzung der EG-Wasserrahmenrichtlinie, TMAP und FFH - Abschlussbericht; http://www.brockmann-consult.de/english/flyers/pdf/PotenzialFEKueste_Abschlussbericht.pdf

Stelzer K, Brockmann C (2006) Optische Fernerkundung für die Küstenzone. In: Traub KP, Kohlus J (Eds.): GIS im Küstenzonen Management. - Grundlagen und Anwendungen. - Heidelberg

Thomson AG, Fuller RM, Sparks TH, Yates MG, Eastwood JA (1998) Ground and airborne radiometry over intertidal surfaces: waveband selection for cover classification. Int. J. Remote Sensing, Vol. 19, No. 6, pp 1189 –1205

Visible and Infrared Remote Sensing of the White Sea Bio-Geo-Chemistry and Hydrology

Dmitry Pozdnyakov[1,2], Anton Korosov[1], and Lasse Pettersson[2]

[1] Nansen International Environmental and Remote Sensing Centre,
 St.- Petersburg, Russia
[2] Nansen Environmental and Remote Sensing Centre, Bergen, Norway

Abstract. A new operational non-satellite-specific algorithm for a simultaneous retrieval of contents of phytoplankton chlorophyll (*chl*), suspended minerals (*sm*) and dissolved organic carbon (*doc*) from space sensor data, was employed to monitor the surface expressions of some biotic and abiotic processes in the White Sea (WS). A special technique has been developed to reconstruct the seasonal variations of the above substances in pixels occasionally masked by cloudiness. The developed software package provided a means to obtain the series of intra-annual spatial and temporal variations of *chl*, *sm*, *doc* and sea surface temperature throughout the WS from SeaWiFS and AVHRR, respectively. The observed variations are controlled by (a) the dynamics of water turbidity and opacity due to seasonal variations in the content of *sm* and *doc* driven by the river discharge varying influence, and (b) thermo-hydrodynamic processes encompassing water density currents, tides, upwellings, fronts, etc.

1. Introduction

The White Sea (WS) is attracting growing attention from both scientists and society. This is due to a new stage in the development of natural resources of the WS and its catchment area. These are extensive fishing, steadily developing farming of mariculture, vast engineering activities aiming to transport natural gas from the deposits in the Barents Sea shelf zone to west Europe. In addition, rapidly unfolding exploitation of diamond and gold quarries, escalating timber cutting/deforestation and agricultural development along with many other activities are inevitably affecting the WS ecosystem. The ongoing climate change is an extra forcing factor that should not be overlooked.

V. Barale, M. Gade (eds.), *Remote Sensing of the European Seas.*
© Springer Science+Business Media B.V. 2008

The WS waters are mesotrophic (Filatov *et al.* 2005). The content of dissolved organic carbon (*doc*) is high across the entire sea, whereas the bays are also rich in suspended, first and foremost, mineral matter (*sm*), due to river discharge but also land runoff. Hence, the optical properties of the WS waters qualify as Case 2 waters (Morel and Prieur 1997).

In turn, this indicates that the National Aeronautics and Space Administration (NASA) and European Space Agency (ESA) standard algorithms used for the retrieval of chlorophyll (*chl*) in Case 1 waters cannot be applied for the WS. Besides, for a comprehensive characterization of ecological processes in mesotrophic water bodies, solely *chl* data are insufficient and information about other water constituents such as *sm* and *doc* is also mandatory. Therefore, non-standard retrieval algorithms are required for processing satellite water colour data from the WS.

Due to cloud screening and imperfect atmospheric correction, there are generally some lacunas in water colour and Sea Surface Temperature (SST) data for certain time-intervals and areas of the WS. To overcome this impediment, some special technique had to be developed and applied to processing remote sensing data on the WS water colour and SST.

Data collected by the Sea-viewing Wide Field-of-view Sensor (SeaWiFS) and the Advanced Very High Resolution Radiometer (AVHRR) were used to obtain time and space distributions of *chl*, *sm*, *doc* and SST, respectively.

2.   General characterization of the White Sea

The WS (Figure 1) is a marginal shelf sea, with a total area of 90,800 km^2, including islands. The greatest width, which is between the cities of Arkhangelsk and Kandalaksha, reaches 450 km. The WS has a watershed area of about 715, 000 km^2. Assessed at 2,200 km^3, the average volume water exchange between the White and Barents Seas (Filatov *et al.* 2005) is highly essential for the WS hydrology and biogeochemistry.

The northern part is connected to the Barents Sea through a strait called Voronka (Funnel). The southern and central parts of the WS, called Bassein (Pool) are the largest and deepest regions of the sea. There are also several large bays in the area, namely Dvinskiy, Onezhskiy, Mezenskiy and Kandalakshskiy Bays, called (with the exception of the latter one) after the respective inflowing rivers (Severnaya Dvina, Onega, and Mezen). It is the Niva River that flows into Kandalakshskiy Bay.

The bottom relief of the sea is irregular: the mean depth is 67 m and the maximum depth is 350 m (Figure 1). Located within the Bassein, the central depression, with depths exceeding 100 m, extends from Kandalashskiy Bay to Dvinskiy Bay.

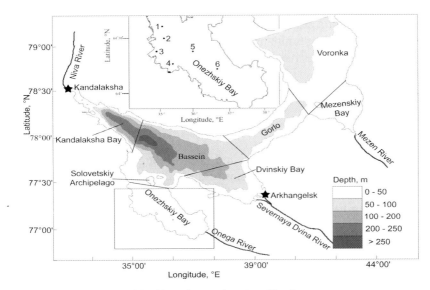

Fig. 1. The WS geographical location and regionalization.

The southern bays are shallow. Onezhskiy Bay, the shallowest region with the depth ranging within 5–25 m, is separated from the central part of the sea by the Solovetskiy Archipelago. The outer part of Kandalaksha Bay is the deepest (~350 m) region of the WS.

As compared to the WS total water volume, the freshwater discharge per year is small (only 5%), but has major consequences for the marine thermo-hydrodynamics and biogeochemistry. The cumulative share of the aforementioned rivers in the annual freshwater input into the sea reaches 88%. The biggest annual discharge comes from the Dvina River (122.3 km^3), the lowest one (21.7 km^3) from the Niva River. The river discharge is highest in May with a secondary enhancement in October.

The water of the tributaries is rich in humic substances: the concentration of *doc* can be as high as 18.3 mgC dm^{-3}. In the open parts of the WS, the maximum concentration of *doc* is much lower (4–5 mgC dm^{-3}).

The WS water is mostly greenish in colour. In river mouths, the colour varies from yellow to brownish. The Secchi depth is 7–8 m in the open sea and 2–3 m in the delta areas. In summer, when riverine waters propagate over larger offshore areas, the seawater transparency is reduced, and the prevailing colour is yellow-green.

The WS is characterized by both the presence of a well-pronounced vertical stratification and fairly large seasonal fluctuations of surface and bulk water temperature. The mean SST reaches 14–18°C in August but declines rapidly in October.

The thermo-hydrodynamics of the WS is fairly intricate although there are some persistent features, *i.e. coastal* currents running along the western coastline from north to south and further back to north in the east, as well as permanent eddies residing mostly in the Bassein.

Semi-diurnal and diurnal tides affect significantly the hydrodynamics of the WS by influencing first of all sea water level (in Mezenskiy Bay, the syzygial tide is 8 m but gradually declines down to 0.7 m in Onezskiy Bay), but also the rate of water mass exchange between the WS and the Barents Sea.

A variety of frontal zones are found in the WS. Some are formed under the influence of river runoff (the runoff and estuarine fronts), whereas the others are driven by tides, seasonal heating as well as by specific features of the bottom relief and shoreline contours.

3. Methodology

3.1 Water quality retrieval algorithms

The developed fast operating algorithm for a simultaneous retrieval of the concentration vector $C = (C_{chl}, C_{sm}, C_{doc})$ is based on two segments - the Levenberg–Marquardt (L–M) multivariate optimization approach in combination with a Neural Network (NN) emulation technique. As inputs, the algorithm uses spaceborne spectral values of subsurface remote sensing reflectance, R_{rsw} which by definition is the upwelling spectral radiance just beneath the water-air interface, $L(-0, \lambda)$ normalized by the downwelling spectral irradiance, $E(-0, \lambda)$ at the same level. Initially, the NN segment of the algorithm rapidly determines the *ranges* of the concentration vector C of the water Colour Producing Agents (CPA), *i.e. chl, sm* and *doc*, which are then used as inputs for the L–M procedure as the *starting* CPA concentrations to determine for each sensor's wavelength, i the absolute minimum of the functional $f(C) = \sum_i g_i^2(C)$ where the term $g_i = (R_{rsw} - R_{rsw,cal})/R_{rsw}$ is the weighted squared residuals between the measured, R_{rsw} and calculated, $R_{rsw,cal}(\lambda, C, a, b_b)$ subsurface remoter sensing reflectance. The $R_{rsw}(\lambda, C, a, b_b)$ parameterization used in this algorithm has been suggested by Jerome *et al.* (1996) and requires spectral values of the CPA absorption and backscattering coefficients (*i.e.* a hydro-optical model). The employed combination of the L–M and NN segments increases manifold the speed of the algorithm determining simultaneously all three CPA components. Importantly, the algorithm is not sensor-specific and can be used to infer the CPA concentrations in Case 2 waters from any ocean colour mission satellite sensors. It is neither area-specific given an adequate hydro-optical model of the targeted water body. The algorithm is outfitted with two *quality assessment units* eliminating *a*) pixels with imprecise atmospheric correction, and *b*) pixels corresponding to waters, whose hydro-optical properties cannot be accurately described by the employed hydro-optical model. A detailed description of the employed advanced bio-optical algorithm is given in (Pozdnyakov *et al.* 2005a).

Since there is no hydro-optical model specifically developed for the WS, the employed hydro-optical model was the one for Lake Ladoga (Pozdnyakov and Grassl 2003). This is justified by the fact that both water bodies (*i*) are within close latitudinal zones, and (*ii*) have geomorphologicallly similar catchment areas. It implies that *doc* and *sm* composition and optical properties can be akin for both water bodies. The phytoplankton in the WS *estuaries* are mostly of riverine origin, and hence are rather similar to the phytoplankton indigenous to Lake Ladoga (Sapozhnikov 1994). In pelagic waters they are maritime and, strictly speaking, differ from those inherent in Lake Ladoga, but given close latitudinal zones, they encompass by and large the same main algal taxa such as Bacillariophyceae, Chrysophyceae, and Cryptophyceae (Sapozhnikov 1994).

The developed bio-optical algorithm was validated for the conditions of the WS. Table 1 illustrates a comparison of water quality variables retrieved from SeaWiFS images over Onezhskiy Bay with the respective *in situ* data (Pozdnyakov *et al.* 2005). The comparison indicates that the algorithm retrieves well *chl* and *doc* concentrations, but is less accurate for *sm*. One of the reasons is undersampling. Indeed, *in situ* measurements refer to a point on the surface of the bay where the sample was taken, whereas the SeaWiFS pixel measures 1.1×1.1 km^2. Thus, the remote sensing CPA values in Table 1 are in reality spatially averaged. Given the well-known spatial inhomogeneity of CPA distributions, such inconsistencies are generally inevitable unless the sampling is frequent enough to account for the above effect. Based on numerous validations (Pozdnyakov *et al.* 2005), the assessed CPA retrieval error is ±50%.

The AVHRR SST data were obtained as ready-to-use products provided by NASA (Kidwell 1997) using the window-split algorithm. The SST assessed accuracy is ±0.5°C.

Table 1. Comparison of *chl* (μg dm^{-3}), *sm* (mg dm^{-3}) and *doc* (mgC dm^{-3}) concentrations retrieved from SeaWiFS images over Onezhskiy Bay with the *in situ* data both obtained on 10 July 2002 for several locations in the bay (Figure 1, inset).

Stations		1	2	3	4	5	6
in situ	*chl*	1.6	1.1	1.1	1.5	1.8	1.8
	sm	8.3	5.8	5.9	5.8	6.3	5.3
	doc	0.25	0.25	0.65	0.7	0.3	0.1
Remote	*chl*	1.3	1.2	1.1	1.5	1.6	1.7
sensing	*sm*	6.5	5.5	4.5	4	4	3.9
	doc	0.8	0.7	1.1	1	1	0.9

3.2 Interpolation algorithm

In order to generate a sequence of images in the cases with frequent cloudiness, a special interpolation technique has been developed. Based on statistically ample data on the WS area, for each pixel with coordinates x and y, established were polynomials relating the retrieved parameter Ψ with the respective date, *i.e.* yielding the temporal variations of the parameter Ψ : Ψ $[x, y] = a_0 + a_1 \cdot date + a_2 \cdot date^2 + a_3 \cdot date^3 + ... a_n \cdot date^n$, where $a_j (j = 0...n)$ are the expansion coefficients, n is the polynomial order. The latter was tested through assessing the significance of the polynomial expansion coefficients. The resulting polynomials were then used for the reconstruction of temporal variations in CPAs and SST in each pixel across the WS for the days with missing/incomplete satellite observations. Importantly, the resulting time resolution of the image sequence can be as high as a few days. This technique is also very useful for producing computer-based animations displaying the spatial/temporal variations in CPAs and SST. Since this approach is not variable-specific, it can be applied to any other satellite-derived product. The algorithm is described in detail in (Korosov *et al.* 2006).

3.3 *In situ* measurements

Shipborne determinations of *sm* and *doc* were performed following the methodologies described in Greengerg *et al.* 1992. For *sm*, the membrane Synpor filters of a 0.85 μm nominal pore size were used. The applied sample incineration temperature was 500°C. *doc* was determined by means of bichromatic oxidation with eventual transition to units of mg C dm^{-3} via a specific coefficient of proportionality (0.375). For *chl* Vladipore filters (with 1 μm pores) and the UNESCO spectrometric method were used

(UNESCO/SCOR 1966). At each station, sampling was performed at three depths: 0.5 m, 2 m and 3 m. However, the preceding high winds resulted in enhanced vertical mixing, so that vertically averaged CPA concentrations differed but insignificantly from subsurface counter values. Located within the area of a narrow depth range (20–40 m), all stations (Figure 1) were performed on the same day of the satellite overflight. The errors of *in situ* determinations of C_{chl}, C_{sm} and C_{doc} are ±5, 1.5 and 10%, respectively.

4.   Remote sensing results

4.1   Spatial and temporal variations of *chl, sm, doc* and SST

The methodology described in section 3.1 was employed to process SeaWiFS images over the WS for the entire period of the sensor operation, *i.e.* 1998–2004. AVHRR data were processed for the same period. Distributions of *chl, sm, doc* and SST were obtained with a time interval of five days. Based on these data, monthly averages were calculated for the period April–September (Figure 2).

The spring phytoplankton are mostly diatoms, which amply develop in cool water. In *April*, while some parts of the WS still remain covered by ice (Filatov *et al.* 2005), the phytoplankton start developing in the southern part of Onezhskiy Bay, and, to a lesser degree, upward to the north, in Dvinskiy Bay in the area immediately neighbouring the delta of the Severnaya Dvina River (Figure 2). There is an area of enhanced *chl* concentrations to the south of the Solovetskiy Archipelago, this area being known as a stable upwelling zone. These distributions are governed by both the nutrients and light availability. In the bays, the nutrients are brought in with river discharge, while in the vicinity of the Solovetskiy Archipelago, they are entrained upward by the upwelling movements. Enhanced *chl* concentration can be observed along the southern coastline of the Gorlo. The concentrations of *doc* and *sm* are still very low throughout the sea.

In *May*, as SST gradually increases, but mostly in shallow Onezhskiy and Dvinskiy bays (however, not yet in Mezenskiy Bay, which is more to the north), and the river discharge is the highest (see Section 2), the pattern of CPA and SST distributions change accordingly (Figure 2).

The phytoplankton strongly develop in Onezhskiy and Dvinskiy Bays and also in the pelagic region of the WS. At the same time, the phytoplankton outburst starts also in Mezenskiy Bay, mainly cold water diatoms that have already bloomed in the two southern bays.

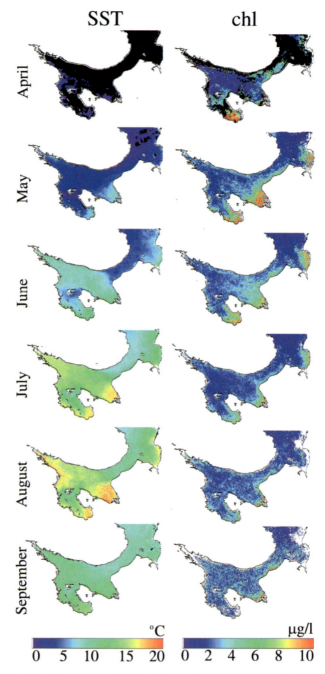

Fig. 2. Panel 1. Monthly averaged SST (°C) and concentration of *chl* (µg dm^{-3}), over the White Sea, for the time period 1998–2004, as obtained from AVHRR and SeaWiFS data.

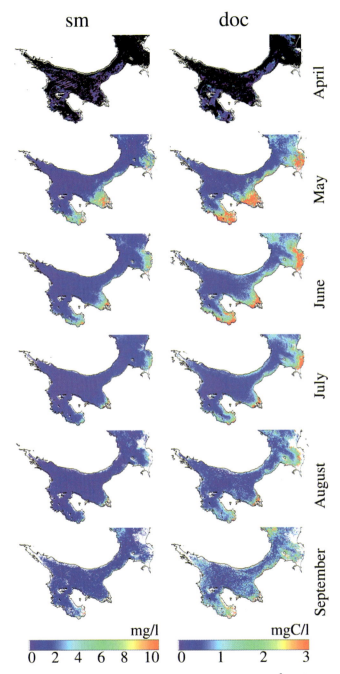

Fig. 2. Panel 2. Monthly averaged concentrations of *sm* (mg dm^{-3}) and *doc* (mgC dm^{-3}), over the White Sea, for the time period 1998–2004, as obtained from SeaWiFS data.

The observed appreciable increase of concentrations of *sm* and *doc* in the bays is definitely due to the peaking river discharge, which is rich in both *sm* and dissolved and suspended soil humus fractions. Phytoplankton are abundant in the Gorlo, mostly along its southern coastline, the algae being advected from Dvinskiy Bay by the coastal current (see Section 2).

Two frontal zones due to river discharge in Onezhskiy and Dvinskiy Bays are distinctly discernible in the SST distribution (Figure 2).

In *June*, the concentrations of CPAs start somewhat receding, which is due to the developing depletion of nutrients for phytoplankton and lower river discharge for *sm* and *doc*. At the same time, the frontal zone separating the Bassein and Gorlo as well as the tide-driven upwelling zone around the Solovetskiy Archipelago become more obvious in the SST distribution (Figure 2). Warm water extends northwards, reaching the deep parts of the WS, *i.e.* Kadalakshskiy Bay and the Bassein. As a result, the thermal gradients across the frontal zones that are due to river discharge temporarily diminish.

In *July*, the SST in the main shallow bays continues to increase and reaches 17–18°C in the innermost parts of Onezhskiy and Dvinskiy Bays. The thermal gradients within the river discharge frontal zones become again enhanced, whereas the spatial distribution of SST becomes rather homogeneous over vast areas covering Kandalakshskiy Bay and the Bassein but also Mezenskiy Bay. However, there is a strip of relatively cool water extending from the Voronka through the Gorlo along its northern coast (known as the Derugin current), which is the residual of the frontal zone caused by the tides.

The phytoplankton concentration continues receding throughout the WS as the nutrient availability becomes more and more scarce. Low levels of river discharge result in shrinking of the associated frontal zones in Onezhskiy and Dvinskiy Bays as well as in a substantial decrease of content of *sm* and *doc* there.

The changes in SST observed in thermal tendencies during July continue into *August*: the water temperature increases further, reaching 20°C not only in bays but also along the north–western coastline of the WS. The thermal fronts persist in the bays, and the tidal front separating the Gorlo and Bassein becomes again expressed somewhat stronger.

At the same time, the biogeochemical cycle in the WS reaches the phase when the nutrient concentrations increase slightly again after the mid-summer minimum. A second, although far less strong phytoplankton bloom begins. Concentrations of *sm* and *doc* remain at low levels.

The phytoplankton growth continues till *mid September*. The autumnal/secondary maximum is not characterized by a well-developed algal biomass peak, as it is influenced by the declining water temperature and lower solar irradiance, although the nutrients might remain fairly abundant during autumn (Filatov *et al.* 2005). Concentrations of *sm* begin to increase slightly, which is thought to be due to the second/autumnal increase in river discharge, peaking in October. However, apparently this does not entail a substantial enhancement of *doc* concentrations.

The surface thermal expressions of the upwelling front in Onezhskiy Bay as well as the tidal front, constituting the Bassein/Gorlo abutment, become further accentuated.

In *October*, the SST is rather homogeneously distributed throughout the sea. The phytoplankton activity comes to an end indicated by the seawater transparency (Secchi depth) that increases to eventually reach its maximum of 7–8 m.

Thus, for the first time the major biogeochemical and thermohydrodynamic features have been visualized from the collected satellite data in the visible and thermal infrared.

4.2 Comparison with historical *in situ* data

In the absence of *in situ* data covering the entire White Sea synchronously (or at least, quasi-synchronously) with spatial resolution comparable to satellite sensors, historical *in situ* data on CPA concentration or SST value ranges have to be used.

Regarding CPAs, historical *in situ* data exist only for *chl* and *doc* concentrations. In the middle and outer parts of Dvinskiy Bay the *chl* concentration in surface waters was registered between 21 June and 8 July 1991 at 8–10 μg dm^{-3} and 3.0–10 μg dm^{-3}, respectively, whereas in the inner part of Kandalakshskiy Bay it was at 3–4 μg dm^{-3}. In the Gorlo, the surface water *chl* concentration was 3 μg dm^{-3}. The concentration of *doc*, varies across the sea remaining, however, below about 5 mg dm^{-3} in pelagic areas, but can reach 18 mg dm^{-3} in the innermost parts of the bays. A comparison with remotely sensed concentrations in Figure 2 shows good general agreement, although, of course, it would be unreasonable to expect a better match because the historical in situ data are scarce and date back to the early 1990s.

For SST, the available in situ data indicate that in late June-early July, it reaches about 4°C and 7°C along the northern and southern coast of the Gorlo. In the Bassein, it ranges between 4°C at the boundary with the Gorlo and 14°C in the areas neighbouring the outer parts of the principal bays. In the bays, it increases with the approach to river estuaries, even reaching 19–21°C (Filatov *et al.* 2005). These SST ranges compare well the range of SST obtained from the satellite data.

5.   Concluding remarks

Thus, for the first time, the WS, has been comprehensively studied from space via combining satellite visible and infrared data. The developed advanced methodologies allowed revealing the spatial and temporal variations SST and CPAs and causally relate ecologically important biogeochemical processes with the sea's thermo-hydrodynamics. The documented perennial life cycles of the phytoplankton community, entrainment and dispersion of phytoplankton by currents, establishment, weakening and disappearance of fronts of various nature provide a much deeper insight into the essence, dynamics and interactions between marine biotic and abiotic processes.

References

Filatov N, Pozdnyakov D, Johannessen O, Pettersson L, Bobylev L, eds (2005) White Sea: its marine environment and ecosystem dynamics influenced by global change. Springer-Praxis, Chichester

Greengerg A, Clesceri L, Eaton A, eds (1992) Standard methods for the examination of marine water and freshwater. Elsevier Pub., Amsterdam

Jerome J, Bukata R, Miller J (1996) Remote sensing reflectance and its relationship to optical properties of natural water. Int. J. Rem. Sens., 17: 43–52

Kidwell KB (1997) Noaa Polar Orbiter Data Users Guide. U.S. Department of Commerce, Suitland USA

Korosov AA, Pozdnyakov DV, Filatov NN, Grassl H, Mazourov AA, Loupyan EA, Ionov VV (2006) A satellite data-based study of seasonal and spatial variations of some ecoparameters in Lake Ladoga. Earth Obs. Rem Sens. 5: 76–85 (in Russian)

Morel A, Prieur L (1977) Analysis of variations in ocean colour. Limnol. Oceanogr. 22: 709–722.

Pozdnyakov D, Grassl H (2003) Colour of Inland and Coastal Waters: A methodology for its Interpretation. Springer-Praxis, Chichester

Pozdnyakov D, Korosov A, Grassl H, Pettersson L (2005) An advanced algorithm for operational retrieval of water quality from satellite data in the visible. Int. J. Rem. Sens. 26: 2669–2688

Sapozhnikov V, ed (1994) Comprehensive Studies of the White Sea Ecosystem. Russian Seas Ecology Series, Publ. All-Russia Res. Inst. f Fishery and Oceanogr, Moscow (in Russian)

Remote Sensing of Coastal Upwelling in the North-Eastern Atlantic Ocean

Isabel Ambar and Joaquim Dias

Instituto de Oceanografia, Faculdade de Ciências, Universidade de Lisboa, Portugal

Abstract. The Sea Surface Temperature (SST) distribution of the sub-tropical North Atlantic shows low SSTs off Iberia and NW Africa due not only to cold water advection by the southward branch of the subtropical gyre but also to the upwelled cold subsurface waters associated to the wind-induced coastal divergence. The upwelling off the western coast of Iberia has a seasonal character associated with the northerly winds during summer whereas, further south (20–25°N), the upwelling off NW Africa is a permanent phenomenon driven by the trade winds. Since low SST and high productivity of the upper ocean characterize upwelling, remote sensing is a widely used tool for its study. The present paper reviews the main aspects, based in satellite observations, related to the physical and biological characteristics of the upwelling off Iberia and to their seasonal and interannual changes.

1. Introduction: coastal upwelling and remote sensing

1.1 Coastal upwelling

Coastal upwelling is a wind-driven phenomenon with strong repercussions on the temperature and biology of the ocean upper layer, occurring at appropriate places and times of the year along the eastern sides of the major ocean basins (Wooster *et al.* 1976). Examples of places where upwelling can occur are the coastal regions off the west coasts of America (California, Peru), of Africa, of Europe (Iberia) and all of them are associated with high rates of primary production and important coastal fisheries.

Coastal upwelling is a consequence of the Ekman transport associated with surface wind stress along a coast on the left (right) of the wind in the Northern Hemisphere (Southern Hemisphere). The wind-driven current decreases exponentially with depth and the resulting horizontal mass

V. Barale, M. Gade (eds.), *Remote Sensing of the European Seas.*
© Springer Science+Business Media B.V. 2008

transport - Ekman transport - is directed to the right of the wind in the Northern Hemisphere. When the Ekman transport is directed offshore, the resulting water depletion in the upper layer creates a surface divergence near the coast which, by mass continuity, forces an ascending motion (up-welling) of subsurface water coming from depths not greater than 200–300 m. The coastal divergence creates also an alongshore sea surface depression which results, by geostrophic adjustment, in an equatorward jet parallel to the coast.

In general, the temperature of the ocean decreases with depth, and so the ascending subsurface water is usually colder than the surface water. If the upwelling-favourable wind is blowing for sufficient time, the colder waters from below will eventually be exposed to the surface, forming a thermal and density front separating cold coastal water from warmer offshore water. This near-surface lateral density gradient - upwelling front - is a region of horizontal velocity shear causing barotropic and baroclinic instabilities.

Usually, the ascending subsurface waters, besides being colder are also richer in nutrients. Thus, the occurrence of upwelling brings new nutrients into the photic zone allowing phytoplankton growth on the inshore side of the upwelling front, while oceanic waters offshore of the front remain poor in phytoplankton due to nutrient depletion. The intensification of plankton production during upwelling is biologically important as can be proved by the fact that upwelling regions (which are about 1% of the ocean surface) account for roughly 50% of the world's fisheries landings.

1.2   Use of remote sensing

Taking into consideration the main sea surface features associated with coastal upwelling, namely the lateral gradients of temperature and of phytoplankton biomass across the fronts, one can understand the reason why remote sensing in both infrared and visible wavelength ranges is so widely used to study the physical and biological aspects of coastal upwelling and their space and time evolution.

Oceanic infrared remote sensing, based on the measurement of the thermal radiance emitted by the ocean, allows retrieving the Sea Surface Temperature (SST) corresponding to the temperature of an infinitesimally thin layer of the ocean.

The remote sensing in the visible wavelengths ("ocean colour") is based on the effect of underwater scattering and absorption of light by seawater molecules, pigments in dissolved organic matter or in phytoplankton cells and suspended particles. The ocean colour corresponds to the variations in the visible spectrum due to the relative concentrations of the absorbers and scatterers in the water (Robinson 2004). Phytoplankton cells contain pigments that strongly absorb part of the sunlight visible spectrum and the photosynthetic pigment chlorophyll-*a* is usually the dominant. As phytoplankton constitute the base of the marine food web by converting CO_2 into organic carbon, the chlorophyll-*a* concentrations give an estimate of the primary production (phytoplankton rate of growth and carbon fixation).

The potentialities of using SST and ocean colour data in upwelling studies can be further explored by combining them. An example of this is the estimate of the upwelling vertical velocities proposed by Ruiz and Navarro (2006) in a study for the Gulf of Cadiz.

The present paper focus the main role of remote sensing on the studies of the physical and biological characteristics of coastal upwelling systems in the northeastern Atlantic, namely off Iberia.

2.   Upwelling regimes along the eastern North Atlantic

One of the main differences between the coastal upwelling systems off NW Africa and off Iberia (Figure 1) is their seasonal cycle, this behaviour being directly connected with the trade winds migration. In general, large scale upwelling favourable trade wind patterns migrate northward during spring and summer, reaching their most northerly position in August followed by a southward displacement until the most southern position is reached in February (Nykjaer and Van Camp 1994). The boundary of the trade winds is approximately between 32° and 20°N in summer and between 25° and 10°N in winter. Therefore, from 25° to 20°N, fairly steady and strong trade winds exist all year round.

Using merchant ship observations (1850–1970), Wooster *et al.* (1976) found a general pattern of variation of the upwelling with latitude which was in accordance with the trade winds migration: (i) 12–20°N, upwelling from January to May; (ii) 20–25°N, strong upwelling throughout the year; (iii) 25–43°N, strongest upwelling from June to October.

The increasing use of remote sensing, especially in the infrared range, has proved how fruitful it could be for the study of the onset and evolution of the upwelling phenomenon on a synoptic scale that oceanographic cruises never reach. Before the availability of satellite remote sensing data, airborne surveys of SST had already been used in several studies of upwelling.

A set of SST satellite images (1981–1989) was used by Nykjaer and Van Camp (1994) to evaluate the seasonal variability of SST and coastal upwelling along the NW Africa: (i) south of 20°N, upwelling during winter;

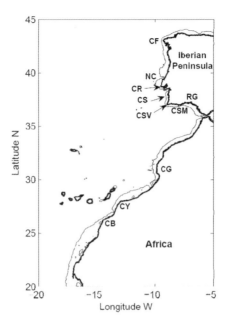

Fig. 1. Coastline configuration along the eastern boundary of N Atlantic (CB – Cape Blanc, CY – Cape Yubi, RG – River Guadiana, CSM – Cape Santa Maria, CSV – Cape S. Vicente, CR – Cape Roca, CF – Cape Finisterre). The 200-m bathymetric contour is also shown.

(ii) 20–26°N, upwelling throughout the year; (iii) 30–32°N, upwelling strongest in summer; (iv) 32°–37°N, upwelling not evident; (v) north of 37°N, upwelling during summer.

Figure 2 shows monthly average values (August 2002 and February 2003) of remote sensed SST and Chlorophyll-*a* concentration for the eastern border of the North Atlantic. The occurrence of coastal upwelling is associated with colder water nearshore which sometimes extends offshore locally as filaments. The comparison between the SST distributions in summer (Figure 2a) and in winter (Figure 2b) illustrates the main features associated with the upwelling patterns and their seasonal dependence. Southward of Cape Blanc (Figure 1) at about 21°N and until 12°N, the upwelling is only present in winter (Figure 2b) whereas northward of about 37°N only in summer (Figure 2a). In the region between about 20°N and 28°N (Cape Yubi), we find strong upwelling in summer and also in winter. Between 28°N and 33°N, there is strong upwelling in summer but very light in winter and between 33°N and 37°N there is no indication of upwelling either in summer or winter. To complement these observations, we can look at the chlorophyll-*a* concentration (Figures 2c and 2d) and point out the main differences between summer and winter distributions. In fact, a large offshore extension of high concentrations southward of about 21°N in winter (Figure 2d) confirms the occurrence of strong upwelling in that region at that time of the year.

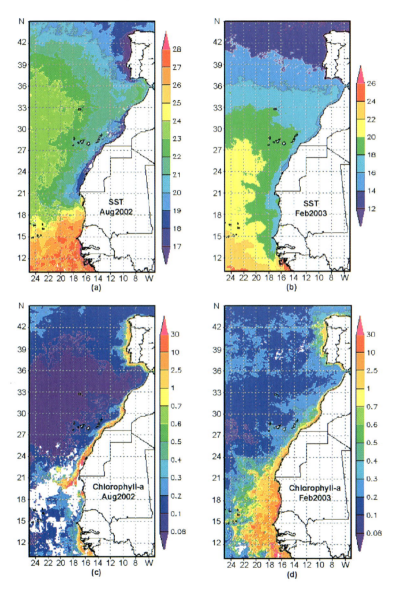

Fig. 2. Monthly average values of remote sensed SST (°C) and chlorophyll-*a* concentration (mg m^{-3}) for the border of NE Atlantic: (a) SST (Aug 2002); (b) SST (Feb 2003); (c) chlorophyll-*a* (Aug 2002) and (d) chlorophyll-*a* (Feb 2003). NASA imagery.

On the other hand, the presence of a broad region of high chlorophyll concentrations with well developed filaments extending offshore in several points off Iberia in summer (Figure 2c) shows the effects of the upwelling on the productivity of these waters. It must be emphasized that the use of ocean colour is not as conclusive as SST images for detecting coastal upwelling occurrence, since there are processes other than upwelling (*e.g.*, rainfall, river discharges) affecting chlorophyll concentrations along the coast (Peliz and Fiúza 1999; Ribeiro *et al.* 2005; Navarro and Ruiz 2006).

3. Coastal upwelling system off Iberia

A few years after the intensive observational program (in the 1970's) on the upwelling off NW Africa, a systematic study started of the coastal upwelling system off Iberia with the pioneer work of Fiúza.

3.1 Seasonal cycle of the upwelling off Iberia

The seasonal evolution of the Iberian upwelling is related to the large-scale climatology of the NE Atlantic. The main aspects of the seasonal cycle of the shelf-slope circulation off Iberia were present in the results of the model of Stevens *et al.* (2000), namely the winter Portugal Coastal Countercurrent flowing poleward over the upper slope and the summer coastal upwelling with the associated equatorward flow.

Fiúza *et al.* (1982) analyzed wind and SST data concluding that the average regime of upwelling-favourable winds presents a maximum in summer with a phase lag of 1 month relative to the temperature anomalies. In the work of Nykjaer and Van Camp (1994), the cross-correlation between calculated Ekman transport and upwelling SST anomalies has shown at 38°N (off Portugal) lags of approximately 2 months.

Superimposed on the seasonal cycle, interannual and long-term variability of the upwelling off Iberia can be detected in association with large-scale processes such as the North Atlantic Oscillation (NAO). King *et al.* (2006) used SST images (1995–2005) off W Iberia to compute an upwelling index, and detected a NAO influence in this index.

A study by Borges *et al.* (2003), using sea level pressure data and Portuguese sardine catches (1946–1991), had already shown a statistically significant correlation between NAO, the alongshore winds off Portugal in winter months and the annual catch of sardine.

Fiúza *et al.* (1982) pointed out the sporadic appearance of upwelling in December–January along the Portuguese coast, less intense than in summer but also locally wind-induced. Brief episodes of upwelling in winter are also reported by Haynes *et al.* (1993). The upwelling winter episodes seem to have increased in the last decades, with the increase in frequency and intensity of equatorward winds (Borges *et al.* 2003; Santos *et al.* 2004).

3.2 Upwelling patterns by infrared remote sensing

Remote sensed SST data and coastal wind data were used by Fiúza (1983) to characterize Portuguese coastal upwelling patterns off the west and south coasts. He found that the upwelled water extent pulsates onshore-offshore in response to cycles of northerly winds. The cold waters reached 30–50 km offshore under calm conditions and 100–200 km during or shortly after strong winds. Figure 3a shows a typical SST distribution off Iberia in summer, with the region off the western coast occupied by cold waters extending offshore in some places as far as about 200 km. Different upwelling patterns off the west coast were mostly related to characteristic topographic constraints.

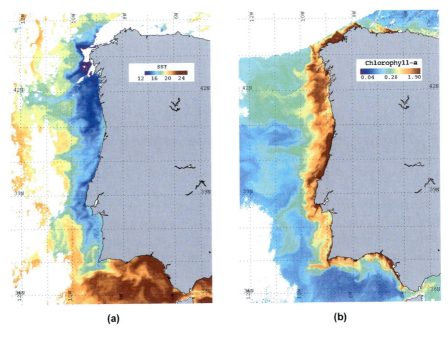

(a)   (b)

Fig. 3. Satellite remote sensed images during upwelling off Iberia: (a) NOAA-AVHRR SST (°C) for 19 August 1998 (from the Instituto de Oceanografia of University of Lisbon) and (b) SeaWiFS chlorophyll-*a* pigment concentration (mg m^{-3}) for 21 August 1998. NASA imagery.

In what respects the southern coast of Portugal, Fiúza (1983), by analyzing infrared images from summer 1979, has noticed that the coastal region is affected directly by upwelling under westerly winds, whereas during cycles of moderate to strong northerly winds, the cold upwelled waters of the west coast turn around Cape S. Vicente (Figure 1) and seem to flow eastward along the shelf break. During upwelling spin-down, Fiúza (1983) has detected a coastal countercurrent carrying warm surface water to the west and, eventually, reaching S. Vicente and proceeding northward. This same aspect was detected by Relvas and Barton (2002, 2005) in their work based on the analysis of long series (1990–1995) of SST images and *in situ* observations. They have noticed that the warm coastal countercurrent flowing westward interacts with the equatorward jet after the relaxation of the north winds, separating the upwelled water from the coast.

Vargas *et al.* (2003) used SST images (1993–1999) to perform a spatial EOF analysis of the variability of SST patterns in the Gulf of Cadiz. The first mode, explaining 60% of SST variance, showed the upwelling zone around Cape S. Vicente. The higher order modes revealed the response of this upwelling zone to local zonal winds.

Similarly to what happens off Cape S. Vicente, the coast orientation changes have influence on the local upwelling also off Cape Finisterre (Figure 1) in Galicia. The analysis of 2-year SST data in this region, by Torres *et al.* (2003), has evidenced persistent upwelling off Cape Finisterre even when not present farther south and an alternation of upwelling between the northern and the western coasts of Galicia.

3.3   Fronts, jets, filaments by infrared/visible remote sensing

One of the dominant features of the coastal upwelling off Iberia is the occurrence of cool and chlorophyll pigment-rich filaments, which have an important role on the shelf-offshore transport of plankton and organic carbon. Due to their characteristics, they can be traced by satellite infrared and visible images (see Figures 3a and 3b).

Sousa and Bricaud (1992) analysed a set of CZCS images (July 1981–September 1983), and identified patterns of chlorophyll pigment concentrations. The analysis for summer and early fall revealed a band along the west coast with high concentrations from which plumes or filaments extended offshore (sometimes as far as 150 km), some of them with a mushroom-like termination. They were located in several fixed points (around 41°N, 40°N, 38.5°N, 38°N and 37°N), which seemed to be associated with ridges between submarine canyons. In all images, part of the Cape Roca (Figure 1) filament extended southward, this being related by the authors to the equatorward coastal jet current (Fiúza 1984). In the southern coast during summer, a coastal band was identified with high phytoplankton concentration as far as Gibraltar, and a recurrent plume spreading eastward from Cape Santa Maria and leaving the coast off river Guadiana (Figure 1). Figure 3b, showing a SeaWiFS image off Iberia in summer, illustrates the presence of most of the just described features.

The development, persistence and variability of upwelling filaments off the Iberian Peninsula was also studied by Haynes *et al.* (1993), using infrared images (1982–1990). They concluded that major filaments start developing in July–August as bulges in the upwelling front, then grow offshore until September (maximum length of 200–250 km) and gradually decrease and become relatively rare in late October. The authors deduced that the main mechanisms of filament formation were topographically forced.

The evidence of a systematic association of upwelling filaments with the irregularities of the bathymetry or of the coastline (Fiúza 1982; Sousa and Bricaud 1992; Haynes *et al.* 1993), led to the investigation of the dynamics of these persistent filaments by Røed and Shi (1999). They conducted numerical experiments combined with an energy diagnosis and identified successive phases in the filament development, concluding that the filaments do not inherently depend on irregularities in the topography or coastline geometry but on barotropic and baroclinic instabilities.

3.4 Biological aspects by visible remote sensing

The variability of surface phytoplankton pigment concentrations off western Iberia was investigated by Peliz and Fiúza (1999) by analysing CZCS images (1979–1985). They presented typical patterns of concentrations along the year and analyzed the observed variability. The authors concluded that the pigment distribution is basically controlled by the stratification cycle due to seasonal heating-cooling and to runoff and by mesoscale and submesoscale processes associated with eddy and frontal activity.

The usual association between coastal upwelling and important fish stocks applies to the region off western Iberia, where the highest biomass in the fisheries results from sardine, which can feed directly on phytoplankton. Based on statistics of sardine landings at Portuguese fishing ports (1939–1974), Fiúza *et al.* (1982) correlated the monthly average catches with climatological upwelling indices obtained from the differences between coastal and oceanic SST measurements. They found a strong correlation, with maximum catches 2 months after the maxima of SST anomalies and 3 months after north wind maxima. Fiúza (1990) and Santos and Fiúza (1992) found that sardine availability is confined to the inner shelf along thin filaments of moderately cool, "old" upwelled waters at the end of the upwelling season. The occurrence of upwelling events off Iberia during the winter spawning season can reduce sardine and horse mackerel recruitment and catches in the following summer (Santos *et al.* 2001). However, as larvae transport is strongly dependent on local coastal flows and buoyant discharges, the winter upwelling events do not always have those negative effects (Santos *et al.* 2004; Ribeiro *et al.* 2005).

The other relevant fisheries in the area are related to tuna and swordfish that migrate seasonally during summer (from the subtropics to the Bay of Biscay) and aggregate in the vicinity of relatively "old" upwelling waters that reach far offshore through filaments and jets (Fiúza 1990; Santos and Fiúza 1992). By comparing SST data (1990–1992) with contemporaneous fisheries (swordfish and tuna species) information, Santos *et al.* (2006) found an apparent decrease of the swordfish catches and an increase of the tuna catches associated with the progressive increase of the coastal upwelling intensity during the study period.

4.   Concluding remarks

Satellite remote sensing is obviously a privileged tool for the study of coastal upwelling, especially when infrared and visible sensors are used. Its capacity of reaching large areas in a synoptic way and with regular sampling rates allows detecting coastal upwelling mesoscale structures, such as fronts, filaments, eddies and jets, and following their evolution. Furthermore, the observation of these structures can play an important role in providing valuable information for research and efficient management of fisheries.

Satellite remote sensing data bases are continuously growing and increasing their potentialities for the long term monitoring of coastal ocean processes, such as upwelling, thus contributing to a better characterization of their climate and variability. In recent years, the access to satellite data sets has become remarkably easier due to their publication in websites, often with free download capability, and this has led to a widespread use of this valuable tool among oceanographers.

The more accurate, frequent and extended measurements of the oceanic phytoplankton concentrations with the new generation of ocean colour sensors, together with the development of new algorithms and models, will improve the estimation of the primary production using satellite remote sensing. Taking all this in consideration, one can hope that the knowledge on coastal upwelling and its decadal or longer time scales fluctuations will be strongly developed in a near future.

Acknowledgements

N. Serra has helped with the construction of Figure 1 and D. Boutov has prepared Figure 3. The images used in Figure 2 were acquired using the GES-DISC Interactive Online Visualization And aNalysis Infrastructure (Giovanni) from NASA's Goddard Earth Sciences Data and Information Center.

References

Borges MF, Santos AMP, Crato N, Mendes H, Mota B (2003) Sardine regime shifts off Portugal: a time series analysis of catches and wind conditions. Scientia Marina, 67(Suppl. 1):235–244

Fiúza AFG (1982) The Portuguese coastal upwelling system. In: Actual problems of Oceanography in Portugal. Junta Nacional de Investigação Científica e Tecnológica, Lisbon, Portugal, pp 45–71

Fiúza AFG (1983) Upwelling patterns off Portugal. In: Suess E, Thiede J (eds) Coastal Upwelling: its sediment record. Plenum Press, pp 85–98

Fiúza AFG (1984) Hidrologia e dinâmica das águas costeiras de Portugal (Hydrology and dynamics of the Portuguese coastal waters). PhD thesis, University of Lisbon, Portugal

Fiúza AFG (1990) Applications of satellite remote sensing to fisheries. In: Rodrigues AG (ed) Operations research and management in fishing. Kluwer, pp 257–279

Fiúza AFG, Macedo ME, Guerreiro R (1982) Climatological space and time variations of the Portuguese coastal upwelling. Oceanol. Acta 5:31–40

Haynes R, Barton ED, Pilling I (1993) Development, persistence, and variability of upwelling filaments off the Atlantic coast of the Iberian Peninsula. J Geophys Res 98(C12):22681–22692

King GP, Yang J, Dias J, Serra N (2006) EOF analysis of seasonal and interannual variability of the surface circulation along the west Iberian coast from 1995–2005. Geophys Res Abs 8:1607–7962/gra/EGU06-A-03127

Navarro G, Ruiz J (2006) Spatial and temporal variability of phytoplankton in the Gulf of Cádiz through remote sensing images. Deep-Sea Res II 53:1241–1260

Nykjaer L, Van Camp L (1994) Seasonal and interannual variability of coastal upwelling along northwest Africa and Portugal from 1981 to 1991. J Geophys Res 99(C7):14197–14207

Peliz A, Fiúza AFG (1999) Temporal and spatial variability of CZCS-derived phytoplankton pigment concentrations off the western Iberian Peninsula. Int J Remote Sens 20(7):1363–1403

Relvas P, Barton ED (2002) Mesoscale patterns in the Cape São Vicente (Iberian Peninsula) upwelling region. J Geophys Res 107(C10), 3164, doi:10.1029/2000JC000456

Relvas P, Barton ED (2005) A separated jet and coastal counterflow during upwelling relaxation off Cape São Vicente (Iberian Peninsula). Cont Shelf Res 25:29–49

Ribeiro AC, Peliz A, Santos AMP (2005) A study of the response of chlorophyll-*a* biomass to a winter upwelling event off western Iberia using SeaWiFS and *in situ* data. J Marine Syst 53:87–107

Robinson IS (2004) Measuring the oceans from space. The principles and methods of satellite oceanography. Springer-Praxis, Chichester

Røed LP, Shi XB (1999) A numerical study of the dynamics and energetics of cool filaments, jets, and eddies off the Iberian Peninsula. J Geophys Res 104(C12):29817–29841

Ruiz J, Navarro G (2006) Upwelling spots and vertical velocities in the Gulf of Cádiz: an approach for their diagnose by combining temperature and ocean colour remote sensing. Deep-Sea Res II 53:1282–1293

Santos AMP, Fiúza AFG (1992) Supporting the Portuguese fisheries with satellites. In: Proceedings of the European International Space Year Conference, Munich, Germany, ISY-1, vol 2 (Paris: ESA), pp 663–668

Santos AMP, Borges MF, Groom S (2001) Sardine and horse mackerel recruitment and upwelling off Portugal. ICES J Marine Sci 58:589–596

Santos AMP, Peliz A, Dubert J, Oliveira PB, Angélico MM, Ré P (2004) Impact of a winter upwelling event on the distribution and transport of sardine (*Sardina pilchardus*) eggs and larvae off western Iberia: a retention mechanism. Cont Shelf Res 24:149–165

Santos AMP, Fiúza AFG, Laurs RM (2006) Influence of SST on catches of swordfish and tuna in the Portuguese domestic longline fishery. Int J Remote Sens 27(15):3131–3152

Sousa F, Bricaud A (1992) Satellite-derived phytoplankton pigment structures in the Portuguese upwelling area. J Geophys Res 97(C7):11343–11356

Stevens I, Hamann M, Johnson JA, Fiúza AFG (2000) Comparisons between a fine resolution model and observations in the Iberian shelf-slope region. J Marine Syst, 26:53–74

Torres R, Barton ED, Miller P, Fanjul E (2003) Spatial patterns of wind and sea surface temperature in the Galician upwelling region. J Geophys Res 108(C4), 3130, doi:10.1029/2002JC001361

Vargas JM, García-Lafuente J, Delgado J, Criado F (2003) Seasonal and wind-induced variability of sea surface temperature patterns in the Gulf of Cádiz. J Marine Syst 38:205–219

Wooster WS, Bakun A, McLain DR (1976) The seasonal upwelling cycle along the eastern boundary of the North Atlantic. J Marine Res 34(2):131–141

On the Use of Thermal Images for Circulation Studies: Applications to the Eastern Mediterranean Basin

Isabelle Taupier-Letage

Centre National de la Recherche Scientifique et Université de la Méditerranée, Laboratoire d'Océanographie et Biogéochimie, La Seyne/Mer, France

Abstract. The use of satellite thermal infrared images to infer marine circulation features is presented, here in the case of the eastern basin of the Mediterranean Sea. Although the first schema of the surface circulation for Atlantic Water in the Mediterranean Sea is one century old, its real path in the Eastern basin is still debated nowadays. Does it flow along the Libyan and Egyptian slopes in a counterclockwise circuit at basin scale, or as an off-shore jet that crosses the basin in its central part (the "Mid-Mediterranean Jet")? In this paper we describe the use and contribution of the thermal images for the study of the surface circulation in the Eastern basin, currently underway within the framework of the programmes EGYPT and EGITTO (2005–2008).

1. Introduction

The water lost by evaporation in the Mediterranean is compensated by light Atlantic Water (AW). AW flows over the saltier – hence denser – Mediterranean Waters (MWs), and determines the surface circulation (for a review of the general circulation in the Mediterranean see Millot and Taupier-Letage (2005a) and references therein). Overall, and due to the earth rotation, AW and MWs are expected to follow the bathymetry and describe quasi-permanent counterclockwise circuits, in both the western and the eastern basins. The first schema of the surface circulation dates back to the early 1910s: it showed counterclockwise circuits in both basins (Figure 1a). In the 1990s, the experiment Physical Oceanography of the Eastern Mediterranean (POEM) issued a schema (Figure 1b) depicting the AW path as an offshore jet, crossing the basin in its central part (the so-called Mid-Mediterranean Jet, MMJ), and feeding mesoscale circulations

153

V. Barale, M. Gade (eds.), *Remote Sensing of the European Seas.*
© Springer Science+Business Media B.V. 2008

on both sides (Robinson *et al.* 1991). That latter representation has been questioned (Millot 1992). Firstly because satellite thermal images display features analogous in both basins (Figure 2), and secondly because no *in situ* observations were available in the southernmost part of the Eastern basin. The thermal signature of AW (warmer in winter; Figure 2a) can be tracked along the southern continental slopes of both basins. In both basins the flow of AW in the South forms currents that are unstable at mesoscale (Figure 2c–f). They meander and generate anticyclonic eddies that often extend down to the bottom and have lifetimes of several months. Eddies propagate usually downstream or in the basins interior at a few km per day, and are thus likely to interact with their parent current or other eddies.

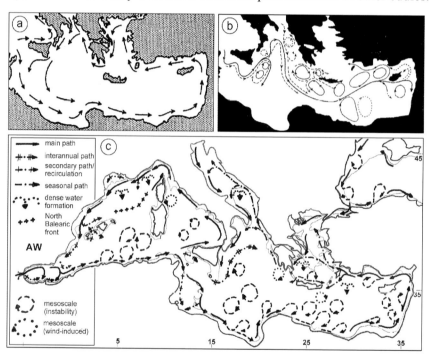

Fig. 1. Schema of the surface circulation in the (eastern) Mediterranean Sea, according to: Nielsen (1912), upper left panel (1a); Robinson *et al.* (1991), upper right panel (1b); Millot and Taupier-Letage (2005a), lower panel.

This intense mesoscale activity induces a very high variability in both space and time, which impairs interpreting the observations that do not re-solve mesoscale properly. By combining thermal (mainly) satellite images and *in situ* observations, Millot *et al.* (1997) and Taupier-Letage *et al.* (2003) showed the role of such mesoscale eddies in entraining AW and

MWs from the periphery towards the central part of the Algerian subbasin, and thus in diverting potentially (part of) the flow from its expected path. This allowed solving a similar controversy about the path of the Levantine Intermediate Water in the southern part of the Western basin (Millot and Taupier-Letage, 2005b). Hamad *et al.* (2005) have analysed a 4-year time series of thermal images covering the Eastern basin, and confronted the *in situ* observations (among which the POEM data and repeated XBT transects) with their thermal signatures. They observed a permanent eastward circulation in the south, that they named the Libyo-Egyptian Current (Figure 1c, Figure 2f). Zervakis *et al.* (2003) and Fusco *et al.* (2003) also concluded that there was a permanent coastal flow (the MMJ in the latter case, *cf* their Figure 10). Given the presence of coastal libyo-egyptian eddies at the time of POEM, Hamad *et al.* (2005, 2006) suggested (*i*) that these eddies were responsible for spreading the AW found offshore, and (*ii*) that the meandering MMJ was a misinterpretation of the AW found on the northern edges of successive eddies. The joint programmes EGYPT[1] and EGITTO[2] are currently (2005–2008) collecting *in situ* observations to study the circulation in the Eastern basin, with focus on its southernmost part. The methodology to infer circulation features from the thermal imagery is described in section 2. The first results of the confrontation between *in situ* and satellite observations are described in section 3 and discussed in section 4, together with the conclusion.

2.   Inferring circulation features from thermal imagery

The retrieval of the sea surface temperature (SST) from remote sensing in the thermal infrared is routine work. The detailed principles are beyond the scope of this paper; therefore the reader is referred to other chapters of this book or for instance to the site of the Medspiration[3] project.

2.1   The eligible products and selection criteria

The most adequate products to track (mesoscale) circulation features are infrared thermal images derived from the Advanced Very High Resolution Radiometer (AVHRR) sensor. The temporal coverage is ensured by two (at least) satellites flying simultaneously and providing at least 4 images/day over

[1] http://www.ifremer.fr/EGYPT
[2] http://doga.ogs.trieste.it/sire/drifter/egitto_main.html
[3] http://www.medspiration.org/science

the Mediterranean, the synoptic view by a swath >2000 km, the spatial resolution by a pixel of ~1 km, and the detection of the features by a ~0.12°C thermal resolution (Figure 2–5). The typical processing includes the retrieval of the SST from the combination of 2 to 3 infrared channels, the geographical registration, with possibly the generation of cloud and land masks. Images can be then composed into daily to monthly products. SST images are routinely generated and made available on numerous data centers for free.

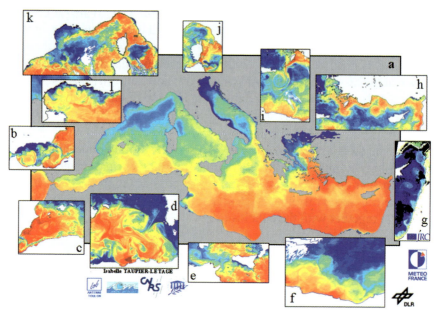

Fig. 2. The surface circulation in the Mediterranean Sea seen with thermal images. (*a*) Atlantic Water (warmer) can be tracked all around the continental slopes in a counterclockwise circuit; (*b–l*) the mesoscale dynamics and associated variability. Temperature increases from blue to red. Image credits: (*a, h*) monthly SST composite for January 1998 (DLR imagery); (*b–f* and *i–l*) single AVHRR images (SATMOS imagery); (*g*) ocean colour from SeaWiFS (JRC imagery).

Although the circulation features can be detected from SST images (see Figure 2a, h and Figure 4e–f), a tailored product has been determined and generated[4] to meet the needs for a fine resolution in both space (1 km) and time (single pass, every day) and for near-real time availability, as required to sample *in situ* mesoscale features (that may be propagating). This product is

---

[4] By the SATMOS/INSU-Météo France facility (http://www.satmos.meteo.fr/), and used throughout this paper unless otherwise stated.

the image of the channel 4[5] (brightness temperatures, hereafter called thermal image). The image is extracted from of a single pass, at a 1-km resolution, and geographically registered. The size of the file is decreased by recoding the image over 8 bits (10 bits originally); the original thermal dynamics is preserved by keeping only the numerical counts that correspond to marine temperatures (usually from 5 to 32°C). To decrease the size further the land is masked, the larger clouds too (flagged by a simple threshold, usually temperatures <5°C), and the file is compacted with a lossless algorithm. If the file is an image a grayscale is used to preserve the possibility to adjust the contrast specifically. The generation of such products is fully automated, cheap, and easily completed within one hour after the satellite pass.

The infrared signal only comes from the upper few microns (the "skin" temperature; see *e.g.* Buongiorno Nardelli *et al.* 2005 for more details). Therefore the thermal signatures cannot be, *a priori*, related to the dynamics of the mixed layer[6]. This is the case when solar heating creates a shallow (and temporary) thermocline, which caps the thermal patterns of the mixed layer. Nighttime passes are thus preferred. On the other hand the wind blows very often and mixes the upper layer, so that the surface temperature is representative of that of the mixed layer[7].

2.2   The link between thermal signature and ocean dynamics

In the Mediterranean, the surface circulation can be tracked most generally by tracking the higher temperatures, which correspond to the lower salinity – and thus lighter- AW[8]. Such conditions are optimum in winter, as seen on Figure 2a. Note that this is also true independently of the latitude, as shown by the AW warmer current flowing along the northernmost parts of both basins on Figure 2a, h and k.

The tracking of the mesoscale features from their thermal signature relies on the fact that there must be coherence between the temporal and the spatial scales. Indeed, the infrared signature of a shallow phenomenon will have a transient lifetime, of the order of hour(s) or day(s). This is the case

---

[5] A single channel is preferred for delineating features since the multi-channel combination increases the noise in the SST image.

[6] Ocean colour images are better since the signal comes from a thicker layer. But their processing is no real routine, and the temporal coverage is much less.

[7] Possibly even under summertime conditions at noontime, as shown by Figure 4a: the northerly winds mix and cool the surface layer on both sides of Crete, clearly revealing anticyclones.

[8] Isopycnals and isotherms mostly co-vary.

for diurnal heating or some wind-induced phenomena. Conversely, the thermal signature that can be tracked for months -up to years- necessarily corresponds to a structure having deep vertical extent, a condition required to maintain the signature over time, especially to survive winter mixing (see *e.g.* the 3-year tracking of algerian eddies in Puillat *et al.* 2002, and of libyo-egyptian eddies in Hamad *et al.* 2006). But there is no direct relation between the intensity (*i.e.* the thermal gradient) of a signature and the intensity of the structure itself (vertical extent or current speed).

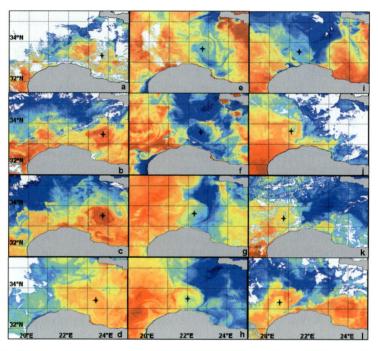

Fig. 3. Time series used to track the Libyan eddy (+) during 2006: *a*) 08 Jan.; *b*) 10 Feb.; *c*) 06 Mar.; *d*) 26 Apr.; *e*) 28 May; *f*) 17 Jun.; *g*) 23 Jul.; *h*) 18 Aug.; *i*) 14 Sep.; *j*) 26 Oct.; *k*) 08 Nov.; *l*) 19 Dec. 2006. Δ: another Libyan eddy.

For instance a cyclonic eddy is associated with a doming of the isopycnals, so that deeper/colder water intersect the surface in the centre and delimit an intense thermal gradient: the resulting small cold spots will always be detected (Figure 2d, 3i). However such cyclonic eddies[9] are known to be shallow and transient. Conversely, an anticyclonic eddy is associated with a depression of the isopycnals, and its generic signature is a warmer (lighter water) central area. While this is always verified during wintertime

---

[9] Cyclonic eddies are secondary phenomena, induced by the shear.

because of mixing (*e.g.* Figures 3b–c, 5b), under stratified conditions the isotherms may not intersect the surface, and hence anticyclones may not exhibit any intrinsic signature. They are detected then indirectly by the superficial water they entrain, up to displaying a colder central area. The resulting variability of an eddy's signature is illustrated by the Figures 3 and 4[10]. Finally, the temperature difference between AW and MW reverses on a yearly cycle: thus, spring and fall are less favourable periods for tracking thermal patterns, and even a difference of ~0.1°C (thermal resolution) can sometimes trace a structure and be significant (see eddy Δ in Figure 3d).

Most often the current is parallel to the isotherms, and generally the inference of the currents is intuitive. One image allows deducing the current direction associated with the eddies, since the isotherms always spiral inside. One can estimate their diameter and centre location, even from segments of isotherms on images partially cloud-covered (Figures 3a, k). But since the eddy's signature may vary according to the water its entrains, estimates of its size can hardly be precise, but in winter when the mixed layer is thick. Time series of images allow deducing the trajectory and the propagation speed from the successive positions of the features (isotherms are then perpendicular to the propagation direction).

During cloudy periods the dynamical features can be tracked using the anomaly they generate on the sea level with altimetric tracks (*e.g.* Pujol and Larnicol 2005), and/or using composite thermal images. But care must be taken when using interpolated or composite images that the longer the time span the smoother the signature of a propagating structure, up to potentially yielding a misleading picture. Indeed, where successive eddies propagating along the coast mainly induce cross-shore gradients (*cf* Figure 2d, f), the image resulting from a long-time average will display a smooth band parallel to the coast, with gradients mainly oriented along-shore.

2.3   The interpretation of the thermal signatures

The first step is to discriminate between oceanic and atmospheric signatures. This is relatively easy because both types of phenomena have different space and time scales. Clouds are changing and moving more rapidly than any oceanic phenomena, so that comparing two images prevents from mistaking the limits of a cloud or haze for those of an eddy. Patterns too are characteristic: isotherms associated with oceanic phenomena are smoother and less patchy, and are mostly tangential: one should suspect atmospheric contamination wherever isotherms intersect.

---

[10] Symmetrically, 2 cyclonic shear eddies display a warmer signature in Figure 2d.

The second step is to detect and evidence the circulation features. As shown above the temperature is not a sufficient criterion, as its value and the gradient it determines vary with its environment, the season, the meteorological conditions and the time of day. This requires stretching the colour scale to adjust the contrast for each image, sometimes even differently for the same image to evidence circulation features at the Mediterranean (Figure 2a) and sub-basin (Figure 2h) scales. If the retrieval of accurate temperature (SST) is crucial for climatological studies, it is not an issue for process studies so that colour-temperature scales are not relevant.

The third step is to check whether the thermal signature corresponds to an actual dynamical structure. Its presence and lifetime must be verified on several images, possibly using other satellite information too (*e.g.* visible imagery, or altimetry to establish the continuity during cloudy periods).

The final step is to characterize the circulation features: one considers the shape, location and consistency of isotherms. The most common analysis consists in reporting the isotherms (possibly segments only) delineating a circulation feature, and in superimposing its successive signatures (as illustrated by the Figure 3 of Marullo *et al.* 2003). The recurrence of contours ends up in delineating the whole feature, its centre in the case of an eddy, and, should it move, its direction and propagation speed.

3.   New results on the circulation in the Eastern Basin

The EGYPT and EGITTO experiments provide a new insight in the surface circulation. The confrontation of concurrent *in situ* with remotely sensed observations is focused here on the year 2006. Note that the data sets analysis is currently underway and related paper still in preparation[11].

3.1   A newly observed drift for libyan eddies

In the previous study spanning 4 years, eddies along the Libyan slope had only been observed drifting eastward (Hamad *et al.* 2006). Figure 3 (d–l) shows 2 libyan eddies drifting westward: one (+) from April during 9 months at least (detailed tracking interrupted after December 2006), the other one upstream (Δ) during 3 months at least (no specific tracking).

---

[11] References of papers will be available on the EGYPT and EGITTO web sites.

3.2 Few-days temporal variability: the merging of eddies

The northerly Etesians winds blowing in summer generate eddies. The anticyclone induced Southwest of Crete is called Ierapetra. Several authors have reported its persistence after the decay of the Etesians, and the fates of successive generations of Ierapetra are described in Hamad *et al.* (2006). The situation of summertime 2006 is detailed in Figure 4.

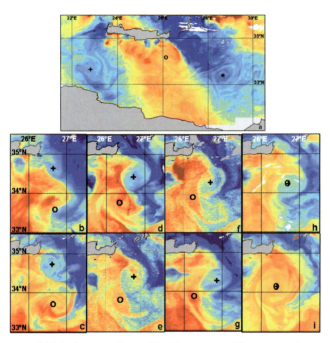

Fig. 4. Summer 2006: the merging of the Ierapetra eddies created respectively in 2005 (I05: **o**) and 2006 (I06: **+**): *a*) situation on 28 May 11:59; *b*) 05 Jul. 20:52; *c*) 10 Jul. 20:38; *d*) 19 Jul. 20:31; *e*) 20 Jul. 16:09; *f*) 21 Jul. 15:30; *g*) 22 Jul. 01:07; *h*) 23 Jul. 20:39; *i*) 27 Jul. 11:48 2006. *e* and *f*: SST from Remote Sensing Group of the Department of Oceanography (OGS)[12].

In May the Ierapetra generated in summer 2005 (I05, o) appears as a large anticyclone. By 5 July the signature of Ierapetra 2006 (I06, +) is definitely established, and both signatures co-exist. The size of I06 doubles in 5 days, and by 19 July I06 is larger than I05. I06 moves to the South and begins interacting strongly with I05: their merging takes less than 4 days, and by 23 July there is only one anticyclone signature. So that by 27 July

---

[12] http://poseidon.ogs.trieste.it/sire/satellite/

(Figure 4j) the situation appears similar to that of late May (Figure 4a), but the Ierapetra eddy is different.

3.3   The sampling strategy dedicated to mesoscale processes

The analysis in near-real time of thermal images transmitted on board allowed crossing several eddies (Figure 5a), and seeding them with surface drifters. The 3 drifters seeded in April in the Libyan eddy (+) remained trapped inside at least till early October, and their trajectories materialize its eastward drift, as already inferred from the thermal images (Figure 3).

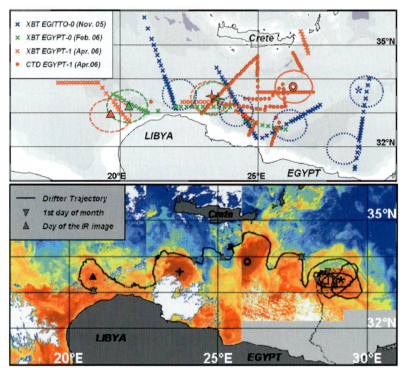

Fig. 5. Upper panel: the strategy of the EGYPT and EGITTO campaigns and positions of the eddies sampled. Lower panel: trajectory of a surface drifter, traced from 1 February to 18 May 2006, superimposed on images from 10 February (left), 10 March (middle) and 1 April 2006 (right).

The CTD transect that crossed it from the Libyan shelf to the Cretan one (Figure 5a) showed[13] that the minimum of salinity (indicating the most

---

[13] It also showed a vertical extent >1000 m

recent/less modified AW) was found on -and limited to- its northern edge. The trajectory (Figure 5b) of the drifter released upstream on the southern periphery of the Libyan eddy (Δ) in February demonstrates how successive eddies act as paddle-wheels to transport water offshore.

4. Discussion and conclusion

The good correlation between *in situ* and satellite observations has been demonstrated again. The Libyan eddy (+) has been tracked for 9 months with drifters, up to 1 year with thermal images. In February and March (Figure 3b–c) the continuity of the warmer signature from the eddy (Δ) to the eddy (+) shows that AW is first flowing alongslope. Then the paddle-wheel effect illustrates the mechanism of its offshore spreading, and explains why, on the CTD transect, recent AW has only been found on the northern edge of the eddy (+). Although the resulting drifter trajectory (Figure 5b) evokes a MMJ, the underlying processes are different.

Knowing the history of the mesoscale phenomena with a fine spatio-temporal interval is also important since situations looking similar can be achieved with different eddies, and since merging of eddies can be completed within few days. This also concerns other disciplines, as for instance the distribution of nutrients not only differs in and out of an eddy, but also in a one-year old eddy (as I05 in 2006) and in a newly formed one.

Provided some precautions are taken, the study of the circulation based on thermal images is efficient in the eastern basin of the Mediterranean too. However there are no constant criteria to characterise the eddies signatures, which are difficult to detect and track automatically. Such studies remain thus rather descriptive, but they are easy to carry, cheap, especially cost-efficient for *in situ* sampling –and indispensable for data interpretation wherever the mesoscale dynamics plays a pivotal role.

Acknowledgments

Thanks are due to J-L. Fuda, R. Gerin, C. Millot, G. Notarstefano, P. Poulain and G. Rougier for their contribution, to the SATMOS (CNRS/ INSU/MétéoFrance) for the thermal images, and to the crews for the campaigns at sea. EGYPT received funding from the CNRS/INSU, programs PATOM/IDAO and GMMC, and Région Provence-Alpes-Côte d'Azur.

References

Buongiorno Nardelli B, Marullo S, Santoleri R (2005) Diurnal variations in AVHRR SST fields: a strategy for removing warm layer effects from daily images. Rem Sens Environ 95 (1): 47–56

Fusco G, Manzella GMR, Cruzado A, Gacic M, Gasparini GP, Kovacevic V, Millot C, Tziavos C, Velasquez Z, Walne A, Zervakis V, Zodiatis G (2003) Variability of mesoscale features in the Mediterranean Sea from XBT data analysis. Ann Geophys 21: 21–32

Hamad N, Millot C, Taupier-Letage I (2005) A new hypothesis about the surface circulation in the eastern basin of the Mediterranean Sea. Progr Oceanogr 66: 287–298

Hamad N, Millot C, Taupier-Letage I (2006) The surface circulation in the eastern basin of the Mediterranean Sea. Sci Mar 70 (3): 457–503

Marullo S, Napolitano E, Santoleri R, Manca BB, Evans R (2003) The variability of Rhodes and Ierapetra gyres studied by remote sensing observation, hydrographic data and model simulations during LIWEX (october 1994–april 1995). J Geophys Res 108 (C9): 8119, doi:10.1029/2002JC001393

Millot C, Benzohra B, Taupier-Letage I (1997) Circulation off Algeria inferred from the Mediprod-5 current meters, Deep Sea Res 44 (9–10): 1467–1495

Millot C (1992) Are there major differences between the largest Mediterranean Seas? A preliminary investigation. Bull Inst Oceanogr Monaco 11: 3–25

Millot C, Taupier-Letage I (2005a) Circulation in the Mediterranean Sea. In: Saliot A (ed) The Mediterranean Sea, Handbook of Environmental Chemistry, vol 5, Part K. Springer-Verlag, Berlin Heidelberg, pp 29–66

Millot C, Taupier-Letage I (2005b) Additional evidence of LIW entrainment across the Algerian Basin by mesoscale eddies and not by a permanent westward-flowing vein. Progr Oceanogr 66: 231–250

Nielsen JN (1912) Hydrography of the Mediterranean and adjacent waters. Rep Dan Oceanogr Exp Medit 1: 77–192

Puillat I, Taupier-Letage I, Millot C (2002) Algerian Eddies lifetime can near 3 years. J Mar Sys 31: 245–259

Pujol MJ, Larnicol G (2005) Mediterranean sea eddy kinetic energy variability from 11 years of altimetric data. J Mar Sys 58: 121–142

Robinson AR, Golnaraghi M, Leslie WG, Artegiani A, Hecht A, Lazzoni E, Michelato A, Sansone E, Theocharis A, Ünlüata Ü (1991) The Eastern Mediterranean general circulation: features, structure and variability. Dyn Atm Oceans 15: 215–240

Taupier-Letage I, Puillat I, Raimbault P, Millot C (2003) Biological response to mesoscale eddies in the Algerian Basin. J Geophys Res 108 (C8): 3245–3267

Zervakis V, Papadoniou G, Tziavos C, Lascaratos A (2003) Seasonal variability and geostrophic circulation in the Eastern Mediterranean as revealed through a repeated XBT transect. Ann Geophys 21: 33–47

Current Tracking in the Mediterranean Sea Using Thermal Satellite Imagery

Steffen Dransfeld

Institut für Meereskunde, Universität Hamburg, Germany

Abstract. Infrared images representing the thermal state of the ocean surface may be used to track ocean currents with feature-tracking algorithms that identify and follow temperature gradients and hence represent advective surface motion. One of these techniques, the Maximum Cross Correlation technique, is based on a comparison of individual subscenes of sequential images, to estimate where a feature has moved from one image to the next. It has distinct advantages compared to alternative feature tracking algorithms such as its simplicity and robustness. Previous research has shown it to achieve a precision of 0.08 to 0.20 m/s rms. This study focuses on surface currents in the central Mediterranean Sea by analysing sequential Advanced Very High Resolution Radiometer local area coverage 1.1 km resolution images from June 2003. Most attention is placed onto the presentation of the method, the results and the main ad/disadvantages of using a Maximum Cross Correlation approach.

1. Introduction

Frequent coverage of oceans by satellite radiometers gave rise to the idea of using successive infra red images from the same region to track ocean currents, by trying to follow displacements of temperature patterns (Emery *et al.* 1986; Kelly 1989; Cote and Tatnall 1995; Bowen *et al.* 2002). In image processing terms this approach is called feature tracking, implemented by different algorithms, such as neural networks or correlation analysis between successive images. Frequently used methods include wavelet techniques (Liu 1997), gradient thresholds (Holyer and Peckinpaugh 1989), neural networks (Cote and Tatnall 1995), an inversion of a heat equation (Vigan *et al.* 2000a) and MCC techniques (Emery *et al.* 1986). Of these, the Maximum Cross Correlation (MCC) technique has proved itself to be easily implementable yet effective. It requires little user input and is suitable for routine processing of large image archives. Bowen *et al.* (2002)

V. Barale, M. Gade (eds.), *Remote Sensing of the European Seas.*
© Springer Science+Business Media B.V. 2008

carried out the most complete study of the MCC method, by applying it to a seven year archive of Advanced Very High Resolution Radiometer (AVHRR) local area coverage (LAC) images of the East Australia Current. They estimated computed current vectors to have a precision of 0.08 to 0.2 m/s, depending on MCC parameterisation. Tokmakian *et al.* (1990) applied MCC to a region off California and estimated the root mean square (rms) error of the method to lie between 0.14 and 0.25 m/s. They compared MCC vectors with *in situ* current vectors obtained from Acoustic Doppler Current Profiler (ADCP) measurements. Kelly and Strub (1992) also compared MCC vectors with *in situ* data based on surface drifter and ADCP sampling and found that MCC currents are typically 35% weaker than ADCP and 55% weaker than surface drifter velocities.

In this paper the MCC is applied to the Strait of Sicily for a short sequence of images to show its potential and the theoretical principle it is based on. The described work is meant to serve as an introduction to the method and to provide the interested readers with a literature and theoretical background to apply the method. Section 2 describes the general theory of the MCC and section 3 some preprocessing steps images have to undergo. Section 4 gives a description of the region of interest (the strait of Sicily) and expected current behaviour. Section 5 presents the used data images, applies the MCC algorithm to them to compute current vectors, and discusses the results. Section 6 concludes this work by stating the method advantages and disadvantages, and giving an outlook of the applicability of MCC to alternative data sources such as ocean colour images.

2.   Theoretical background of MCC

The essence of the algorithm lies in locating a small subscene from a first image inside the next image from the same region. It is done by computing a cross correlation between the subscene and a correspondingly sized area in the second image moved over a predefined search area, until a maximum correlation is found (for a full description see Garcia and Robinson 1989). The algorithm divides the first image into a number of template windows (see Figure 1) sized according to the oceanic flow structures to be resolved. Each of these templates is searched for in a second image inside a search window (dashed line in Figure 1) of size depending on maximum current speed expected between two sequential images. The search window is hence larger than its corresponding template. Once an area that corresponds closest to a template has been found inside the second image, as illustrated by the dotted rectangle in Figure 1, a vector is computed from the template centre to the centre of the area of maximum cross correlation. The two main parameters of the MCC technique are the sizes of the template window and of the search area (Bowen *et al.* 2002). The size of the template window has to be chosen carefully, as it represents a balance between having sufficient features to track in the template and smoothing the resultant vector flow field. Larger template windows result in a smoother flow field, whereas small template sizes will allow tracking of finer features, giving a less homogeneous flow field.

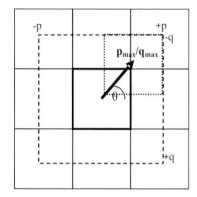

Fig. 1. Division of first image into templates (solid lines) and search area (dashed line) corresponding to central template; vector indicates where in the second image the centre of the window (dotted line) corresponding closest to the original middle template (thicker solid line) lies.

The underlying mathematical expression to compute cross correlation coefficients is given by (1). Pixel values of the template window are $A(x, y)$ and the corresponding pixels in the second window $B(x, y)$ at no lag and $B(x + p, y + q)$ for a lag of (p, q). (1) is the normalized spatial cross correlation function between the template and search window at lag (p, q). P and q hence determine the position, where in the search area a cross correlation is calculated. At the position p_{max} and q_{max} of maximum correlation, the subscene centre corresponding to the search template in image 1 is found.

$$r(p,q) = \frac{1}{\sigma_A \sigma_B'} \sum \sum \left\{ \left[A(x,y) - \bar{A}(x,y) \right] \cdot \left[B(x+p, y+q) - \bar{B}(x+p, y+q) \right] \right\} = \frac{Cov(p,q)}{\sigma_A \sigma_B'} \quad (1)$$

The summation is performed over all pixel values of the template indexed by x and y coordinates. $\bar{A}(x,y)$ and $\bar{B}(x + p, y + q)$ correspond to the average of all template pixels and the equivalent window in the search area. $Cov(p,q)$ is the covariance of A and B at lag (p, q) and σ_A and σ_B' are the standard deviations of the template and lagged window, entering the correlation calculation. Displacement (p, q) at the maximum cross correlation determines the advective velocity c as given by (2).

$$c = \frac{\left[(p_{max} \Delta x)^2 + (q_{max} \Delta y)^2 \right]^{1/2}}{\Delta t} \quad (2) \qquad \theta = \arctan \left(q_{max} \Delta y / p_{max} \Delta x \right) \quad (3)$$

Spatial pixel interval and hence pixel size is represented by Δx and Δy, p_{max} and q_{max} represent lag values for the maximum cross correlation and Δt the time interval between both images. Direction of propagation of the current vector is given by θ as shown in (3). Before the MCC algorithm is applied to image sequences, they need to undergo a preprocessing to map all images onto an identical geographic grid.

3. Pre-processing performed on images

As well as mapped to a geographic grid the AVHRR LAC data used for this study had to be calibrated and corrected for variations in satellite attitude, or the roll, pitch and yaw angle. Fluctuations in attitude can slightly shift an image position from its corresponding geographic position. These fluctuations need to be taken into account and corrected for, since the mapping algorithm assumes a constant attitude and may otherwise produce artificial image offsets that would cause spurious current vectors. To correct for attitude errors an algorithm was used for geo-locating the images to an estimated accuracy of about 1 km for all coastal oceans.

Satellite radiances were converted to brightness temperatures and not actual Sea Surface Temperature (SST) values, as SST results out of a combination of several bands and their noise terms. By using the brightness temperature of one channel (normally channel 4 at 10.6 μm), the image noise reduces to noise of a single channel and therefore thermal gradient structures are better conserved. Before the MCC algorithm is applied, the images are gridded onto a geographic grid using an approach called indirect navigation. It combines a geographic grid with an orbital model to associate scan elements (line-spot values) from the satellite field of vision with corresponding pixels of the geographic grid. More specifically, a line-spot value is computed every eight pixels in latitude and longitude directions and bilinear interpolation used to calculate the remaining line-spot values. These are used to index the field of vision and retrieve radiance values. Following Emery et al. (1992) correlation coefficients were computed only for windows containing at least 60% of ocean pixels with a brightness temperature value to preserve statistical significance.

4. Description of region of interest

The Strait of Sicily corresponds to a very dynamic part of the Mediterranean, connecting the Eastern and Western Mediterranean basins. It therefore represents a suitable area for this study. The strait is characterised by a deep interior basin in the middle of two sills east and west at depths of 400–500 m, whereas the basin reaches as far down as 1400 m. Minimum width of the strait is about 140 km and its predominant current is a surface current flowing south eastward, carrying fresher Atlantic Water, transported there by the Algerian current.

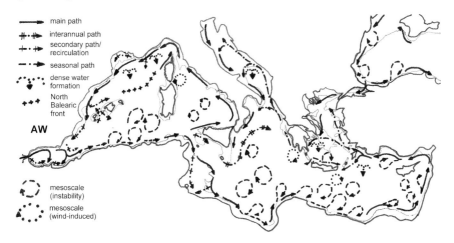

Fig. 2. Predominant current and mesoscale activity throughout the Mediterranean Basin (from Millot and Taupier-Letage 2005).

Observations (Béthoux 1980) show that two-thirds of the Atlantic water in the Algerian current flow through the strait, while one third continues into the Tyrrhenian Sea (Figure 2). A second dominant current flowing through the Strait of Sicily is a denser and saltier current flowing at depth in the reverse northwest direction into the western basin and transporting the so-called Levantine Intermediate Water. By applying the MCC to the images, we expect to find vectors that follow general circulation patterns of the fresher Atlantic water surface current. Current speeds measured by moored current meters (Gasparini *et al.* 2004) show zonal velocities reaching up to 40 cm/s and meridional components of up to a 100 cm/s.

5.   Presentation of data and application of MCC to it

AVHRR images used for this study are from NOAA's satellite active archive. Three pairs of cloudless images (see Figures 3a–f for images and acquisition times) from June 2003 were selected. As can be seen the first image of the third pair is also the last image of the second pair. The last two pairs may thus be seen as a sequence of three images.

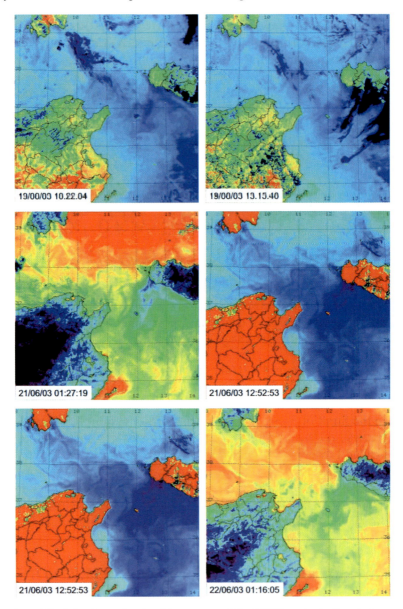

Fig. 3. a–f Each horizontal pair of images used for MCC analysis (3 pairs).

Vector fields computed by MCC are shown in Figures 4a–d. The vector field depicted in Figure 4a is the unfiltered version of 4b. It demonstrates the filtering process of vectors generated by the MCC algorithm which removes vectors with lower than 0.7 cross correlation coefficient and low spatial homogeneity. This value was chosen in coherence with previous publications that carried out statistical significance testing of correlation coefficients for increasing search ranges (Bowen *et al.* 2002). The spatial filter compares central vectors with neighbouring vectors. At least three of the neighbouring vectors have to be made up of components that indicate a pixel displacement of two or less pixels relative to the central vector. Non-compliant vectors are discarded.

Parameters used for the MCC algorithm included a 22 pixel template window size and a search area in the second image of 66 pixels. Seen from the centre of the search area, the search range was hence equivalent to the template size in both directions.

All vector fields indicate a splitting of the Algerian current into a branch flowing through the Strait of Sicily and a branch continuing eastwards into the Tyrrhenian Sea. Figure 4b indicates a relatively strong flow through the strait of up to 85 cm/s. Flows up to this magnitude were measured by Gasparini *et al.* (2004). 4b also shows a flow that orientates more along the Tunisian coastline, whereas vectors in 4c show a flow through the middle and more along the Sicilian coastline. In fact there have been two major currents identified into which the Atlantic water flowing through the strait separates (Gasparini *et al.* 2004). The stream along the Tunisian coastline is generally fresher. It is an important stream of the strait (Onken *et al.* 2003) known as African Modified Atlantic Water (MAW). The other major current, defined by Robinson *et al.* (1991) as the Atlantic Ionian Stream AIS, is an energetic and meandering flow along southern Sicily around generally three thermal surface features; the Adventure Bank Vortex, the Maltese Channel Crest and the Ionian Shelf Break Vortex (Ollta *et al.* 2006). The velocity of the AIS seen in Figure 4c reaches up to 76 cm/s at the entrance of the strait over the Adventure bank on the Eastern side of the strait. Similar velocities were measured by Gasparini *et al.* (2004) during the same period of the year 2000 by ADCP cruise transects. Figure 4c shows part of the Atlantic water transported via the Algerian Current continuing towards the Tyrrhenian Sea. The current forecast for November 2004 by the MFSTEP (Mediterranean Forecasting System toward Environmental Predictions) forecast system (see Figure 5), based on *in situ* measurements and a General Circulation Model (GCM), focuses on the Tyrrhenian Sea and Strait of Sicily. It corresponds to a large extent in direction as well as dimension with MCC vectors. The more energetic regions of Figure 5 can also be found in Figures 4b–d, and a southward current flow east of Sardinia, part of an anticyclonic gyre eventually formed by the Tyrrhenian current, can be located south of Sardinia in 4d.

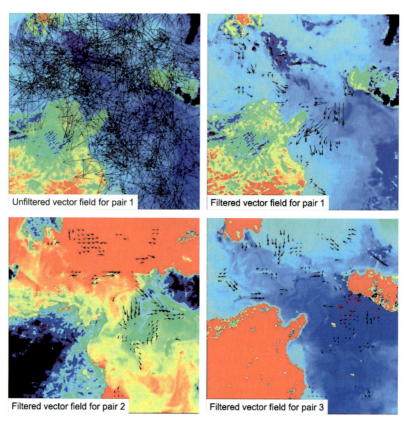

Fig. 4. a–d (a) Raw and unfiltered vector image showing all vectors before some are removed by correlation and nearest neighbour filter; (b) Vectors indicate a strong flow through the centre of the Strait of Sicily and along the Tunisian coastline; (c) Vectors indicate a flow through the centre of the Strait of Sicily and along the Sicilian coastline; (d) Vectors indicate an anticyclonic flow south of Sardinia.

6. Summary

The MCC algorithm is an effective tool for mapping currents using thermal satellite imagery. A prerequisite for a successful application of the MCC routine is that cloud free images with strong thermal gradients are present providing features for tracking. One main advantage of currents derived by feature tracking compared to currents derived from altimeters is their indication of the total advective current combining geostrophic and ageostrophic flow components, as opposed to altimetry only providing geostrophic flow. Furthermore MCC vectors have a much higher spatial resolution since altimeters only provide flow components along the altimeter tracks. Coastal zones are also problem areas for altimeters and MCC provides current information for these regions to allow capturing the flow kinematics of a boundary current (Wilkin *et al.* 2002).

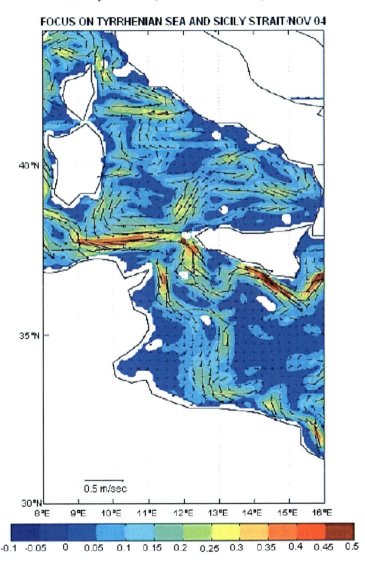

Fig. 5. MFSTEP Bulletin for 11/2004 focussing on the Tyrrhenian Sea and Strait of Sicily, and illustrating the main current features of the region[1].

---

[1] Source: http://www.bo.ingv.it/mfstep/ WP8/Doc/MFSTEP_bulletin_112004.pdf

There have been attempts to apply MCC analysis to data other than AVHRR LAC data, with 1.1 km resolution. Dransfeld *et al.* (2006) have used it on AVHRR Global Area Coverage (GAC) data, with 4.4 km resolution, not to be restricted only to areas covered by a receiving station. This study found that degradation of vector precision, due to the lower image resolution and the averaging procedure used for producing GAC images, is too severe for reliable current fields to be extracted routinely. The averaging scheme of GAC data remains problematic as it does not correspond to a simple average in both horizontal and vertical directions. For GAC production four pixels along a line are averaged and only every third line is used. This implies a real resolution of 4.4 km × 1.1 km and introduces uncertainties into the feature tracking.

A limitation of the MCC method is that by computing discrete pixel offsets to measure continuous velocities, it is not able to represent a continuous velocity spectrum of ocean currents. Two LAC AVHRR images separated by 6 hours indicating a feature displacement of 2 pixels would correspond to a current velocity of 10.2 cm/s and a displacement of 3 pixels to a current velocity of 15.3 cm/s. A possible velocity of *e.g.* 12.7 cm/s could therefore not be calculated by the MCC. Due to this discretisation of velocities each vector component has an associated error that for the mentioned example may attain 20%. That error diminishes for more intense currents, as for displacements of 7 to 8 pixels at a 6 hour separation it may reach only about 7%.

Furthermore MCC is limited to purely horizontal advection processes. It is unable to account for SST gradient changes based on local heating or cooling of the surface due to transient heat fluxes at the air-sea interface. The method relies on the SST purely as a passive tracer for tracking advection processes. This however means that diffusive and vertical mixing processes are neglected (Domingues *et al.* 2000). It is therefore essential to have a detailed knowledge of the predominant physical processes of areas undergoing an MCC analysis.

Emery *et al.* (2006) applied MCC to ocean colour imagery and showed that it provides very similar information to brightness temperature images. For areas of weak thermal gradients, ocean colour hence provides an alternative tracer for advective motion. The information provided by ocean colour is more depth-penetrating, so that ocean colour sensors have potentially a better representation of the upper ocean. However, for winds between 5–10 m/s, the SST seen by the AVHRR instrument gives a reasonable representation (Tokmakian *et al.* 1990). Because wind has less effect on the ocean colour patterns seen from a satellite, compared to thermal patterns, the time separation between ocean colour images may be longer to get similar information about currents as from thermal images with a shorter temporal separation. An interesting research area, within this context, is to analyse differences in vectors obtained from ocean colour and SST imagery in terms of pelagic processes that may have caused them.

The MCC method is a reliable method for researchers concerned with monitoring approximate current strengths in coastal areas. If higher accuracy however is vital, alternative measurement methods such as coastal radars could be used that typically have errors of 5 cm/s or less. The advantage of the MCC method is that it uses freely available satellite data and thus allows gaining a quick insight into surface current behaviour of any coastal area in the global ocean.

Acknowledgements

The author would like to thank Prof. William Emery and his research group at Colorado Centre for Astrodynamical Research for the provision of and help with the navigational software and MCC routines.

References

Bowen MM, Emery WJ, Wilkin JL, Tildesley PC, Barton IJ, Knewtson R (2002) Extracting Multiyear Surface Currents from Sequential Thermal Imagery Using the Maximum Cross-Correlation Technique. Journal of Atmospheric and Oceanic Technology 19: 1665–1676

Béthoux JP (1980) Mean water fluxes across sections in the Mediterranean Sea, evaluated on the basis of water and salt budgets and of observed salinities. Oceanol Acta 3: 79–88

Cote S, Tatnall ARL (1995) The Hopfield neural network as a tool for feature tracking and recognition from satellite sensor images. Inte J Rem Sens 18: 871–885

Domingues CM, Goncalves GA, Ghisolfi RD, Garcia CAE (2000) Advective Surface Velocities Derived from Sequential Infrared Images in the Southwestern Atlantic Ocean. Rem Sens Environ 73 (2): 218–226

Dransfeld S, Larnicol G, LeTraon PY (2006) The potential of maximum cross-correlation technique to estimate surface currents from thermal AVHRR global coverage data. IEEE Geoscience and Rem Sens Letters 3 (4): 508–511

Emery WJ, Thomas AC, Collins MJ, Crawford WR, Mackas DL (1986) An Objective Method for Computing Advective Surface Velocities From Sequential Infrared Satellite Images. J Geophys Res 91 (C11): 12,865–12,878

Emery WJ, Fowler C, Clayson CA (1992) Satellite-image-derived Gulf Stream currents compared with numerical model results. Journal of Atmospheric and Oceanic Technology 19: 286–304

Emery WJ, Crocker I, Matthews D, Baldwin D (2006) Computing Ocean Surface Currents from Infrared and Ocean Color Imagery. Trans Geosci Rem Sens (in press)

Garcia, AEC, Robinson IS (1989) Sea Surface Velocities in Shallow Seas Extracted from Sequential Coastal Zone Color Scanner Satellite Data. J Geophys Res 94 (C9): 12,681–12,691

Gasparini GP, Smeed DA, Alderson S, Sparnocchia S, Vetrano A, Mazzola S (2004) Tidal and subtidal currents in the Strait of Sicily. J Geophys Res 109 (C2), C02011, doi: 10.1029/2003JC002011

Holyer RJ, Peckinpaugh SH (1989) Edge detection applied to satellite imagery of the oceans. IEEE Trans Geosci Rem Sens 27: 46–56

Kelly KA (1989) An inverse model for near-surface velocity from infrared images. J Phys Oceanogr 19: 1845–1864

Kelly KA, Strub PT (1992) Comparison of velocity estimates from Advanced Very High Resolution Radiometer in the coastal transition zone. J Geophys Res 97 (C6): 9653–9668

Liu AK, Martin S, Kwok R (1997) Tracking of Ice Edges and Ice Floes by Wavelet Analysis of SAR Images. Journal of Atmospheric and Oceanic Technology 14: 1187–1198

Millot C, Taupier-Letage I (2005) Circulation in the Mediterranean Sea. In: Saliot A (ed) The Mediterranean Sea, Handbook of Environmental Chemistry, vol 5, Water Pollution, Part K. Springer-Verlag, Berlin Heidelberg, pp 29–66

Ollta A, Sorgente R, Rlbottl A, Natale S, Gabersek S (2006) Effects of the 2003 European heatwave on the Central Mediterranean Sea surface layer: a numerical simulation. Ocean Science Discussions 3: 85–125

Onken R, Robinson AR, Lermusiaux PFJ, Haley Jr. PJ, Anderson LA (2003) Data-driven simulations of synoptic circulation and transports in the Tunisia-Sardinia-Sicily region. J Geophis Res 108 (C9): 8123–8136

Robinson AR, Golnaraghi M, Leslie WG, Artegiani A, Hecht A, Lazzoni E, Michelato A, Sansone E, Teocharis A, Unluata U (1991) The Eastern Mediterranean general circulation: features, structure and variability. Dynamics of Atmospheres and Oceans 15: 215–240

Tokmakian RT, Strub PT, McClean-Padman J (1990) Evaluation of the maximum cross-correlation method of estimating sea surface velocities from sequential satellite images. J Atmos Ocean Tech 7: 852–865

Vigan X, Provost C, Bleck R, Courtier P (2000a) Sea surface velocities from sea surface temperature image sequences. 1, Method and validation using primitive equation model output. J Geophys Res 105: 19499–19514

Wilkin JL, Bowen M, Emery WJ (2002) Mapping mesoscale currents by optimal interpolation of satellite radiometer and altimeter data. Ocean Dynamics 52: 95–103

The Next Generation of Multi-Sensor Merged Sea Surface Temperature Data Sets for Europe

Craig Donlon

The Met Office Hadley Centre, Exeter, United Kingdom

Abstract. Sea Surface Temperature (SST) measured from satellites in considerable spatial detail and at high frequency is required for operational monitoring and forecasting of the ocean, assimilation into coupled ocean-atmosphere models, numerical weather prediction, seasonal forecasting and climate change applications. Currently, many different SST data sets derived from satellite systems are available, and European scientists and agencies alike are presented with a bewildering set of options in terms of SST product content, coverage, spatial resolution, timeliness, format and accuracy. In response, a new generation of integrated SST data products and services are being provided by the Global Ocean Data Assimilation Experiment (GODAE) High resolution SST Pilot Project (GHRSST-PP). An international distributed framework, called the Regional/Global Task Sharing Framework (R/GTS), has been implemented, in which L2 satellite SST data products are processed, following a common Detailed Processing Specification agreed by the GHRSST-PP International Science Team. Output products contain dynamic flags and uncertainty estimates for each SST measurement, for quality control of SST data prior to direct use or to analysis/assimilation. Gap-free L4 analysis systems have also been developed, to combine complementary satellite and *in situ* SST observations in real time, to improve spatial coverage, temporal resolution, cross-sensor calibration stability and SST product accuracy. European systems provide daily 2-km Mediterranean Sea maps and 1/20 degree (~6 km) global maps. Using GHRSST-PP data, SST anomalies for the North Sea and U waters are presented for 2006, clearly showing strong (>3°C) anomalies for July.

1. Introduction

Oceanographic research has grown significantly in response to concerns over the role of the oceans within the climate system and, as more industries move offshore, the growing demand for MetOcean data. The Global

V. Barale, M. Gade (eds.), *Remote Sensing of the European Seas.*

Ocean Data Assimilation Experiment (GODAE) is an international project convened to develop operational global ocean analysis and prediction systems to serve these industrial sectors, to save lives and property and to further oceanographic research and development. GODAE was convened to demonstrate societal and economic advantages of oceanography by implementing the concept of:

"a global system of observations, communications, modeling and assimilation, that will deliver regular, comprehensive information on the state of the oceans in a way that will promote and engender wide utility and availability of this resource for maximum benefit to society."

European operational ocean forecast centres and research teams alike have made a significant contribution to the development and implementation of GODAE resulting in a clear focus for the marine component of the emerging European Global Monitoring for Environment and Security (GMES) program. One of the most important dependencies for ocean forecasting systems is on sea surface temperature (SST) products that are required to properly constrain the upper ocean circulation and thermal structure. SST data products need to be accurate (better than 0.5 K), be available in near real time and have high spatial (<10 km) and temporal (6–12 hours) resolution. In addition, they should properly address the difficult issue of SST at the sea ice edge and diurnal variability. In 2000, no SST products could satisfy these requirements for the European and global domain systems. As a direct response GODAE initiated a GODAE High Resolution SST Pilot Project (GHRSST-PP).

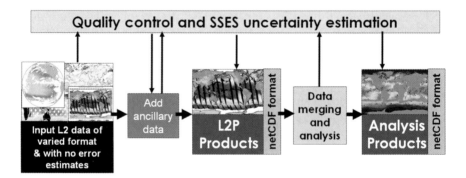

Fig. 1. The data processing strategy of the GHRSST-PP. Complementary satellite and *in situ* data of varied format are merged together to provide analysis products. Data are first quality controlled and error estimates added along with several other auxiliary fields to provide observational data products in a common netCDF data format.

The primary aim of the GHRSST-PP is to develop and demonstrate a system that will deliver high-resolution global coverage SST data products operationally in near real time according to GODAE specifications. The project has worked to establish a common internationally distributed system that quality controls different origin satellite SST data sets into a common format, adds uncertainty estimation and supporting data to each pixel value and ultimately provides merged data products building in the synergy between complementary satellite and *in situ* SST data as summarized in Figure 1.

This paper first reviews current definitions of SST, it then summarizes the international distributed SST processing system developed by the GHRSST-PP including a review of the main SST outputs from European analysis systems currently available. GHRSST-PP data products are then used to review SST anomalies in the North Sea and surrounding the British Isles in 2006. Finally a summary and conclusions are presented.

2.   Definitions of sea surface temperature

What exactly *is* the sea surface temperature? SST is a difficult parameter to define exactly because the upper ocean (~10 m) has a complex and variable vertical temperature structure that is related to ocean turbulence and the air-sea fluxes of heat, moisture and momentum. A framework is required to understand the information content and relationships between measurements of SST made by different satellite and *in situ* instruments, especially if these are to be merged together as in the case of GHRSST-PP analysis systems. The definitions of SST developed by the GHRSST-PP SST Science Team achieve the closest possible coincidence between what is *defined* and what *can be measured operationally*, bearing in mind current scientific knowledge and understanding of how the near surface thermal structure of the ocean behaves in nature.

Two hypothetical vertical profiles of temperature in low wind speed conditions during the night and day are shown in Figure 2 and these encapsulate the effects of the dominant heat transport processes and time scales of variability associated with distinct vertical and volume regimes. At the exact air-sea interface a hypothetical temperature called the interface temperature (SSTint) is defined although this is of no practical use because it cannot be measured using current technology. The skin temperature (SSTskin) is defined as the temperature measured by an infrared radiometer typically operating at wavelengths 3.7–12 μm (chosen for consistency with the majority of infrared satellite measurements) and represents the temperature within the conductive diffusion-dominated sub-layer at a depth of ~15–20 μm. SSTskin measurements are subject to a large potential diurnal cycle including cool skin layer effects (especially at night under clear skies and low wind speed conditions) and warm layer effects in the daytime with a time-scale of ~10 s (Jessup *et al.* 1997). The mean difference between the SSTskin and the SST at depth is typically stable having a value of –0.17 K ± 0.3 K in moderate (>6 ms^{-2}) wind speed (*e.g.* Donlon *et al.* 2002).

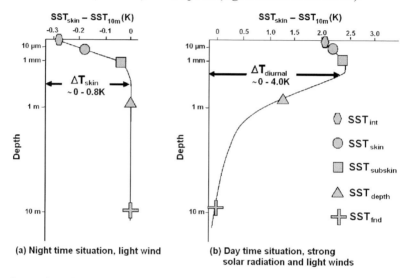

(a) Night time situation, light wind   **(b) Day time situation, strong solar radiation and light winds**

Fig. 2. Modern definitions of sea surface temperature used within the framework of the GHRSST-PP. SSTint is the interface temperature, SSTskin the temperature measured by an infrared radiometer, SSTsub-skin the temperature assumed to be measured by a microwave radiometer, SSTdepth the temperature measured by conventional contact thermometers at depth (z). The foundation temperature (SSTfnd) is the temperature without any diurnal variability typically obtained using analysis techniques that blend complementary SST data together.

The subskin temperature (SSTsubskin) represents the temperature at the base of the conductive laminar sub-layer of the ocean surface. For practical purposes, SSTsubskin can be well approximated to the measurement of surface temperature by a microwave radiometer operating in the 6–11 GHz frequency range, but the relationship is not direct and varies with changing physical conditions and the geometry of the microwave measurement. All measurements of water temperature beneath the SSTsubskin are referred to as depth temperatures (SSTdepth) and are measured using a wide variety of platforms and sensors such as drifting buoys, vertical profiling floats, or deep thermistor chains at depths at depths ranging from 10^{-2}–10^{3} m. These temperature observations are distinct from those obtained using remote sensing techniques (SSTskin and SSTsubskin) and must be qualified by a measurement depth in meters (or SST(z), *e.g.* SST5 m). The foundation SST, SSTfnd, is defined as the temperature of the water column free of diurnal temperature variability (daytime warming or nocturnal cooling) and is considered equivalent to the SSTsubskin in the absence of any diurnal signal. It is named to indicate that it is the foundation temperature from which the growth of the diurnal thermocline develops each day (noting that on some occasions with a deep mixed layer there is no clear SSTfnd profile in the surface layer). Only *in situ* contact thermometry is able to measure SSTfnd and analysis procedures must be used to estimate the SSTfnd from radiometric satellite measurements of SSTskin and SSTsubskin. SSTfnd provides a connection with the historical concept of a "bulk" SST considered representative of the oceanic mixed layer temperature and represented by any SSTdepth measurement within the upper ocean over a depth range of ~1–20 m. In the surface waters, diurnal variability may introduce significant variability within a 24 hour period and it is impossible to define exactly what an analysis of such disparate data would mean without observation times and vertical location of each observation. Such data sets are therefore best considered a blend of SST (SSTblend). SSTfnd provides a more precise, well-defined quantity and in general, SSTfnd will be similar to a night time minimum or pre-dawn value at depths of ~1–5 m, but some differences could exist. Note that SSTfnd does not imply a constant depth mixed layer, but rather a surface layer of variable depth depending on the balance between stratification and turbulent energy and is expected to change slowly over the course of a day.

3. The GHRSST-PP Regional Global Task Sharing (R/GTS) framework

Figure 3 provides a schematic overview of the distributed GHRSST-PP Regional/Global Task Sharing (R/GTS) framework in which satellite and *in situ* data processing operations are shared by international partners who generate and distribute global coverage high resolution SST data sets.

Regional Data Assembly Centres (RDAC) first ingest, quality control and merge complementary satellite and *in situ* SST data sources, to generate consistent-format regional (and in some cases global) coverage quality-controlled SST data products in real-time. Each RDAC delivers data products to the same specification in terms of quality control methods applied to input data based on the GHRSST-PP Data Processing Specification

(GDS, Donlon *et al.* 2006). Observation data products (denoted as L2P) are generated for every swath of polar orbiting satellite instruments and for each geostationary satellite instrument scene (or the temporal mean of several scenes). Multi-sensor combined analysis products (denoted as L4) build on the synergy of different but complementary satellite and *in situ* observations to produce combined gap-free gridded SST analysis product.

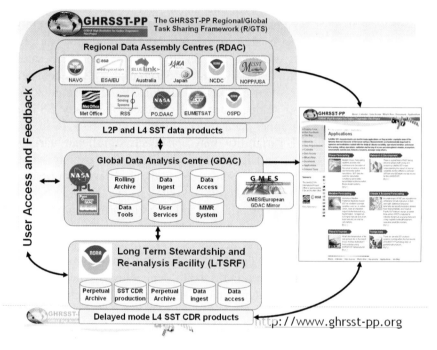

Fig. 3. The Regional/Global Task sharing (RTGS) distributed implementation of the GHRSST-PP. Arrows indicate the flow of data.

GHRSST-PP RDAC data products are all assembled together in near real time at Global Data Analysis Centres (GDAC) to facilitate easy on-line access over a period of 30 days. After this time, they are swept away to a Long Term Stewardship and Reanalysis Facility (LTSRF) located at the NOAA National Oceanographic Data Center. The LTSRF provides both the GHRSST-PP long term archive and forms the central hub of the distributed GHRSST-PP re-analysis system. The re-analysis project is working towards a combined 4 km global coverage data sets spanning 1983-present with the first basic set of re-analysis products available in 2007. As the name LTSRF implies, complete stewardship rather than a simple archive in the traditional sense, is required to enable a successful

RAN system which protects the significant investment made by RDAC and GDAC teams that have generated GHRSST-PP data products.

4.   A new generation of SST products for Europe

The Medspiration Project[1] is a European initiative funded by the European Space Agency (ESA) to develop a GHRSST-PP RDAC for Europe providing L2P and L4 products. Figure 4 shows an example L2P data set and associated auxiliary fields, generated by the Medspiration project, using the EUMETSAT Spinning Enhanced Visible Infra-Red Imager (SEVIRI) instrument flown on the Meteosat second generation geostationary satellite.

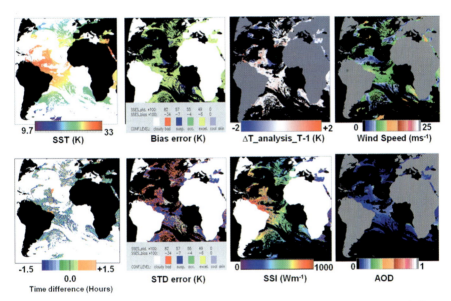

Fig. 4. Example GHRSST-PP L2P data set derived using Meteosat second Generation SEVIRI SST data collected over a 3 hour period. Shown (from top left to bottom right) are SST, SST bias error, difference from previous reference SST analysis field, surface wind speed (ECMWF), time of observation within a 3 hour window (SEVIRI makes observations every 15 minutes), standard deviation of SST error, surface solar irradiance (from SEVIRI visible channels) and aerosol optical depth (NEDSIS).

L2P products do not alter the SST data provided by an instrument operator but complement these observations by (a) providing a extensible common

---

[1] see http://www.medspiration.org

interface format data set that is internet 'aware' and (b) adding value through metadata and auxiliary fields that greatly aid the application and interpretation of the SST observations. Specifically, a L2P data record, provided for every pixel in a L2P data product, includes the native SST data, an estimate of the bias error and standard deviation of error, surface wind speed, aerosol optical depth (AOD), surface solar radiation (SSI), sea ice (SI) concentration, time of observation and a set of quality control flags. The auxiliary geophysical fields are provided as dynamic flags that can be used by different user communities to filter data that are not suitable for a given application. For example, wind speed and surface solar radiation data can be used as filters prior to combined analysis to determine the likelihood of thermal stratification and estimate the extent of any diurnal temperature variability in the afternoon. Aerosol data may be used to filter contaminated data out by setting an appropriate threshold above which infrared SST data from satellite are considered unreliable.

A number of ultra-high resolution (UHR, <5 km resolution) regional SST L4 analysis products are also being generated by the Medspiration project. A L4 SST product at a resolution of 2 km covering the Mediterranean Sea is generated each day based on optimum interpolation which is available every day since 1.2.2005. The Mediterranean is a particularly challenging area for developing L4 SST products due to strong diurnal variability, North African atmospheric aerosol and strong surface ocean dynamics. L4 maps are produced daily based on all available satellite and *in situ* SST data and made available via ftp from both RDAC and GDAC server systems. Figure 5(a) provides a typical example L4 Medspiration product including a map of uncertainty estimation for each SST grid point.

Figure 5(b) shows a global L4 SST analysis produced by the Met Office, based on available L2P and *in situ* observations for the 20.11.2005. The Operational SST and Sea Ice Analysis (OSTIA[2]) is produced each day based on optimal interpolation providing SST on a 1/20° global grid (~6 km). At present OSTIA uses fixed x and y correlation length scales at 100km representing synoptic scales and 10km representing meso scales in a two pass process. Figure 5(c) shows the corresponding SST anomaly computed against a Pathfinder SST weekly climatology derived from AVHRR SST data covering the period 1985–2001. These types of products are extremely useful for monitoring seasonal forecast products[3] that have strong dependencies on tropical region SST patterns and magnitudes.

Figure 6 shows the SST anomaly computed from monthly mean OSTIA SST analysis outputs for 2006. Anomalies have been computed relative to

---

[2] see http://ghrsst-pp.metoffice.com/pages/latest_analysis/ostia.html
[3] e.g., http://www.metoffice.gov.uk/research/seasonal/index.html

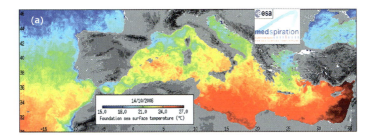

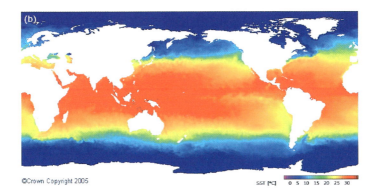

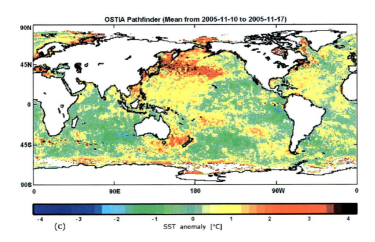

Fig. 5. (a) L4 2 km SST Analysis product developed by the ESA Medspiration project for 14/10/2006. (b) SST foundation L4 product (OSTIA) produced by the Met Office based on GHRSST-PP data products for 20/11/2005. (c) OSTIA – NCEP OIv2.0 climatology (1985–2001) for 10-17/11/2005. (© Met Office 2005).

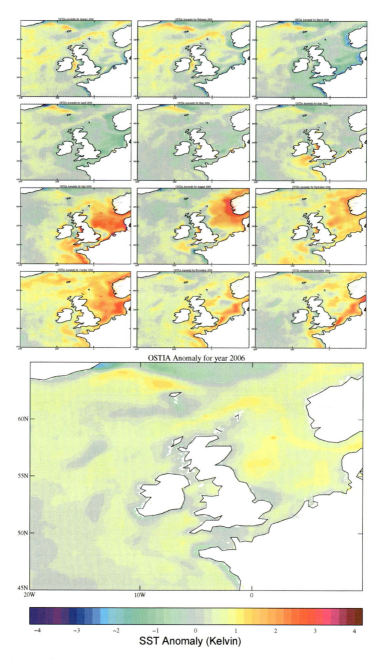

Fig. 6. Monthly average SST anomaly for January–December 2006 computed from the MetOffice OSTIA system referenced to a multi-year NCEP NCEP OIv2.0 analysis. The reference period used for the climatology is 1985 to 2001. The lower figure shows the annual mean SST anomaly for 2006.

the NCEP OIv2.0 for the period 1985–2002 (Reynolds and Smith 1994). The North Sea and Cost of Norway were more than 3°C warmer than expected. The lower panel of Figure 6 shows the mean SST anomaly for the North Sea and UK coastal waters for 2006 which is 0.8 K. A new set of OSTIA anomaly maps are being prepared based on the HadSST data set (Rayner *et al.* 2006) and building on a new high-resolution version of the HadISST climate reference (Rayner *et al.* 2003).

In 2006 European summer temperatures were exceptionally warm and the mean average temperature for July in the region 10W–30E, 35N–75N was 2.7°C ± 0.2°C above normal. Throughout the summer the surface of the North Atlantic averaged about 0.8°C warmer than normal but coastal waters around the UK and were 1–2°C warmer than normal during July. During July 2006, atmospheric pressure over the UK and Northern Europe was anomalously high and such persistent anti-cyclonic conditions and associated clear skies. Much of the 2006 warming is the local response to increased sunshine, light winds and warm overlying air characteristic of the high atmospheric pressure systems that predominated in July and in August, the impact of the warm NE Atlantic temperatures and climate change. In August a series of low pressure systems developed which would normally bring cold air from the northwest and cool the surface waters but in this case, the impact of a warmer than normal Atlantic temperatures moderated such an event and anomaly temperatures remained high.

5. Conclusions

The GHRSST-PP[4] is a large international project that provides a framework for the exchange, processing and application of satellite SST products. It has over €16 million invested by projects in Europe, Japan, Australia and the USA. Operational sea ice concentration, surface solar irradiance, aerosol optical depth, and wind speeds constitute the core data sets used by the GHRSST-PP. European SST users and producers are now in transition towards a fully operational distributed SST system contributing to the international GHRSST-PP within the Global Monitoring for Environment and Security (GMES) program. For studies of diurnal SST variability, hourly products of SST and SSI at full SEVIRI native resolution are now becoming available via Medspiration. A dedicated re-analysis program is now developing within the GHRSST-PP and a set of on-line diagnostic

---

[4] Full details of the GHRSST-PP, its data products and services are available at http://www.ghrsst-pp.org.

services are available. For the future, the GHRSST-PP International Science Team requests that SST user and producer communities consider the GHRSST-PP L2P format and methods as a baseline standard for the present and next generation of satellite SST data products and services which represents the best international scientific consensus opinion on SST data format and quality control procedures.

Acknowledgements

The Author would like to thank John Stark, John Kennedy, David Parker and Chris Folland at the Met Office for providing inputs to this paper.

References

Donlon, C J and the GHRSST-PP Science Team, (2006) The GHRSST-PP Data processing Specification version 1.7, available form the GHRSST-PP International Project Office, Met Office, Exeter, UK or at http://www.ghrsst-pp.org, pp 241

Donlon CJ, Minnett P, Gentemann C, Nightingale TJ, Barton IJ, Ward B, Murray J (2002) Towards Improved Validation of Satellite Sea Surface Skin Temperature Measurements for Climate Research. J Climate 15: 353–369

Jessup AT, Zappa CJ, Loewen MR, Hesany V (1997) Infrared remote sensing of breaking waves. Nature 385: 52–55

Rayner NA, Parker DE, Horton EB, Folland CK, Alexander LV, Rowell DP, Kent EC, Kaplan A (2003) Global analyses of sea surface temperature, sea ice, and night marine air temperature since the late nineteenth century. J Geophys Res 108 (D14), 4407, doi:10.1029/2002JD002670.

Rayner NA, Brohan P, Parker DE, Folland CK, Kennedy JJ, Vanicek M, Ansell T, Tett SFB (2006) Improved analyses of changes and uncertainties in sea surface temperature measured *in situ* since the mid-nineteenth century: the HadSST2 data set. J Climate 19: 446–469

Reynolds RW, Smith TM (1994) Improved global sea surface temperature analyses. J Climate 7: 929–948

Laser Remote Sensing of the European Marine Environment: LIF Technology and Applications

Sergey Babichenko

Laser Diagnostic Instruments AS, Tallinn, Estonia

Abstract. Laser remote sensing is an efficient, proven tool capable of providing quantitative, spatially-resolved, *Real-Time* data for chemical pollution, eutrophication, biomass, and hydrographic processes over large water surface areas with high spatial resolution; and is often the only solution for many environmental marine applications. Various types of Light Detection and Ranging systems (LIDAR) utilize Laser Induced Fluorescence (LIF) and light backscattering to analyze bodies of water remotely. LIDAR systems are installed as a payload on airborne, shipboard or stationary platforms for operational purposes and scientific research. This paper provides an overview of and references to LIF LIDAR technology, together with a brief insight to applications development in Europe.

1. Introduction

The technology of Laser Induced Fluorescence (LIF) is widely applied to remote sensing of the marine environment. LIF and light backscattering (elastic and inelastic) provide radar-like measurements utilized by Laser remote sensing systems – LIDAR (LIght Detection And Ranging). One of the principal merits of Laser remote sensing is its direct analytical capability. LIDAR can measure hydrographic parameters and substances in water, which could otherwise (*e.g.* with ocean color) only be evaluated indirectly using theoretical or empirical bio-optical algorithms. Such systems can be used as a payload for airborne, shipboard or stationary platforms and is capable of providing quantitative, spatially resolved, *Real-Time* data.

Reduction of water quality caused by pollution and eutrophication is a typical problem for coastal waters and inner seas. Due to the complex nature of, and multiple interactions between physical, chemical, and biotic factors in water ecosystems, the *Real-Time* monitoring of water quality is a most effective approach to diagnose unwanted deviations in water characteristics. For this application, the LIDAR's primary objective is rapid

V. Barale, M. Gade (eds.), *Remote Sensing of the European Seas.*

screening of large water areas to detect oil pollution (Hoge and Swift 1980; Hengstermann and Reuter 1990) and to measure phytoplankton abundance (Babichenko *et al.* 1993; Chekaljuk *et al.* 1995).

Furthermore, a LIDAR provides in-depth analytical information on the type of chemical pollution (Dudelzak *et al.* 1991; Brown *et al.* 1997), pigment composition of phytoplankton (Babichenko *et al.* 1999), spectral characterization of Chromophoric Dissolved Organic Matter (CDOM) (Poryvkina *et al.* 1992; Determann *et al.* 1994, Patsayeva 1995). LIDAR is also able to provide data on water transparency and turbidity (Bristow *et al.* 1985), and to measure hydrographic fronts (Hengstermann *et al.* 1992). The technology can be equally effective in the detection of various underwater targets, such as industrial waste, sunken cargo, pipelines etc., benthic habitat and sediment characterization, and other underwater studies related to oceanography, marine biology and ecology.

LIF LIDAR is based on the detection and analysis of the spectra of fluorescence, induced in the target object by illumination with monochromatic laser radiation (Figure 1). A secondary emission of the sensed object constitutes an echo-signal of Laser remote sensing as a composite of the light scattering and LIF responses of the individual compounds present in water. This signal, collected by a receiving telescope and recorded by spectral selective detectors, is analyzed. The intensity of Raman (inelastic) scattering of a laser emission on water molecules is often used to normalize fluorescence in marine applications. This method was introduced by Klyshko and Fadeev in 1978, and Hoge and Swift in 1981, references and description can be found in (Determann *et al.* 1994).

Due to the pulsed emission of the laser, it is possible to either integrate the echo-signal through the water column or to take distance-resolved measurements by time-gating the received echo-signal. The time delay between the laser pulse and the gate pulse defines the sensing distance, while the duration of the gate pulse determines the thickness of the sensed water layer. The characteristic penetration distance of laser emission in water depends on water transparency and laser wavelength. As LIF LIDAR systems typically employ a laser with the ultraviolet (UV) or blue/green sensing wavelengths, in most natural waters a penetration depth from a few meters up to tens of meters can be achieved.

Compared to passive remote sensing: selective excitation by monochromatic laser emission; directionality; controlled light source; the use of an inner spectroscopic bench mark; and the possibility of time-(distance)-resolved measurement by time-gating the echo-signal are all advantages of the LIF LIDAR technique. These capabilities encourage and suit the application of this technology to monitoring natural waters. This paper is mainly focused on LIF LIDAR applications and development in Europe.

2. LIF LIDAR techniques

While LIF LIDAR systems are similar in their basic operational principles, they differ in layout depending on application and intended use. A LIDAR can be defined according to its key technical parameters. These include: sensing wavelength(s); the number of spectral channels of the detector; time-resolving capabilities, etc. It is also possible to classify a LIDAR according to its operational features, such as sensing distance (short or long range); spot or scanning or imaging mode of operation.

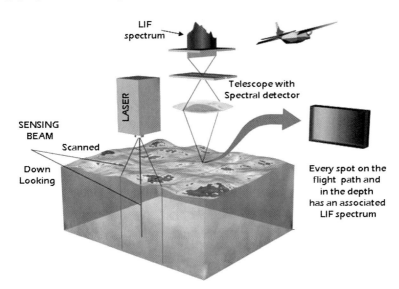

Fig. 1. The schematics of LIF LIDAR operation.

2.1 Laser Fluorosensor

Laser Fluorosensor, is a widely used term specifying a LIF LIDAR operating with an echo-signal spectrum recorded in a discrete, predefined number of spectral ranges, spectrally separated and having a certain spectral width (spectral channels). In the optical layout of such systems, the echo-signal is split into spectral portions and registered by a set of photomultipliers (PMT). The spectral channels are selected according to the main features of a water emission spectrum and, typically, correspond to Raman scattering, CDOM and chlorophyll fluorescence, and to the characteristic fluorescence ranges of oil or other substances of interest. The theory behind, and first results of the application of a Laser Fluorosensor to measurement of oil-film thickness, were introduced in (Hoge and Swift 1980).

The ability of an airborne Laser Fluorosensor to detect and classify small discharges of mineral oils in the sea was experimentally well demonstrated in the North Sea in 1986 (Hengstermann and Reuter 1990). An oceanographic LIDAR system (OLS), used for the experiment, was developed for installation in a Dornier Do 28/228 research aircraft. An XeCl excimer laser (308 nm, 10 MW peak power, 10 Hz) was the main sensing source and used to pump the dye-laser (emission at 450 nm). Spectral channels with a spectral width of 10 nm were centered at 344, 366, 380, 450/500, 533, 650, 685 nm to record LIF spectrum (Figure 2a). A significant result of the experiment was the detection and classification of oil pollution in water (at 200 *ppm – part per million* - concentration), which was neither visible on the water surface nor detected by Side-Looking Airborne Radar (SLAR), UV and Infra-Red (IR) onboard sensors.

Laser Fluorosensors were enhanced in time by adding depth-resolved features to measure vertical profiles of bio-optical parameters of sea-water (Ohm *et al.* 1998, Barbini *et al.* 2001), more spectral channels for better LIF spectrum analysis (Zielinski *et al.* 2001), and scanning the laser beam to produce a swath on the sensed water (Hengstermann at al. 1992).

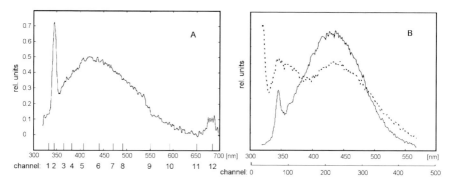

Fig. 2. (A) - Typical emission spectrum of seawater, measured in a lab at 308 nm excitation with indicated LFS detection channels; the peak at 344 nm is water Raman scattering; the bands with maxima at 420 and 680 nm are due to CDOM and phytoplankton fluorescence; (B) - Typical LIF spectra of clean (bold line) and polluted water (dotted line) in the Gulf of Finland, measured with FLS-A(M) LIDAR onboard a helicopter at the distance of 200 m.

2.2 Hyperspectral LIDAR

Hyperspectral refers to a LIF LIDAR operated with an echo-signal recorded as a continuous emission spectrum. The main difference between such systems and a Laser Fluorosensor is that there are no predefined spectral ranges to detect echo-signal, and a comprehensive shape of water emission spectrum is recorded at every laser shot for consequent analysis. The hyperspectral detector of such LIDAR consists typically of a polychromator, defining detection spectral range and resolution, and a multichannel gated optical detector to read-out an emission spectrum.

The analytical capabilities of LIDAR operated with high spectral resolution were studied in (Babichenko *et al.* 1993, Pantani *et al.* 1995). Initially, such systems were developed mainly for shipboard operation. Late 1980s, a multi-channel shipboard LIDAR with several sensing wavelengths was used to measure the horizontal profile of phytoplankton distribution, CDOM and organic pollution (Dudelzak *et al.* 1991). An excimer laser served as a sensing source and to pump the dye-laser. An intensified, gated, linear, CCD-camera was used as a detector.

The analytical capabilities of hyperspectral LIDAR for detecting oil pollution in water from onboard an airplane were studied in the Gulf of Finland and Estonian lakes in 1994. Based on an XeCl excimer laser (308 nm, 90 mJ pulse energy, 20 Hz pulse repetition rate), 28 cm aperture telescope, and gated intensified CCD detector (500 channels, 300–550 nm spectral range), an FLS-A LIDAR was designed as a mono-block to be installed onboard an "aircraft of opportunity." This airborne survey, accompanied by water sampling in the area of interest, confirmed that water pollution down to a concentration level of 1 ppm can be identified by hyperspectral LIDAR at distances of 200 m (Babichenko *et al.* 1995). Modern intensified CCD cameras are sensitive enough (2–3 times less sensitive than PMT) to serve as a LIF detector (see Figure 2b). Optional signal accumulation is typically used to increase the signal-to-noise ratio (SNR).

2.3   Imaging LIDAR

Laser Fluorosensor and hyperspectral LIDAR with beam-scanning capabilities are considered to be imaging LIDAR. Indeed, a map of the water surface scanned by a laser beam can be presented as a pixel image of the area plotted in various spectral ranges, providing information on the spatial distribution of different fluorophores in water.

Another group of systems referred to as Imaging LIDAR are based on a different concept. The object is illuminated by laser beam, and its image is registered through the optical filter. A compact imaging LIDAR aimed at oil-slick detection is described in (Hitomi *et al.* 2002). The system is designed to display the pollution spot in an image of a wide-field-of-view. Reportedly, the ability of the LIDAR to detect an oil film on surface water was proved through airborne tests at an altitude of 1000 ft.

3.   Airborne LIF LIDAR in Europe

One of the main assumptions behind the use of a LIDAR is that it can de-liver information in near *Real-Time* to meet the clients' needs, rather than raw data for research studies. Environmental and emergency response agencies require in-time information on oil spills: location, type and vol-ume, potential movement and spread, and other data necessary to plan and optimize clean-up. In order to satisfy these requirements and fulfill expec-tations of pollution diagnostics, the following measures should be targeted:

- Reliable detection of oil spill; 24 hours a day;
- Identification of oil type for physical and chemical properties;
- Evaluation of pollution volume;
- Mapping polluted area and oil concentration.

Information on hydrographical influences in the coastal zone is also im-portant for effective marine monitoring. The detection of events such as harmful algal blooms, or transport of chemicals and other pollutants by water masses is within the capabilities of an airborne LIDAR.

As for other remote sensing techniques, the integration of LIDAR data with that of other technologies helps to fully address the customers' infor-mation requirements. Accordingly, LIDAR systems work typically either as a part of a multi-sensor platform or are integrated at the data level as one of multiple sources of information. Airborne remote sensing equipment may be divided into two categories: far-range detection (SLAR) and near-range detection (IR/UV line scanner). Microwave radiometers (MWR) and air-borne LIF LIDAR can be specified as advanced sensors for near-range de-tection, facilitating measurement of oil-film thickness.

The distinguishing features of LIF LIDAR is its capability to classify oil remotely; detect oil emulsion and dissolved fractions in water; reveal sub-mersed oil in sub-surface layer; and locate oil pollution in iced water.

Two German maritime surveillance Do228-212LM aircraft have been equipped with sensor suites, including a scanning Laser Fluorosensor Sys-tem (LFS). The LFS is designed with two high energy pulse lasers in the UV spectral range. An XeCl excimer laser (308 nm, 150 mJ, 110 Hz) is used for analysis of oils, fulvic acid (CDOM), and organic pollutants. A dye-laser (383 nm, 20 mJ) serves for chlorophyll fluorescence excitation. Induced fluorescence and scattering are detected with a 20 cm aperture telescope and a 12-channel optical spectrograph. The spectrograph consists of identical modules with gated PMT detectors, spectrally divided by di-chroic and interference filters. Detection channels are located at wavelengths of 332, 344. 365, 382, 407, 441, 471, 492, 551, 592, 650, and 684 nm, and have typical optical bandwidth of 10 nm (Reuter *et al.* 1995).

Two-dimensional mapping of the sea surface is carried out with a conical scanner providing typically 10 m pixel-to-pixel resolution at 300 m flight altitude. Maps of oil-film thickness, and finally, released volume, are derived from these measurements. For hydrographic and biological applications, the sensor measures the concentration of CDOM and phytoplankton by their fluorescence. Additionally, turbidity of water can be determined through analyzing the backscatter signal.

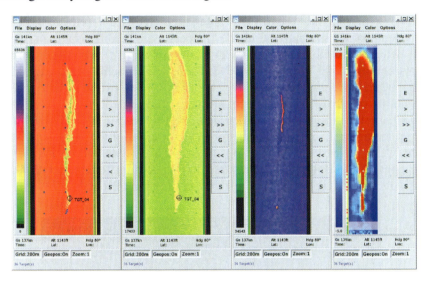

Fig. 3. Four windows of the MEDUSA software module for oil spill analysis. From left to right: IR image, UV image, MWR image (89 GHz channel) and LFS thickness image of an oil spill (Robbe and Hengstermann 2006).

LFS is part of the onboard Multispectral Environmental Data Unit for Surveillance Applications (MEDUSA), managing data acquisition and pollution information (Robbe and Hengstermann 2006). The MEDUSA system provides data and sensors fusion for *Real-Time* on-line visualization of information and reliable data storage (Figure 3). It combines 6 sensors - SLAR, IR, UV, MWR, LFS, FLIR (Forward-Looking Infra-Red camera) - in one environment and includes telemetry links, and geo-correction. The central mission computer stores the data, provides time and visual information, and operational controls. Sensor management, on-line visualization and data analysis is facilitated by a graphical user interface (Robbe and Zielinski 2004). Reportedly, four new Spanish maritime surveillance aircraft shall be equipped with a similar sensor suite, including a nadir looking LFS. The most recent review of a multi-sensor suite, airborne, oil-spill surveillance in Europe can be found in (Zielinski *et al.* 2006).

Hyperspectral airborne FLS-A(M) LIDAR pre-operational tests were recently undertaken in Estonia (Babichenko *et al.* 2006). The tests were the culmination of several years of development, preceded by a number of pilot projects including the combined use of LIF LIDAR with other onboard sensors (Lennon *et al.* 2006). The FLS-A(M) LIDAR is designed as a down-looking-and-scanning airborne system with self-adjusted, 28 cm aperture telescope, acting as a transmitting and receiving system. A ruggedized XeCl excimer laser (308 nm, 150 mJ, 150 Hz) serves as a sensing source with optional use of two integrated dye-lasers (typical sensing wavelengths 360 nm and 460 nm). LIF spectra are recorded by hyperspectral detector (gated intensified CCD camera) having 500 channels in the spectral ranges 300–600 nm or 450–750 nm. The scanning of the expanded laser beam enables a ground level corridor to be observed with spatial resolution of 10 m (at 250 m flight altitude). The LIDAR has an integrated video targeting system with two outputs. One serves for the pilot screen to target the aircraft, and another to store video frames of the underlying surface. Data processing is carried out in *Real-Time* mode and visualized on the operator console. Each measured LIF spectrum, together with GPS position and corresponding video frame, is stored in the database.

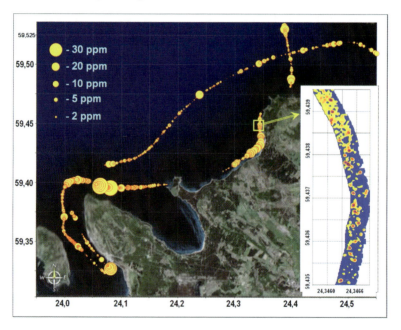

Fig. 4. Example of oil pollution in iced water (Feb 2006) in the Gulf of Finland along the flight path of FLS–A(M) LIDAR. Zoomed area on the right side shows pollution distribution inside the observation corridor of the LIDAR.

Laser remote sensing by hyperspectral detector delivers some analytical benefits - especially for coastal-water and shoreline monitoring. Typically, shallow depths, reduced water transparency, and higher amounts of suspended solids cause high variability in LIF spectra and require detailed analysis of the comprehensive shape of the echo-signal spectrum. The same is applicable to the detection of oil pollution on solid ice, in seawater with broken ice, and the location of submersed oil.

The FLS-A(M) LIDAR, operated onboard Estonian rescue helicopter, monitored open water, the coastal zone and shoreline in the Baltic Sea. In the winter-spring period of 2006, it was used to detect several oil spills in iced water in the Gulf of Finland (Figure 4) and to observe oil leaks from the damaged fuel tank of a sunken cargo ship.

A hydrographic airborne LIDAR PAL-1M operates in Russia onboard the research aircraft "Arktika" (AN-26), operated by the Polar Research Institute of Marine Fisheries and Oceanography in Murmansk. The aircraft is equipped with MWR, IR and Synthetic Aperture Radar (SAR). LIDAR PAL-1 is used to measure turbidity and phytoplankton to locate pelagic fish schools in the Norwegian and Barents seas. It is based on a Nd:YAG laser (532 nm, 120 mJ, 140 Hz) and a 3-channel detector. Two channels are used for polarized detection of scattered light and the third for fluorescence of *chlorophyll a*. It is reported that PAL-1M is effective up to a sensing depth of 45 m (Norwegian Sea) at a flight altitude of 150–200 m and flight speed of 300 kmph (Zabavnikov *et al.* 2005).

4.   LIF-LIDARs under development and for scientific application

The development of applications for LIF LIDAR technology is constant. Scientific research is focused on the enhancement of the analytical capabilities of Laser remote sensing. Technical developments for marine applications are characterized by two major directions. The first, considers LIF LIDAR as a payload for various platforms, including Remote Operated Vehicles (ROV) and Unmanned Aerial Vehicles (UAV). The second addresses the combined use of LIF LIDAR with other remote sensors.

4.1 Shipboard LIDAR

While operational systems are mainly airborne, the use of shipboard LIDAR still delivers some benefits in gathering analytical information from water bodies. Lower speed of the vehicle, shorter distance to the water surface, potential ability to introduce the laser beam directly under the water surface – all these factors provide the possibility to carry out range-resolved measurements, and finally, 3D spatial distribution of marine organics and phytoplankton.

The possible layout of a range-resolved shipboard LIDAR is described in (Ohm *et al.* 1998). The LIDAR was installed onboard the research vessel "Polarstern" at 11 m depth below the sea surface to investigate subsurface seawater layers. Such installation reduces background sunlight and enhances the detection limit for signals coming from the water mass. The system operated with a Nd:YAG laser at three sensing wavelengths (532 nm, 355 nm, and 283 nm) and an 8-channel detector. The sensing depth at open sea was 40 m, but only 10–20 m depth was sensed in coastal waters.

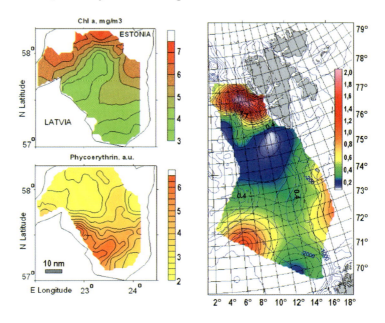

Fig. 5. (a): Map of chlorophyll a and phycoerythrin distribution in the Gulf of Riga; (b): The fluorescence factor of chlorophyll a in the water column normalized with surface concentration of chlorophyll a measured with FLS lidar, sensing wavelength 440 nm (Drozdowska *et al.* 2004).

The hyperspectral LIF LIDAR onboard the research vessel were widely used in regional studies of phytoplankton abundance and taxonomic composition based on measuring *chlorophyll a* and other pigments in the subsurface water layer. Comparison of spatial distribution of different pigments indicates the structure of the phytoplankton community. Fig 5a shows the difference in *chlorophyll a* and *phycoerythrin* maps measured with FLS shipboard LIDAR – indicative of the distribution of blue-green phytoplankton in the Gulf of Riga (Babichenko *et al.* 1999).

In 2002, the inhomogeneous spatial distribution of *chlorophyll a* in the water column was studied in the Greenland and Norwegian Seas with FLS shipboard LIDAR. The area is characterized by the mixing of the salty, warm water of the Spitsbergen current with cool, fresh glacial water (Drozdowska *et al.* 2004). Investigation of the spatial variability of the phytoplankton biomass in the surface water layer - influenced by salinity, temperature, and water circulation - permitted studies of the patchiness of fresh and ocean water inflow (Figure 5b).

For the last decade, a LIDAR laboratory has operated onboard research vessel "Italica". The ENEA LIDAR fluorosensor (ELF) measures surface *chlorophyll a* and monitors seawater quality. A sub-surface LIF LIDAR - with time-resolved function – is used for cross-calibration with simultaneously operating ELF system. The LIDAR installed on the bottom of the ship is based on a Nd:YAG laser (355 nm, 30 mJ) and a 4-channel detector (355, 403, 450 and 680 nm). Range-resolved data for CDOM and phytoplankton, computed from return signals, are accurate within 10% – 15% at the distances up to 30 m (Barbini *et al.* 2001).

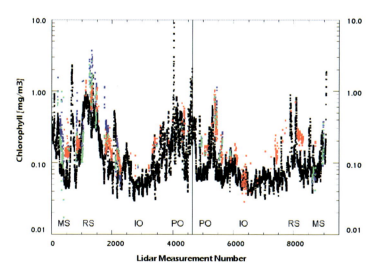

Fig. 6. Surface *chlorophyll a* measured with ELF lidar (black color), SeaWiFS (red) and MODIS (blue and green). MS - Mediterranean Sea, RS - Red Sea, IO - Indian Ocean, PO - Pacific Ocean (Fiorani *et al.* 2004).

Shipboard LIDAR can deliver valuable information for large-scale monitoring. The ELF system was used for inter-calibration of imagery collected by

SeaWiFS and MODIS (Figure 6). This study confirmed the advantages, and possible synergies, of LIDAR fluorosensors and spaceborne radiometers. A potential extension of this application is the correction of satellite data for the effects of CDOM absorbance (Fiorani *et al.* 2004).

4.2   Stationary LIDAR installations

For certain applications, the stationary installation of LIF LIDAR can be considered a cost-efficient alternative to airborne LIDAR. Where there is a need to monitor specific local areas, such as harbors, oil terminals or select sections of waterways, *i.e.* the water area is limited, but continuous observation is required, LIF LIDAR installed on a stationary platform can serve as an effective surveillance tool. To achieve reasonable sampling distances, scans are performed with small sampling angles between the water surface and the optical axes of the laser beam and the telescope.

Studies completed in (Maslov *et al.* 2001) have demonstrated the use of stationary LIF LIDAR in inclined operation on the shores of the Black Sea. The LIDAR consisted of a Nd:YAG laser, 30 cm telescope with polychromator, and UV enhanced gated CCD camera. Sensing was carried out at inclination angles of $78°–83°$. Signal accumulation (up to 2000 laser shots) was used to obtain LIF spectrum of water with acceptable SNR.

Similar sensing geometry was used in (Karpicz *et al.* 2006). A LIDAR was installed on the shore of Lake Skaistis (Lithuania) and on a Baltic Sea beach. The LIDAR included a diode pumped solid-state laser (355 nm, 1 mJ), 21 cm telescope, and 6 spectrally-selective detectors. It was demonstrated that oil-slick on surface water can be detected up to a sensing angle of $85°$ (approximately at 100 m distance, and at 10 m height of LIDAR installation). The results of the study indicated that reconstruction of the pollutant's LIF spectrum and identification is possible only when the pollutant's fluorescence is much stronger than that of the natural water.

4.3   LIF LIDAR onboard ROV and UAV

Probably one of the most promising installations of LIF LIDAR is its use as a payload for underwater ROV. Such applications can address the shortcomings of airborne and shipboard LIDAR. Indeed, due to the limited penetration depth of light into water, chemicals dispersed in the water column or spread on the seafloor are mostly undetectable with airborne or even shipboard LIDAR.

The combined use of a LIF LIDAR and a range-gated imaging device onboard an underwater ROV was proposed in (Harsdorf *et al.* 1999). This system of sensors is capable of inspecting the seafloor, and can detect and locate the release of toxic chemicals in the water column. A short laser pulse of a Nd:YAG laser (532 nm, 160 mJ) is used as the illumination light for a time-gated video camera to locate and inspect underwater objects. Compared with conventional imaging, range-resolved image recording - synchronized with the laser pulse - enhances the contrast of underwater video imagery. Upon detection of an object, the LIF LIDAR records the echo-signals of substances within the water column and on the seafloor. The 3rd harmonic of the Nd:YAG laser (355 nm, 60 mJ) is used as a sensing emission to obtain LIF spectra. Time- and spectrally-resolved echo-signals are detected with variable interference filter (15 nm spectral resolution) and fast PMT (2 ns sampling time). It is reported that recorded LIF spectra provided a distinction between CDOM and pollutants.

An Underwater LIDAR Imaging System (ULIS) with angular-resolved detection was developed to improve underwater imaging (Stute *et al.* 2001). The first sea trials with ULIS demonstrated its potentials for the investigation of fluorescence profiles, even in very turbid estuaries.

The development of a Laser Fluorosensor payload for submarine applications onboard an ROV is reported in (Fantoni *et al.* 2004). The design of the instrument addressed ROV requirements for ruggedness, weight, size, power consumption, and extreme environmental conditions (lower temperature limit from −2°C in seawater down to −40°C outside). The LIDAR is based on a Nd:YAG laser (355 nm, 70 mJ) and range-resolved detector consisting of 4 spectral channels (402, 450, 650, 680 nm). The tests onboard an ROV confirmed the capability of underwater LIDAR to collect data for CDOM and *chlorophyll a* in deep sub-surface waters.

Development of a new lightweight flying payload, LIF LIDAR, for UAV is also reported (Fantoni *et al.* 2004). A compact diode-pumped solid state laser and miniaturized, PMT based, detector are being considered as a system layout. The development is focused on such requirements as weight reduction (below 10 kg), size of LIDAR, and minimal power requirements (not exceeding 600 W). Potential use of a relatively low-cost unmanned aerial vehicle is very attractive due to low operational costs and could expand the applications of LIF LIDAR to survey marine ecosystems.

5. Conclusions

Turning data into useful information is a core part of any application development; based first on scientific research and accumulated knowledge. Therefore, moving from demonstration projects to operational use requires a certain level of technology maturity – maturity that cannot be achieved without feedback to technology developers from operational use. As a result of this time-consuming process, the operational use of LIF LIDAR - while expanding - remains behind the technological achievements.

Over the past decades in the course of applications development, interest in LIDAR technology has varied; beginning with high expectations and intensive scientific achievements (1970s and early 1980s), through understanding the application challenges and enhancing analytical capabilities (during the 1990s) to renewed and expanding attention for operational use in the last decade. The latter was influenced by noticeable improvements in the availability of other remote sensing information and newly emerged spatial data technologies, such as GPS and GIS, allowing precise geo- referencing of remote-sensing data.

For many practical applications, valuable information can be obtained by combining large-scale data with higher-resolution surveys of key locations flown by airborne LIF LIDAR. LIF LIDAR can serve as an effective sensing tool when a survey is either scheduled for a specific purpose and location, or is aimed at periodical data collection over large areas for long periods to enable trend identification. LIF LIDAR operation onboard a travelling ship has its own merits: In terms of application, it can provide near-surface and valuable in-depth range-resolved measurements. In terms of operation, it can be in use for extended periods of time, which is of particular value in long-term monitoring programmes.

New potential for LIF LIDAR is coming from applied research carried out by various LIDAR installations. Stationary installations of LIF LIDAR (sea shore, platforms) can be considered a cost-efficient alternative to monitor specific local areas, such as harbours, oil terminals or select sections of waterways. Extremely promising and effective applications of LIF LIDAR could come as a result of technology developments for various ROV, or in the future for UAV.

References

Babichenko S, Poryvkina L, Arikese V, Kaitala S, Kuosa H (1993) Remote sensing of phytoplankton using laser induced fluorescence. Rem Sens Environ 45: 43–50

Babichenko S, Lapimaa J, Porovkina L, Varlamov V (1995) On-line fluorescent techniques for diagnostics of water environment. In: Russwurm GM (ed) Air Toxics and Water Monitoring, SPIE vol 2503, pp 157–161

Babichenko S, Kaitala S, Leeben A, Poryvkina L, Seppälä J (1999) Phytoplankton pigments and Dissolved Organic Matter distribution in the Gulf of Riga. J Mar Systems 23: 69–82

Babichenko S, Dudelzak A, Lapimaa J, Lisin A, Poryvkina L, Vorobiev A (2006) Locating water pollution and shore discharges in coastal zone and inland waters with FLS lidar. In: EARSeL eProc 5: 32–41

Barbini R, Colao F, Fantoni R, Frassanito C, Palucci1 A, Ribezzo S (2001) Range resolved lidar fluorosensor for marine investigations. In: EARSeL eProc 1: 185–195

Bristow M, Bundy D, Edmonds CM, Ponto PE, Frey BE, Small LF (1985) Airborne laser fluorosensor survey of the Colombia and Snake rivers: simultaneous measurements of chlorophyll, dissolved organics and optical attenuation. Int J Rem Sens 6: 1707–1734

Brown CE, Nelson R, Fingas MF, Mullin JV (1997) Airborne Laser Fluorosensing: Overflights During Lift Opeartions of a Sunken Oil Bargade. In: Proc of IV Thematic Conference Remote Sensing for Marine and Coastal Environments. Orlando, Florida, vol 1, pp 23–30

Chekaljuk AM, Demidov AA, Fadeev VV, Gorbunov MYu (1995) Lidar monitoring of phytoplankton and organic matter in the inner seas of Europe. EARSeL Adv Rem Sens 3: 131–139

Determann S, Reuter R, and Willkomm R (1994) Fluorescent matter in the eastern Atlantic Ocean, Part 1: method of measurement and near surface distribution. Deep Sea Res 41: 659–675

Drozdowska V, Walczowski W, Hapter R, Stoń J, Irczuk M, Zieliński T, and Piskozub J (2004) Fluorescence characteristics of the upper water layer of the Arctic seas based on lidar, spectrophotometric and optical methods. In: EARSeL eProc 3: 136–142

Dudelzak AE, Babichenko SM, Poryvkina LV, Saar KU (1991) Total luminescent spectroscopy for remote laser diagnostics of natural water conditions. Appl Opt 30: 453–458

Fantoni R, Barbini R, Colao F, Ferrante D, Fiorani L, and Palucci A (2004) Integration of two lidar fluorosensor payloads in submarine ROV and flying UAV platforms. In: EARSeL eProc 3: 45–53

Fiorani L, Barbini R, Colao F, De Dominicis L, Fantoni R, Palucci A, and Artamonov ES (2004) Combination of lidar, MODIS and SeaWiFS sensors for simultaneous chlorophyll monitoring. In: EARSeL eProc. 3: 8–17

Harsdorf S, Janssen M, Reuter R, Toeneboen S, Wachowicz B, Willkomm R (1999) Submarine lidar for seafloor inspection. Meas Sci Tech 10: 1178–1184

Hengstermann T and Reuter R (1990) Lidar fluorosensing of mineral oil spills on the sea surface. Appl Opt 29: 3218–3227

Hengstermann T, Loquay K, Reuter R, Wang H, and Willkomm R. (1992) A laser fluorosensor for airborne measurements of maritime pollution and of hydrographic parameters. EARSeL Adv Rem Sens 1: 85–98

Hitomi K, Yamagishi S, Yamanouchi H, Yamaguchi Y (2002) Detection of Spilled Oil Using a Compact Fluorescence Lidar. J Visualization Society of Japan 22: 77–85

Hoge FE, Swift RN (1980) Oil film thickness measurement using airborne laser induced Raman backscatter. Appl Opt 19: 3269–3281

Karpicz R, Dementjev A, Kuprionis Z, Pakalnis S, Westphal R, Reuter R, and Gulbinas V (2006) Oil spill fluorosensing lidar for inclined onshore or shipboard operation. Appl Opt 45: 6620–6625

Lennon M, Babichenko S, Thomas N, Mariette V, Mercier G, and Lisin A (2006) Detection and mapping of oil slicks in the sea by combined use of hyperspectral imagery and laser induced fluorescence. In: EARSeL eProc 5: 120–128

Maslov DV, Fadeev VV, and Lyashenko AI (2001) A shore-based lidar for coastal seawater monitoring. In: EARSeL eProc 1: 46–52

Ohm K, Reuter R, Stolze M, and Willkomm R (1998) Shipboard oceanographic fluorescence lidar development and evaluation based on measurements in Antarctic waters. In: EARSeL Yearbook 1997, Paris, pp 105–113

Pantani L, Cecchi G, and Bazzani M (1995) Remote Sensing of Marine Environments with the High Spectral Resolution Fluorosensor, FLIDAR 3, SPIE, vol 2586, pp 56–64

Patsaeva S (1995) New Methodological aspects of the old problem. Laser Diagnostics of Dissolved Organic Matter. EARSeL Adv Rem Sens 3: 66–70

Poryvkina LV, Babichenko SM, Lapimaa J (1992) Spectral variability of humus substance in marine ecosystems. AMBIO 21: 465–467

Reuter R, Wang H, Willkomm R, Loquay K, Hengstermann T, Braun A (1995) A laser fluorosensor for maritime surveillance: Measurement of oil spills. EARSeL Adv Rem Sens 3: 152–169

Robbe N, Zielinski O (2004) Airborne remote sensing of oil spills: analysis and fusion of multi-spectral near-range data. J Mar Sci Environ C2: 19–27

Robbe N, Hengstermann T (2006) Remote Sensing of marine oil spills from airborne platforms using multi-sensor systems. In: WIT Transactions on Ecology and the Environment, WIT Press, vol 95, pp 347–355

Stute U, LeHaitre M, and Lado-Bordowsky O (2001) Aspects of spatial and temporal ranging for bistatic submarine lidar. In: EARSeL eProc 1: 96–105

Zabavnikov V, Vasiliev A, Lisovsky A, Chernook V (2005) Usc of airborne LIDAR for carrying out research in fisheries oceanography. In: 31st Int. Symposium ISRSE, St.-Petersburg, Russia, pp 201–202

Zielinski O, Andrews R, Göbel J, Hanslik M, Hunsänger T, and Reuter R (2001) Operational Air-borne Hydrographic Laser Fluorosensing. In: EARSeL eProc 1: 53–60

Zielinski O., Hengstermann T. & Robbe N. (2006) Detection of oil spills by airborne sensors. In: Gade M, Hühnerfuss H, Korenowski GM (eds), Marine Surface Films, Springer, Berlin Heidelberg New York, pp 255–27

Section 3:

Microwave
Passive/Active Remote Sensing

Microwave Radiometry and Radiometers for Ocean Applications

Niels Skou

Danish National Space Center, Technical University of Denmark, Lyngby, Denmark

Abstract. The microwave radiometer system measures, within its bandwidth, the naturally emitted radiation - the brightness temperature - of substances within its antenna's field of view. Thus a radiometer is really a sensitive and calibrated microwave receiver. The radiometer can be a basic total power radiometer or a more stable Dicke radiometer. Also correlation receivers play an important role in modern systems. The radiometer system might be single or dual polarized (horizontal and vertical) – or even be polarimetric, *i.e.* measure all 4 Stokes parameters, thus providing additional geophysical information at any given frequency. The radiometer system will very often be configured as an imaging system on a spacecraft for example. This normally implies scanning the antenna. Then there are certain relationships (or even conflicts) between achievable radiometric sensitivity/ground resolution/antenna size, and the problem: scanning antenna/spacecraft stability. In many cases good compromises have been reached, as evident recalling the many successful missions throughout the recent 30 years. But in some cases the situation calls for special solutions, like the push-broom system or the synthetic aperture radiometer technique, both yielding imaging capability without scanning. Typical applications of microwave radiometry concerning oceans are: sea salinity, sea surface temperature, wind speed and direction, sea ice detection and classification. However, in an attempt to measure properties of the sea from space, the intervening atmosphere will disturb the process, and corrections might be required. Also, at some frequencies and for some applications, the Faraday rotation in the Ionosphere must be taken into account.

1.   What is radiometry about?

All bodies at a temperature above the absolute zero (0 K = −273 °C) radiate power, according to Planck's Law. At microwave frequencies the

V. Barale, M. Gade (eds.), *Remote Sensing of the European Seas.*

Rayleigh-Jeans approximation holds, and the radiation is proportional to physical temperature. Actually, this only holds for so-called blackbodies, which are perfect emitters. Natural bodies radiate less, and we introduce the term emissivity (ε) which is a number between 0 and 1 describing how well the body radiates relative to a blackbody. Within radiometry the radiated power is expressed as the so-called brightness temperature, T_B, so that $T_B = \varepsilon \cdot T_{phys}$. The brightness temperature of a blackbody is thus equal to its physical temperature, while all natural bodies will have brightness temperatures lower than that. The brightness temperature of a natural body can also be understood as that physical temperature a blackbody would have to have in order to emit the power in question.

First-year sea ice is close to a blackbody at microwaves and will thus have a T_B close to the physical temperature of its emitting layer ($\approx$260 K). Water is quite far from a blackbody and exhibits low T_B in the range 100 – 150 K very dependent on a range of parameters. This contrast between water and ice is basically why spaceborne radiometry has been used with great success for sea ice monitoring.

The radiated energy can be picked up by an antenna in order to be measured by a radiometer. Unfortunately antennas are not perfect: a perfect antenna would have a sharply defined main lobe so that when this is pointed towards the target only this contributes to the received power. But the main lobe is not with sharp cut off, so some power is received from the surroundings of the target. Furthermore, antennas have side lobes far from the main lobe meaning that some power is also received from other directions. The process can be expressed by the following equation:

$$ T_A = \tfrac{1}{4\pi} \cdot \int \int_{4\pi} T_B(\theta,\phi) \cdot G(\theta,\phi) d\Omega \qquad (1) $$

This convolution integral shows how the received power, denoted as the so-called antenna temperature, T_A, is a gain-weighted summation of the brightness temperatures from each direction (Ulaby *et al.* 1981).

2. Basic radiometers and their sensitivity

As stated above, the task of the radiometer is to measure the power picked up by the antenna. So in fact the radiometer is basically a calibrated microwave receiver. Figure 1 shows some basic radiometer types to be discussed in the following.

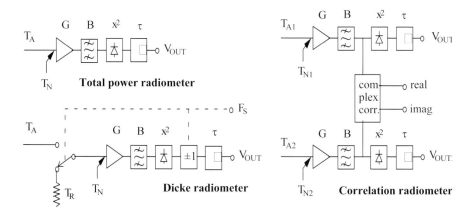

Fig. 1. Three basic radiometer types.

The **Total Power Radiometer** is the direct implementation of what it is all about: the received power is amplified with a gain G, a certain bandwidth B around the center frequency is selected by the filter, the microwave signal is detected by a square law detector, and since we are handling noise-like signals, a certain integration time τ is required. Finally, it is indicated that we cannot make a receiver without introducing an additional noise T_N.

The output will be a power measure expressed as: $P = k \cdot B \cdot G \cdot (T_A + T_N)$ where k is Boltzmann's constant: $1.38 \cdot 10^{-23}$ J/K. As already stated the signals are noise-like, so the output fluctuates, but the fluctuations are smoothed by the integration. The level of fluctuations, the standard deviation of the output to be specific, is expressed as the radiometer sensitivity, ΔT, and it is calculated from equation 2.

$$\Delta T_{TPR} = \frac{T_A + T_N}{\sqrt{B \cdot \tau}} \tag{2}$$

A fundamental problem with this simple and direct implementation of the radiometer is that the output is dependent on stability of the gain G and the receiver noise temperature T_N. Hence stability of total power radiometers may be problematic, and frequent calibration is required.

In 1946 R.H. Dicke found a way of alleviating the stability problems in radiometers (Dicke 1946). By using the radiometer not to measure directly the antenna temperature, but rather the difference between this and some known reference temperature, the sensitivity of the measurement to gain and noise temperature instabilities is greatly reduced. Take note of the **Dicke Radiometer** in Figure 1 where rapid switching between T_A and a reference T_R is indicated.

A price has to be paid, however, for the better immunity to instabilities. Since only half of the measurement time is spent on the antenna signal (the other half is spent on the reference temperature), the sensitivity is poorer than for the total power radiometer by a factor of 2:

$$\Delta T_{DR} = 2 \cdot \frac{T_A + T_N}{\sqrt{B \cdot \tau}} \tag{3}$$

The **Correlation Radiometer** is a multichannel system that finds use in the case where two brightness temperatures are measured as well as the correlation between them. This is the case in the polarimetric radiometer, see Section 3, where the vertical and the horizontal brightness temperatures are measured together with their correlation thus finding the so-called Stokes parameters. This is also the case in interferometric radiometers like the synthetic aperture radiometer, where the outputs of two different antennas pointing in the same direction in space are measured.

The correlation radiometer is illustrated to the right in Figure 1. Two identical receivers, here total power radiometers, are connected to the two output ports of the antenna system. The outputs of the receivers are detected the usual way to yield the normal brightness temperatures. The signals of the receivers are also (before detection) fed into the complex correlator providing the real and the imaginary parts of the cross correlation between the two input signals from the antenna system.

For the 2 normally detected outputs we find the usual sensitivity as expressed by equation 2. Assuming that the input signals are only weakly correlated, which is usually the case, the sensitivity of the correlator outputs can be found to be : $\Delta T_{CORR} = \Delta T_{TPR} / \sqrt{2}$.

3. Polarimetric radiometers

Generally, the radiation from an object is partly polarized meaning that the brightness temperature at vertical polarization T_V is different from the brightness temperature at horizontal polarization T_H. Many radiometer systems measuring T_V and T_H have been built over the years. But measuring these 2 polarizations does not necessarily exhaust the possibilities at any given frequency: in some cases additional information can be gained by measuring the correlation between the two. A well-known example is the sea surface (Yueh *et al.* 1995). To describe scenes with partial polarization it is convenient to use the Stokes parameters (4).

$$\overline{T_B} = \begin{pmatrix} I \\ Q \\ U \\ V \end{pmatrix} = \begin{pmatrix} T_V + T_H \\ T_V - T_H \\ T_{45°} - T_{-45°} \\ T_\ell - T_r \end{pmatrix} = \frac{\lambda^2}{k \cdot z} \begin{pmatrix} \langle E_V^2 \rangle + \langle E_H^2 \rangle \\ \langle E_V^2 \rangle - \langle E_H^2 \rangle \\ 2 \cdot \mathrm{Re} \langle E_V \cdot E_H^* \rangle \\ 2 \cdot \mathrm{Im} \langle E_V \cdot E_H^* \rangle \end{pmatrix} \tag{4}$$

where z is the impedance of the medium in which the wave propagates, and λ is the wavelength. It is seen that I represents the total power, Q represents the difference of the vertical and horizontal power components, while U and V represent the real and imaginary parts of the cross correlation of the electrical fields.

The vertical and horizontal outputs of a conventional dual polarized antenna are connected to the two inputs of the correlation radiometer of Figure 1, $T_V = T_{A1}$ and $T_H = T_{A2}$. The outputs are: $I = V_{OUT1} + V_{OUT2}$, $Q = V_{OUT1} - V_{OUT2}$, U = 2·real, V = 2· imag. Concerning sensitivities, it is noted that the two first Stokes parameters are found by addition and subtraction of similar, independent signals having standard deviations of ΔT_{TPR} resulting in $\Delta I = \Delta Q = \sqrt{2} \cdot \Delta T_{TPR}$. Recalling that $\Delta T_{CORR} = \Delta T_{TPR} / \sqrt{2}$ and the factor 2 in equation (4) it is seen that also $\Delta U = \Delta V = \sqrt{2} \cdot \Delta T_{TPR}$. The polarimetric radiometer can measure wind direction over the sea as demonstrated with success in airborne campaigns over the North Sea (Laursen and Skou 2001).

4.   Spaceborne imaging radiometers

4.1 Antenna and imager considerations

As already discussed the ideal antenna has a certain gain within its field of view and zero gain outside, but a real antenna will receive radiation from many directions. An important antenna property is the so-called beam efficiency η, defined as the ratio between the energy received through the main beam and the total energy received by the antenna (main beam + all sidelobes). An antenna with $\eta = 95\%$ within the -20 dB points is about the best you can get.

For typical spaceborne reflector antennas, where the electrical aperture is roughly equal to the physical aperture, the 3 dB beamwidth is slightly larger than the reciprocal of the aperture measured in wavelengths of the operating frequency:

$$\theta \sim 1.4 \cdot \lambda/D \qquad (5)$$

Many readers will know this rule-of-thumb equation with a factor 1.2 ahead of λ/D. The factor 1.4 is used here to reflect the radiometer's need for high beam efficiency, which generally results by a careful design having low aperture-edge illumination (hence a wider beam compared to other situations).

A 1 m parabolic reflector antenna used at 30 GHz (1 cm wavelength) will thus have a 0.014 (= 0.8°) beamwidth. Looking straight down from a satellite at H = 800 km altitude, the footprint will be FP = $\theta \cdot$H = 11.2 km.

If a radiometer with its antenna is mounted on a satellite, and the antenna points straight down, the forward movement of the satellite (~7.5 km/sec) will facilitate measurements on the ground along a straight line (the nadir path of the satellite). Coverage of the entire Earth by such "profiles" will require an enormous number of orbits! A dramatic increase in mapping efficiency results from scanning the antenna, see Figure 2. The antenna rotates about a vertical axis, and the footprint will cover a wide swath on the Earth dependent on satellite altitude and incidence angle. Other scanning methods are possible, but the rotating scan is attractive due to constant incidence angle on the ground, and the lack of accelerations associated with reciprocating scans.

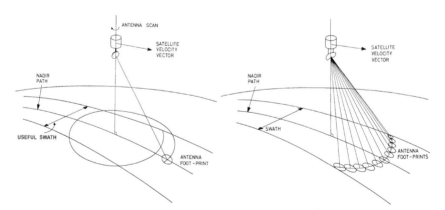

Fig. 2. Conical scan by rotating antenna (left) and push-broom imager (right).

In a scanning system it is obvious that there is only a limited time for the radiometer to carry out its measurement before the footprint moves to another position within the swath. The so-called dwell time per footprint is short. The users in general want small footprints (or high spatial resolution, to put it differently). As technology evolves, high-resolution systems become possible, but a small footprint results in rapid rotation (mechanical problems) and in a very small dwell time per footprint, hence in a short integration time, which, through our radiometer sensitivity formula, directly translate into poor sensitivity! The solution to this fundamental problem is offered by the so-called push-broom concept also illustrated in Figure 2. In the push-broom radiometer system a multiple-beam antenna covers the swath while the satellite moves forward. A host of radiometer receivers are connected to an equal number of antenna feeds, producing individual beams to sense the Earth simultaneously.

The obvious advantages of the push-broom system compared to a scanning system are: no moving antenna to cause problems in the satellite design, and much larger dwell time per footprint, hence better sensitivity.

For the scanning radiometer systems, the requirement for receiver sensitivity is severe. At the same time, frequent calibration is easily achieved: once per scan, while the antenna is anyway looking away from the swath, the receiver is calibrated. Hence the total power radiometer is an obvious candidate for such systems, due to its optimal sensitivity and since potential instabilities are taken care of by frequent calibration.

For the push-broom system, requirements for receiver sensitivity are greatly relaxed due to the much larger dwell time per footprint as compared to the scanner situation. At the same time, frequent calibration is not attractive, as all receivers are always busy sensing the Earth. Hence, the push-broom situation favours a trading of sensitivity for stability and, in conclusion, the Dicke radiometer is preferred.

Following the discussions above, two mission oriented systems will be covered as illustrations of the differences between the scanner and the push-broom system. An 800 km orbit and a 53° incidence angle on ground are assumed being typical values within remote sensing. The swath will typically be 1500 km (for details see Skou and LeVine 2006).

4.2   General purpose multi-frequency mission

Consider the 18.7 GHz channel of a multi-frequency radiometer system, like the Special Sensor Microwave/Imager (SSM/I). A 1 m aperture is assumed (this is a typical size for such a system). The ground resolution will be 35 km.

For the scanner we find: 17 msec integration time, a sensitivity of 0.28 K using a state-of-the-art total power radiometer, and an antenna rotation rate of 12.5 rpm.

For the push-broom system: 4.8 sec integration time, a typical sensitivity of 0.03 K (Dicke radiometer), and we need 79 channels to cover the swath.

A radiometric resolution of 0.28 K will satisfy most users and 12.5 rpm for a 1 m reflector is not frightening. Surely the push-broom system offers better sensitivity but at the expense of complexity (79 receivers and a more complicated antenna system). Hence the trade-off between scanner and push-broom in this case favors the first.

Note that this example represents a very successful concept that has had many representatives operating is space since the launch of the Scanning Multichannel Microwave Radiometer (SMMR) in 1978. This ground-breaking mission has been followed by many others, but foremost the SSM/I series of systems providing continuous data to the user community. The quite established frequencies are around: 7, 11, 19, 22–24, 37, and 90 GHz (not all mission have them all), and they enable retrieval of ocean related parameters like: sea surface temperature and wind speed, sea ice mapping and classification.

4.3 Salinity sensor

The 1.400–1.427 GHz protected band is the generally accepted frequency for the purpose. A radiometric sensitivity of a fraction of a Kelvin (around 0.1 K) is required by the users.

A realistic salinity sensor could assume a 10 m aperture, which in turn results in a ground resolution of 47 km.

For a scanning system we find a 30 msec integration time, a sensitivity of 0.4 K using a state-of-the-art total power radiometer, and a 9.4 rpm antenna rotation rate. Not only are the requirements for sensitivity not fulfilled but the spacecraft designer is left with the problem of having a 10 m dish rotating with 9.4 rpm. Not a very attractive solution.

For a push-broom system we find: 6.4 sec integration time, a sensitivity of 0.08 K, and we shall need 59 channels to cover the swath. 0.08 K sensitivity fulfills all reasonable requirements.

The cost is receiver and antenna system complexity. 59 identical receivers are needed, but integrated techniques can be used, thus greatly facilitating "mass production", and at the same time keeping volume and mass to reasonable levels. The antenna system must include 59 individual feeds and have an unusually shaped and sized reflector. More about such issues in Section 5.

5.   Example of a sea salinity mission

The example to be covered is the push-broom system briefly described above. So we are talking about a 10 m aperture system with $\approx$50 km ground resolution well suited for providing data to the global climate and weather models. As a target value we assume a resolution in the salinity measurement of 0.1 psu (practical salinity unit; 1 psu is in effect equal to 1 part per thousand). More details can be found in Skou and Le Vine 2006.

Note that this example is taken as a comprehensive way of discussing typical aspects of a salinity mission. Also note that the NASA Aquarius mission, presently being developed, is in fact a push-broom system (although a reduced one with 3 beams). Finally, note the forthcoming Soil Moisture and Ocean Salinity (SMOS) mission – the first European Earth sensing space mission having a radiometer as the prime instrument.

5.1   Brightness temperature of the sea

The brightness temperature of the sea surface depends on several parameters: frequency, incidence angle, polarization, sea surface temperature (SST), wind speed (WS) (actually surface roughness), and salinity (S). Disregarding surface roughness, the brightness temperature can in fact be calculated in a relatively straight forward manner using standard electromagnetics procedures involving the dielectric constant and Fresnel reflection (Klein and Swift 1977). The dielectric constant of water is only sensitive to salinity at rather low microwave frequencies – well below 5 GHz, and in principle the lower the better. For practical reasons this means use of the protected radio astronomy band 1400–1427 MHz (L-band). Take note that microwave radiometers only measure surface properties of the ocean, as the penetration depth in saline water even at L-band is only a few millimeters. Figure 3 shows the result of running the Klein & Swift model for the brightness temperature of the sea using vertical polarization and 53° incidence angle.

It is noted that the sensitivity to salinity is best at warm temperatures. It is also noted that some dependence on temperature is evident especially for low salinity water. Fortunately we are in the situation that a large part of the Earth's oceans are in the best possible category: moderate to warm temperatures and high salinity around 35 psu. Arctic oceans and brackish seas are more difficult to deal with. Dealing with the open ocean/high salinity case, the curves in Figure 3 reveal a brightness temperature sensitivity to salinity at best approaching $\Delta T_B/\Delta S = 1$ K/psu (warm water at 30°C), and, more typically, the sensitivity will be better than $\Delta T_B/\Delta S = 0.5$

K/psu for temperatures above 10°C. This latter value will be used in the following. For the same case the sensitivity to sea surface temperature never exceeds $\Delta T_B / \Delta SST = 0.2$ K/°C.

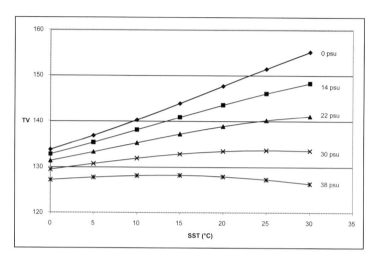

Fig. 3. Vertical brightness temperature as a function of sea surface temperature with salinity as parameter. Wind speed is zero.

But as already alluded to, the brightness temperature also depends on sea surface roughness, which in turn is a result of winds. Using data from for example (Sasaki *et al.* 1987) we note a sensitivity to wind speed of $\Delta T_B / \Delta WS = 0.1$ K per m/sec.

The requirement by users for resolution of the salinity measurement is 0.1 psu.

Assume a 0.08 K radiometer sensitivity as calculated in Section 4.3, and uncertainties in temperature and wind speed of 1°C and 1 m/sec respectively. The temperature uncertainty results in a radiometric uncertainty of 0.2 K and the wind speed uncertainty gives a 0.1 K uncertainty. As the three uncertainties are statistically independent, the total standard deviation of the radiometer measurement will be:

$$\Delta T_{tot} = \sqrt{0.08^2 + 0.2^2 + 0.1^2} = 0.24 K \qquad (6)$$

resulting in a 0.48 psu uncertainty in salinity, which is far from fulfilling the requirements. It is obvious that the dominating factor is the uncertainties in the disturbing geophysical parameters, especially sea temperature, and it would help little to strive for a better radiometer ΔT. Fortunately, the oceans surface temperature is quite well known from satellite sources like infrared radiometers augmented by buoy measurements.

In the following it will be assumed that the radiometer sensitivity is still 0.08 K, the temperature of the sea is known to within 0.3°C, and the wind speed is known to within 1 m/sec. This results in a measurement sensitivity of 0.14 K, which again transforms into a 0.28 psu uncertainty in retrieved salinity. This is still not fulfilling the original requirements, but there is little to do about it, as sea surface temperature and wind speed cannot be known to better accuracy than assumed here. The only remedy is spatial or time averaging as part of post processing of the data once transferred to ground. And in the present case this is a viable solution: ground resolution is not always the greatest issue concerning ocean salinity, and 100 km would serve adequately most users. This means that 4 footprints can be averaged, and with the reasonable assumption that measurement uncertainties are statistically independent from footprint to footprint, the uncertainties are reduced by a square root of 4, *i.e.* to 0.14 psu, which is quite satisfactory. If a user would prefer to retain the 50 km ground resolution, another integration scheme is possible, namely to integrate over 4 consecutive orbits over the area in question. This is possible as sea salinity is generally a slowly varying parameter. Actually, even further time averaging may be possible, which would reduce the salinity uncertainty to below the 0.1 psu target value.

5.2 System considerations

The system discussed in Section 4.3 is still a demanding concept with its 59 channels. A more realistic system having 21 channels is evaluated in the following. The only consequence of the reduced number of channels is a reduced swath of 530 km, hence longer revisit time.

Figure 4 illustrates the system. 21 feeds illuminate the reflector, which is in the shape of a torus in order to satisfy the need for a structure that is rotationally symmetric around a vertical axis and hence is able to cover a wide swath with constant incidence angle. The feeds horns are arranged in two rows, since their physical size exceeds the distance between positions required for proper beam squinting corresponding to correct footprint position on ground. To the right in the figure is shown a photo of a demonstration model scaled to a frequency of 36.5 GHz.

For many years the huge antenna structures as considered here has represented an insurmountable problem! But substantial developments have taken place recently, and a 12 m mesh antenna, having a weight of 85 kg and being able to serve an L-band radiometer, has been flown in space. A communication satellite having two 19 × 17 m antennas has been designed and the antennas demonstrated. Based on these experiences, a 10 × 19 m mesh antenna for the present purpose could have a weight of around 150 kg including feed mast and structure.

Finally, a weight and power budget shall be outlined: the weight of the total system is assumed to be 392 kg (150 kg for the antenna, 210 kg for feeds, 21 kg for the receivers, 11 kg for misc. electronics); the power consumption of the system is assumed to be 112 W (84 W for receivers, 10 W for data handling, 18 W for power supply). Such a system is realistic by today's standards.

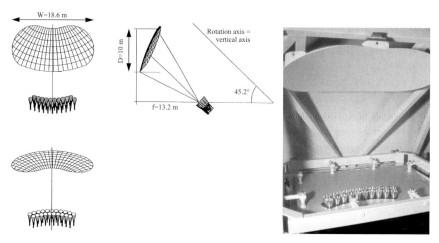

Fig. 4. Push-broom torus antenna with 21 feeds and demo system.

5.3   The Faraday rotation

Usually, within radiometry, ionospheric effects are ignored, but in the present case the frequency is so low and the measurement requirements so stringent that an investigation is necessary.

The plane of polarization of microwave signals propagating from Earth through the ionosphere to a satellite is rotated by an angle θ. The amount of rotation depends on the direction and location of the ray path with respect to the Earth's geomagnetic field, and on the state of the ionosphere. To get a feeling for the magnitude of the polarization rotation angle θ, a mean daytime value can be estimated from: $\theta = 17/F^2$ (F in GHz) taken from (Hollinger and Lo 1984). Hence, the average daytime rotation is found to be θ = 8.7° at 1.4 GHz, and furthermore it is illustrated why the effect generally can be ignored at higher frequencies. A more in-depth analysis of the Faraday rotation reveals monthly averages of the daytime maximum rotation around $25° - 30°$, whereas monthly maximum averages at 6.00 AM are around $5°$. In addition, day-to-day variations can reach values within $+100\%$ to -50% of the averaged values due to the unpredictable nature of the ionosphere. Proposed missions generally have a 6 AM orbit, so from these considerations it is clear that the radiometer system has to cope with at least $5°$, possibly up to $10°$, Faraday rotation.

The polarization rotation will result in a slight mixing of the true vertical (T_{BV}) and horizontal (T_{BH}) brightness temperatures. A dual polarized radiometer will thus measure:

$$T'_{BV} = T_{BV} \cdot \cos^2 \theta + T_{BH} \cdot \sin^2 \theta$$
$$T'_{BH} = T_{BV} \cdot \sin^2 \theta + T_{BH} \cdot \cos^2 \theta \qquad (7)$$

Typical values are T_{BV} = 132 K and T_{BH} = 66 K. Assuming $\theta = 10°$ we find: T'_{BV} =130.0 K and T'_{BH} = 68.0 K. The 2 K error in T_{BV} translates into an error in retrieved salinity of about 2–4 psu (depending on sea temperature) which is totally unacceptable.

The set of equations (7) can be solved with respect to T_{BV} for example. After some reductions the following expression is found:

$$T_{BV} = \frac{T'_{BV} - T'_{BH} \cdot tg^2\theta}{1 - tg^2\theta} \qquad (8)$$

Hence, if both the local vertical (T'$_{BV}$) and horizontal (T'$_{BH}$) brightness temperatures are measured, and θ is known, the true vertical brightness temperature can be found. The rotation angle can be calculated from the Total Electron Contents (TEC) maps available from a variety of sources.

Note, that in order to correct for Faraday rotation, the radiometer system must measure both vertical and horizontal polarizations. This doubles the number of receivers (marginal weight and power consequences already taken into account in the previous section), and requires the feeds to be dual polarized (design complexity but marginal influence on weight). The antenna reflector serves both polarizations without problems.

It is noteworthy that a dual polarized system actually is optimum for a combined sea salinity/soil moisture mission as this solves the problem that the optimum polarization for sea salinity measurements is vertical while the optimum polarization for soil moisture is horizontal.

As it is obvious from the previous discussion, the radiometer needs to measure both the vertical and the horizontal polarizations and by addition of those we in fact have the 1'st Stokes parameter, see equation (4). This parameter represents the total power in the electrical field, and it is totally invariant to Faraday rotation. Hence, if the geophysical parameters in question (here salinity) can be retrieved as well from the 1'st Stokes parameter as from the vertical polarization, the Faraday problem is solved and need no further concern. This is largely the case: salinity is found from the 1'st Stokes parameter with good sensitivity, but the influence of wind is slightly larger. Thus we are faced with a choice: having to correct for both wind and Faraday each with their individual uncertainties, or just correcting for the wind with a slightly larger uncertainty (1'st Stokes). More about this in Skou 2003.

5.4   Other disturbing factors: space and atmosphere

There are other disturbing factors than Faraday rotation that must be taken into account in order to obtain correct measurements of the radiated sea brightness temperature: space radiation and atmospheric effects. They will only be mentioned briefly here. More detailed information can be found in Yueh *et al.* 2001, Le Vine and Abraham 2002 and Le Vine and Abraham 2004.

Microwave power emitted from space will generally enter the antenna main beam through reflection in the sea surface. Three contributions must be considered:

The cosmic radiation is isotropic at a constant level of 2.8 K, and this bias does not affect measurement accuracy.

The galactic noise exhibits a great variation, depending on whether the antenna beam reflected in the sea surface looks toward the galactic pole or the galactic center (0.8 K and 16 K respectively at 1.4 GHz). This effect must be taken into consideration. Corrections must be carried out on the measured data, which is possible as the galactic noise is well mapped.

The Sun is a very intense microwave emitter, with a brightness temperature dependent on solar activity, but always on the order of 100,000 K or more at L-band. Direct Sun reflection in the sea surface must be avoided, which is best done by choosing an early morning, Sun- synchronous, near polar orbit, for example a 6 AM orbit. This, in short, means that the equatorial ascending passing takes place at sunrise, and the Sun will be almost 90° away from the look direction of the radiometer. But this is only a crude assessment, and at high latitudes - additionally bearing in mind that the radiometer has a certain swath width, and that under rough sea surface conditions, scattered solar radiation will be received from directions away from the specular direction - some pixels will be contaminated by reflected Sun radiation. This requires detailed analysis and pixel flagging as part of the data processing.

The atmospheric effects can be summarized as follows:

Losses due to constituents like water vapour, oxygen, clouds, and rain will attenuate the brightness temperature emitted from the sea surface as it passes through the atmosphere to the spaceborne radiometer. When there is loss, there is also re-radiation from the lossy medium according to its physical temperature. To deal with this it is necessary to introduce a layered model of the atmosphere with different parameters as functions of altitude. These parameters are modelled and hence it is possible to calculate the integrated losses and emissions (radiative transfer equation) in order to correct for them.

Absorption by water vapour is generally of great concern within microwave radiometry. However, for the low frequencies considered here, this effect can be neglected. Also oxygen contributes to the atmospheric effects. The loss and re-radiation due to oxygen contributes around 4.5 K to the received brightness temperature, and must be corrected for. Moreover, the contribution is dependent on surface pressure and temperature, but with a low sensitivity, which makes it relatively straight forward to correct for. Typical clouds and rain rates give very low contributions and can be neglected (see Skou and Hoffmann-Bang 2005).

6.   Another example of a sea salinity mission: SMOS

The European Space Agency mission SMOS represents a very different way of overcoming the problem with imagers and large antennas. It is a synthetic aperture radiometer that works as a radio camera and it acquires a two-dimensional image of the ground by interferometric means. SMOS is synthesizing a ground resolution corresponding to a roughly 9 m real aperture antenna by cross correlating signals from all possible pairs of small antenna elements suitably positioned within that aperture.

7.   Conclusions

Following the introduction of basic radiometer concepts and imaging principles, a comprehensive example is discussed in some detail. Through this discussion a variety of considerations are presented ranging from geophysical issues to technical trade-offs and designs. One possible layout of a future sea salinity mission is thus demonstrated, its performance is discussed, and a variety of disturbing factors are evaluated including how to compensate for them.

References

Dicke RH (1946) The Measurement of Thermal Radiation at Microwave Frequencies. Rev Sci Instr 17: 268–279

Hollinger JP, Lo RC (1984) Low-Frequency Microwave Radiometer for N-ROSS. Large Space Antenna Systems Technology NASA Conference Publication 2368 pp. 87–95

Klein LA, Swift CT (1977) An Improved Model for the dielectric constant of Sea Water at Microwave Frequencies. IEEE Trans Antennas and Propagation 25 (1): 104 – 111

Laursen B, Skou N (2001) Wind Direction over the Ocean Determined by an Airborne, Imaging, Polarimetric Radiometer System. IEEE Trans Geosci Remote Sensing 39 (7) 1547–1555

LeVine DM, Abraham S (2002) The Effect of the Ionosphere on Remote Sensing of Sea Surface Salinity from Space: Absorption and Emission at L Band. IEEE Trans Geosci Remote Sensing 40: 771–782

LeVine DM, Abraham S (2004) Galactic Noise and Passive Microwave Remote Sensing From Space at L Band. IEEE Trans Geosci Rem Sens 42: 119–129

Sasaki Y, Asanuma I, Muneyama K, Naito G, Suzuki T (1987) The Dependence of Sea-surface Microwave Emission on Wind Speed, Frequency, Incidence Angle, and Polarization over the Frequency Range from 1 – 40 GHz. IEEE Trans Geosci Rem Sens 25 (2) 138–146

Skou N (2003) Faraday Rotation and L-band Oceanographic Measurements. Radio Science 38 (4) 24–1 to 24–8

Skou N, LeVine D (2006) Microwave Radiometer Systems, Design and Analysis. Artech House

Skou N, Hoffmann-Bang D (2005) L-Band Radiometers Measuring Salinity from Space: Atmospheric Propagation Effects. IEEE Trans Geosci Rem Sens 43 (10):2210–2217

Ulaby FT, Moore RK, Fung AK (1981) Microwave Remote Sensing, Vol.1. Artech House

Yueh SH, Wilson WJ, Li FK, Nghiem SV, Ricketts WB (1995) Polarimetric Measurements of Sea Surface Brightness Temperatures Using an Aircraft K-band Radiometer. IEEE Trans Geosci Rem Sens 33 (1) 85–92

Yueh SH, West R, Wilson WJ, Li FK, Njoku EG, Rahmat-Samii Y (2001) Error Sources and Feasibility for Microwave Remote Sensing of Ocean Surface Salinity. IEEE Trans Geosci Rem Sens 39 (5): 1049–1059

Microwave Aperture Synthesis Radiometry: Paving the Path for Sea Surface Salinity Measurement from Space

Jordi Font[1], Adriano Camps[2], and Joaquim Ballabrera-Poy[1]

[1] Institut de Ciències del Mar, CSIC, Barcelona, Spain
[2] Departament de Teoria del Senyal i Comunicacions, Universitat Politècnica de Catalunya, Barcelona, Spain

Abstract. This chapter summarises the main objectives and characteristics of the ESA's SMOS mission and its remote sensing applications. The SMOS payload is MIRAS, a new type of instrument in Earth observation: the first two-dimensional aperture synthesis interferometric radiometer. It operates at L-band, has multi-angular and multi-look imaging capabilities, and can be operated in dual-polarisation or full-polarimetric modes. Due to its novelty, the principles of operation, imaging characteristics and its main performance parameters (spatial resolution and radiometric sensitivity and accuracy) are described, as well as the approach selected in the retrieval algorithms of sea surface salinity.

1. Introduction

Remote measurement of sea surface salinity (SSS), using microwave radiometry at L-band, was first proposed by Swift and McIntosh (1983). The polarised brightness temperatures (T_h and T_v) measured by the radiometer are linked to salinity through the dielectric constant of sea water. The sensitivity to salinity (conductivity) increases with decreasing frequency and the 1.400–1.427 MHz window, reserved for passive observations, has advantages for SSS remote sensing. This requires special care because of the low sensitivity of brightness temperature to SSS: from 0.8 K to 0.2 K per unit of salinity, which depends on ocean temperature, radiometer incidence angle, and polarisation (Yueh *et al.* 2001). The stringent requirements pose technical challenges to achieve the required radiometric accuracy and stability. Finally, the low frequency involved requires the use of very large antennas to achieve a moderate spatial resolution on ground. For these reasons, only

V. Barale, M. Gade (eds.), *Remote Sensing of the European Seas.*
© Springer Science+Business Media B.V. 2008

two L-band space-borne radiometers, until present, have been flown: in 1968 aboard the Cosmos 243 and in 1973 aboard the Skylab S-194.

In 1995, at the "Soil Moisture and Ocean Salinity" Workshop organised at ESTEC (the European Space Research and Technology Centre, Noordwijk, the Netherlands), microwave radiometry at L-band was still considered as the most adequate technique to remotely measure these geophysical variables. However, instead of the real aperture microwave radiometers that were considered until then, it was concluded that the most promising technique was aperture synthesis radiometry that had successfully been demonstrated a few years earlier (Ruf *et al.* 1988).

Miller *et al.* (1996) published the first SSS map using the airborne Scanning Low Frequency Microwave Radiometer (SLFMR), a 6 beam, real aperture pushbroom radiometer. Le Vine *et al.* (2000) created an SSS map using the Electronically Steered Thinned Array Radiometer (ESTAR), the first 1D synthetic aperture radiometer flown on a plane.

2. The SMOS mission

After the completion of the Microwave Imaging Radiometer by Aperture Synthesis (MIRAS) feasibility study (Martín-Neira and Goutoule 1997), an international team of land and ocean scientists proposed the Soil Moisture and Ocean Salinity (SMOS) mission (Silvestrin *et al.* 2001) to European Space Agency (ESA) first call for Earth Explorer Opportunity Missions. SMOS was selected for feasibility studies in 1999 and for implementation in 2002. Launch is scheduled for 2008. With the technologically challenging approach of SMOS, Europe is taking the lead in attempting to measure salinity, a missing key variable in ocean remote sensing.

The objectives of SMOS are to obtain global sea surface salinity maps (accuracy of 0.1 every 30 days with a 200 km spatial resolution), global soil moisture maps (accuracy of 4%, *i.e.* 0.04 m^3/m^3, every 3 days), and vegetation water content maps (accuracy of 0.2 kg/m^2, with spatial resolution better than 50 km). In order to achieve the required SSS accuracy it will be necessary to average the SMOS pixel (between 30 × 30 km^2 to 50 × 50 km^2) both in space and time to reduce the measurement uncertainties. As a result, the mission will focus only on large-scale oceanography (Font *et al.* 2004). However, several phenomena extremely relevant for large-scale and climatic studies can benefit from the SMOS observational approach: barrier layer effects on tropical Pacific heat flux, halosteric adjustment of heat storage from sea level, North Atlantic (NA) thermohaline circulation, surface freshwater flux balance, etc., which require an accuracy of 0.1–0.4 in salinity over 100×100 km^2 to 300×300 km^2 in 10–30 days (Lagerloef 2000, Koblinsky *et al.* 2003).

3. Applications of salinity remote sensing

3.1 Operational models and process studies

The need for accurate salinity information to reconstruct the state of the ocean was already pointed out by Cooper (1988) in the context of data assimilation in ocean models, as univariate temperature corrections lead to density imbalances that rapidly amplify the errors of the model. However, and despite the fact that salinity has been an active variable in most numerical ocean models, many operational ocean predictions have been performed with a "free-evolving" salinity field obeying only to advection and diffusion without additional constraints imposed through data assimilation or surface forcing. Neglecting the need for realistic salinity fields in operational simulations was justified by the lack of observations and the assumption that salinity plays only a minor role on density changes.

Increases of temperature and salinity have opposite effects on density and, in turn, on dynamic height. Scrutiny of the ocean fields provided by the National Centers for Environmental Prediction (NCEP) data assimilation system identified errors in the near-surface currents (Acero-Schetzer *et al.* 1997) due to lack of salinity corrections. In the context of tropical dynamics, Murtugudde and Busalacchi (1998) compared ocean simulations with and without salinity variability, illustrating the importance of accounting for salinity variability in tropical dynamics. Maes (1999) revealed that, during 1996, the salinity anomalies could have contributed to sea level as much as 5–10 cm. Such an estimate was later verified in the NCEP operational system by Ji *et al.* (2000).

The El Niño/Southern Oscillation (ENSO) is the major source of interannual climate variability. As leading intermediate coupled models have been able to predict ENSO without accounting for salinity effects, the role of salinity in operational prediction of ENSO has often been neglected. This situation has changed due to the proposed missions for remote sensing of SSS and the availability of an unprecedented amount of salinity observations given by the Argo profiles. For example, Ballabrera-Poy *et al.* (2002) showed that SSS observations would play little role in the statistical nowcast of ENSO, but a significant role in the 6–12 month predictions. At these lags, positive SSS anomalies off the equator have the potential to modify the subsurface stratification of the western Pacific as they are subducted westward. In this region, the most prominent feature related to salinity is the existence of a "barrier layer" (Lukas and Lindstrom 1991) that isolates the mixed layer from the entrainment of cold water from below. Thus, salinity stratification helps to preserve a warm anomaly, increases the fetch of westerly winds, and leads to the ocean-atmosphere coupled instability leading to an El Niño event (Maes *et al.* 2002). On the other hand, the eastern edge of the warm pool is distinguished by a sharp SSS gradient, but by a weak Sea Surface Temperature (SST) gradient. Remote sensing of SSS is thus expected to improve our characterisation of the state of the equatorial Pacific previous and during the initial phases of ENSO events.

ENSO teleconnections and other climate signals have an impact on the precipitation over the Atlantic Ocean and its catchment areas, modifying its SSS. Surface salinity anomalies may then be advected to the deep convection regions, modulating the thermocline circulation (Latif *et al.* 2000). One of the largest ocean climate events recorded on the Atlantic Ocean is the Great Salinity anomaly (Dickson *et al.* 1988) which lasted from 1968 to 1982. As such a salinity anomaly propagated, it reached the Labrador Sea and perturbed the thermocline circulation intensity as well. Origin and evolution of these anomalies is still not fully understood because of the lack of enough salinity observations. In such a context, studies focusing on the mechanisms by which these SSS anomalies evolve are usually based on the output of ocean models (Mignot and Frankignoul 2003). Errors on the physics, subscale parameterisation, and surface forcing do introduce uncertainties in their results. Direct data assimilation of SSS observations on such ocean models will reduce the impact of model and forcing errors and provide a better understanding of NA haline variability.

3.2   Assimilation of sea surface salinity data

The awareness of the importance of accounting for salinity variability both in numerical simulations/predictions and in data assimilation algorithms has not been accompanied by the availability of a continuous stream of salinity observations. This has lead to the development of different strategies to palliate the scarcity of subsurface salinity. For example, in the National Oceanic and Atmospheric Administration (NOAA) approach, synthetic salinity profiles are obtained from vertical profiles of temperature based on their statistical relationship derived from historical data (Vossepoel *et al.* 2002). In the European Center for Medium-range Weather Forecast (ECMWF) approach, the assimilation of data must be followed by a post-processing step in which salinity profiles are shifted vertically to maintain the background T-S relationship (Troccoli *et al.* 2002). In a third approach, multivariate statistics from a long simulation of an ocean model are used to derive corrections to the salinity field in response to temperature innovations (Gourdeau *et al.* 2000). All three approaches will benefit from SSS data, as the new data will introduce a new constraint onto the statistical relationships, allow the estimation of actual SSS covariance maps, and provide information about salinity at the mixed layer, the region where the T-S conservation approach has the lowest performance.

4.   Principles of synthetic aperture microwave radiometers

4.1   Basic concepts on synthetic aperture microwave radiometry imaging

As compared to real aperture radiometers, in which the brightness temperature maps are obtained by a mechanical scan of a large antenna or by a large push-broom system, in aperture synthesis radiometers, the brightness temperature image reconstruction is performed through a Fourier synthesis process. Instead of a power measurement, a synthetic aperture radiometer measures the cross-correlations (V_{12}^{pq}) between all signal pairs ($b_1(t)$ and $b_2(t)$ at p and q polarizations), collected by the array elements which are located in the XY plane. According to Corbella *et al.* (2004), the samples of the visibility function are given by:

$$V_{12}^{pq}(u_{12}, v_{12}, w_{12} = 0) \stackrel{\Delta}{=} \frac{1}{k_B \sqrt{B_1 B_2} \sqrt{G_1 G_2}} \cdot \frac{1}{2} \left\langle b_1^p(t) b_2^{q*}(t) \right\rangle =$$

$$= \frac{1}{\sqrt{\Omega_1 \Omega_2}} \iint\limits_{\xi^2 + \eta^2 \leq 1} \frac{\left(T_{pq}(\xi, \eta) - T_{rec}\delta_{pq}\right)}{\sqrt{1 - \xi^2 - \eta^2}} F_{np1}(\xi, \eta) F_{nq2}^*(\xi, \eta) \cdot$$

$$\tilde{r}_{12}\left(-\frac{u_{12}\xi + v_{12}\eta}{f_0}\right) \exp\left(-j2\pi(u_{12}\xi + v_{12}\eta)\right) d\xi d\eta \qquad (1)$$

where k_B is the Boltzmann's constant, $B_{1,2}$ and $G_{1,2}$ are the receiver noise bandwidth and power gain, $\Omega_{1,2}$ is the solid angle of the antennas, $T_{pq}(\xi, \eta)$ is the brightness temperature of the scene, T_{rec} is the physical temperature of the receiver (the "Corbella term"), $\delta_{pq} = 1$ if $p=q$ and 0 if $p \neq q$, $F_{np,q1,2}(\xi, \eta)$ are the normalized antenna co-polar voltage patterns at p and q polarisations, $\tilde{r}_{12}\left(-(u_{12}\xi + v_{12}\eta)/f_0\right)$ is the so-called fringe-washing function that accounts for spatial decorrelation effects and depends on the frequency response of the pair of elements collecting the signals being correlated, (u_{12}, v_{12}, w_{12}) is the spatial frequency (baseline) that depends on the antenna position difference: $(u_{12}, v_{12}, w_{12}) = (x_2 - x_1, y_2 - y_1, z_2 - z_1)/\lambda_0$ (in this case

$z_1=z_2$ and $w_{12}=0$), $\lambda_0=c/f_0$ is the wavelength and f_0 is the central frequency of the receivers, and the direction cosines $(\xi, \eta)=(\sin\theta\cos\varphi,\sin\theta\sin\varphi)$ are defined with respect to the X and Y axes.

In the ideal case, when all antenna patterns are equal (*i.e.*, $F_{np1}(\xi,\eta)=F_{nq2}(\xi,\eta)=F_n(\xi,\eta)$ and $\Omega_1=\Omega_2=\Omega$), spatial decorrelation effects are negligible ($\tilde{r}_{12}\approx 1$), and there are no antenna position errors (the (u,v) points are equal to their nominal values). Then, the relationship between the visibility samples and the T_B reduces to a discrete Fourier transform:

$$V_{12}^{pq}\left(u_{12},v_{12}\right)=F\left[\frac{\left(T_{pq}\left(\xi,\eta\right)-T_{rec}\delta_{pq}\right)}{\sqrt{1-\xi^2-\eta^2}}\frac{\left|F_n\left(\xi,\eta\right)\right|^2}{\Omega}\right] \tag{2}$$

Since the support of the T_B to be retrieved is the unit circle $\xi^2+\eta^2\leq 1$, it can be demonstrated that the optimum sampling strategy of the (u,v) plane is on a hexagonal grid, instead of the usual rectangular one. This sampling strategy allows the increase of the maximum antenna separation from $d=\lambda_0/2$ to $d=\lambda_0/\sqrt{3}$ without suffering from aliasing effects in the image reconstruction process. In addition, if the (ξ,η) are properly selected, the hexagonally sampled visibility function can be processed with standard Fast Fourier Transform (FFT) (Camps 1996; Camps *et al.* 1997).

For a given number of elements, the array structure that provides a hexagonal sampling of the (u,v) plane and the largest (u,v) coverage (the best angular resolution) is a Y structure, which is the shape of the MIRAS instrument (Figure 1a). However, in MIRAS, after an optimisation process of the swath width, the revisit time, the angular resolution, and the range of incidence angles over the Earth being imaged, it was decided to increase the antenna spacing even more from $d=\lambda_0/\sqrt{3}$ to $d=0.875\lambda_0$. Then, the closest six replicas of the T_B supported by the unit circle overlap with the main one and there is "aliasing." The alias-free field-of-view (AF-FOV) limited by the periodic repetition of the unit circle is, in fact, very small (Figure 1b, grey region). However, since a significant part of the aliases correspond to the cold and known sky, the AF-FOV can be extended (E-AF-FOV) up to the region limited by the periodic repetition of the of the Earth's disk (Camps 1996; Camps *et al.* 1997; Camps *et al.* 2006) (Figure 1b, thick irregularly hexagonal line).

In the non-ideal case (antennas not in the nominal positions, and different antenna patterns and receivers' frequency responses), the brightness temperature maps can be retrieved by discretising (1) and solving the resulting system system of equations (Camps *et al.* 2006).

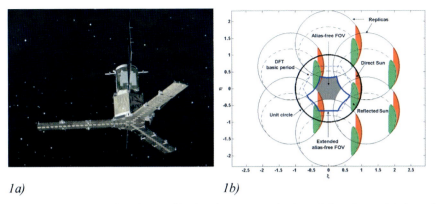

1a)   *1b)*

Fig. 1. a) SMOS artist view with the three arms deployed forming a Y-shaped structure; b) Aliasing in 2D aperture synthesis Y-shaped interferometric radiometers for an antenna spacing of 0.875 wavelengths, 32° array tilt, 30° array steering and 755 km platform height. Representation of the Earth's disk (Earth-sky horizon, dash-dotted), unit circle, DFT (Discrete Fourier Transform) basic period, geometric place of the Sun positions (direct and reflected images), and their six closest replicas to the main DFT period.

4.2 Instrument performance: angular resolution and radiometric errors

Angular resolution

The angular resolution of MIRAS is given by the half-power width of the synthetic beam (Bará *et al.* 1998) or "equivalent array factor" (*AF*), which can be adjusted with the windowing function $W(u_{mn}, v_{mn})$ used to taper the visibility samples:

$$AF(\xi - \xi', \eta - \eta') = \frac{\sqrt{3}}{2} d^2 \sum_m \sum_n W(u_{mn}, v_{mn}) e^{j2\pi(u_{mn}(\xi - \xi') + v_{mn}(\eta - \eta'))} \qquad (3)$$

Figure 2a shows the shape of the array factor corresponding to the Blackmann window (used as default in SMOS data processing) in the direction cosines domain. Figures 2b and 2c show the same factor for different directions in the field-of view (FOV) both in the direction cosines and in cross-track/along-track coordinates, respectively. As it can be appreciated in Figs. 2b and 2c, while the width of the synthetic beam is constant in the direction cosines domain, it enlarges and distorts when projected over the Earth. This effect can be mitigated by using the so-called strip adaptive processing, in which a different window is applied in each direction, at the expense of varying side lobe levels (Anterrieu *et al.* 2004).

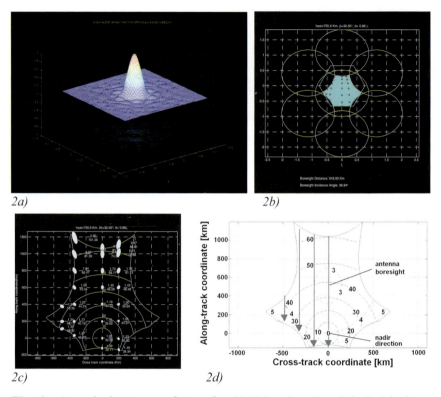

2a)   *2b)*

2c)   *2d)*

Fig. 2. a) equivalent array factor for SMOS using the default Blackmann window; b) half-power beamwidth in different positions of the FOV in direction cosines; c) half-power beamwidth in different positions of the FOV in Earth coordinates; d) Dwell lines shorten as the pixel under observation is far away from the satellite ground-track. Contours of incidence angle (dashed) and radiometric sensitivities (dash-dotted) are shown over the SMOS FOV.

Radiometric errors

In any imaging instrument there are three types of radiometric errors: radiometric sensitivity, radiometric bias and radiometric accuracy. Random errors (noise due to finite integration time) have a zero temporal average, and their temporal standard deviation is called the radiometric sensitivity, which can be computed as:

$$\Delta T(\xi,\eta) = \Omega_{ant} \left(\frac{\sqrt{3}}{2} d^2 \right) \frac{T_A + T_B}{\sqrt{B \tau_{eff}}} \alpha_W \frac{\alpha_{LO}}{\alpha_F} \sqrt{N_V}$$ (4)

where Ω_{ant} is the antenna solid angle, T_A and T_R are the antenna and receiver noise temperatures, τ_{eff} is the effective integration time, which depends on the correlator type, α_W, α_{LO} and α_F are parameters depending on the window, the type of demodulation, and the shape of the frequency response (Camps 1996), and N_V is the total number of (u,v) points sampled by the array.

The spatial average of systematic errors (instrumental errors) appears as a radiometric bias (scene bias) in the whole brightness temperature image, and their spatial standard deviation is the radiometric accuracy (pixel bias). The different instrumental error sources can be grouped as antenna errors, affecting each pixel in the scene in a different way, receiver amplitude and phase errors, that can be assigned to each element forming the baseline, and baseline errors, that can only be assigned to the pair of elements forming the baseline. These errors are calibrated by internal and the external calibration. The first consists of two-level correlated noise sources distributed in the hub and along the arms, and an uncorrelated noise source in each receiver. The second consists of measuring external known targets (the homogeneous galaxy pole, preferably) to calibrate the noise injection radiometers and to measure the so-called Flat Target Response to be applied in the Flat Target Transformation (Martín-Neira 2004).

Finally, Tables 1 and 2 summarise the MIRAS radiometric sensitivity and accuracy requirements and performances at X- and Y-polarisations (in the antenna reference frame), as of March 2007.

Table 1. MIRAS radiometric sensitivity at X- and Y-polarisations (antenna reference frame). TA: average brightness temperature

Radiometric Sensitivity (K)	Requirement	X-Pol	Y-Pol
Ocean Salinity (TA = 150 K) – Boresight	2.5	1.84	1.80
Ocean Salinity (TA = 150 K) – 32°	4.1	3.02	2.95
Soil Moisture (TA = 220 K) – Boresight	3.5	1.99	1.98
Soil Moisture (TA = 220 K) – 32°	5.8	3.26	3.55

Table 2. Boresight MIRAS radiometric accuracy at X- and Y-polarisations (antenna reference frame)

Radiometric Accuracy (K)	Requirement	X-Pol	Y-Pol
Radiometric Systematic Error (TA = 273 K)	3.7	2.26	2.25
Measurement Accuracy (TA = 298 K)	4.1	2.41	2.41

4.3   Multi-look and multi-angular imaging capabilities

The SMOS mission is a challenge in its own for many reasons. Firstly, it carries a new type of instrument and the errors and image reconstruction algorithms to be used are new. Secondly, the type of multi-look and multi-angle observations with different pixel sizes and orientations, and polarisation mixing in the pass from the antenna to the Earth reference frame and vice-versa, is also new. Finally, L-band emission models, covering the whole range of incidence angles from 0° to nearly 60° have had to be developed in recent years, but are still a subject of refinement. The big difference in the SMOS mission is its multi-look capability: each SMOS snapshot provides a two-dimensional image where each pixel is seen under a different angle and as the satellite moves on, the same pixel is being seen under different incidence angles at both polarisations. This wealth of information can be used to infer:

- over the ocean: not only the SSS, but to tune other auxiliary geophysical parameters that also influence L-band emission and have to be used in the retrieval,
- over land: not only the soil moisture, but also the vegetation opacity (related to the vegetation water content), the albedo and the surface roughness parameter.

4.4   Sea surface salinity estimations from SMOS data

The primary algorithm implemented for SMOS SSS retrieval from radiometric information (Font *et al.* 2006) is based on an iterative convergence scheme that compares the measured values with those provided by a L-band forward model (that includes emission from the sea surface plus all other sources and phenomena that modify it). This model uses a guessed salinity that can be adjusted until obtaining an optimal fit with the radiometric measurement.

While a reasonably accurate model for the L-band emissivity of a flat sea exists (Klein and Swift 1977), the different processes affecting the emission of a roughened surface are not fully described or considered in the theoretical formulations available until now. However, it is critical to know the changes in the ocean brightness temperature produced by the sea state as they can be even larger than the salinity-induced change itself.

In a first step, the brightness temperature of the sea, $T_{h,v}^{sea}$, can be decomposed in two terms: the "flat sea" contribution and a deviation with respect to it. The deviation term, $\Delta T_{h,v}(\theta, param)$, depends on the incidence angle and a parameterisation of the surface roughness by means of variables such as wind speed, significant wave height, wave age, atmospheric stability etc. (Gabarró et al. 2004):

$$T_{h,v}^{sea}(\theta, SST, SSS) \approx \left(1 - \Gamma_{h,v}\left(\theta, \varepsilon_r(SST, SSS)\right)\right) \cdot SST + \Delta T_{h,v}(\theta, param) \qquad (5)$$

In (5), $\Gamma_{h,v}\left(\theta, \varepsilon_r(SST, SSS)\right)$ is the electric field Fresnel reflection coefficient (amplitude squared) that through the dielectric constant depends on SSS and SST. At the selected frequency, the measured sea emission comes from the upper 1 cm of the ocean.

In the presence of foam, there is an increase of the brightness temperatures, which is function of the sea surface fraction covered by foam and the brightness temperature of the sea foam. Several controlled measurements (Camps et al. 2004a; Camps et al. 2005) indicate a foam-induced emissivity in good agreement with the Reul-Chapron (2003) model specifically developed for SMOS under some conditions. Other effects that modify sea brightness temperature are those of rain and of oil slicks that can modify the sea surface spectrum, which affects the brightness temperature.

Before comparing the modelled and measured T_B in the iterative process, it is necessary to add other components to the modelled T_B and finally make the geometric transformation from the Earth reference frame (where the forward model has been applied) to the antenna reference frame (where measurements are done). Several atmospheric effects (upwelling atmospheric/ionospheric radiation, downwelling sky/atmospheric/ionospheric radiation scattered over the sea surface, atmospheric/ionospheric losses) are sufficiently well modelled (Skou and Hoffman-Bang 2005), while the Faraday rotation, due to the electromagnetic wave propagation through the ionosphere in the presence of the geomagnetic field, can be either modelled from the knowledge of the ionospheric Total Electron Content (TEC) or avoided by using the first Stokes parameter $I = T_h + T_v$ instead of both polarisations separately (Camps et al. 2003). This presents several other advantages, such as the cancellation of geometric rotation effects, and the minimization of uncertainties in the T_B associated to the sea water dielectric constant model, even though the number of observables is halved.

Radiation by celestial sources illuminating the ocean surface that are further reflected (through scattering produced by the surface roughness) towards the radiometer has to be taken into account. The T_B of the source brightness can be estimated from sky surveys. The surface level scattered signals are estimated through a proper weighting of the sky T_B illuminating the considered Earth target by the rough sea surface bistatic scattering coefficients estimated at that point. Reflected solar radiations are extremely intense at L-band and their contribution needs to be accounted for. However, Sun glint events being rare, after applying the developed Sun cancellation algorithms (Camps *et al.* 2004b) the few affected grid points and angular measurements will be discarded for salinity retrieval instead of attempting a correction (Font *et al.* 2006).

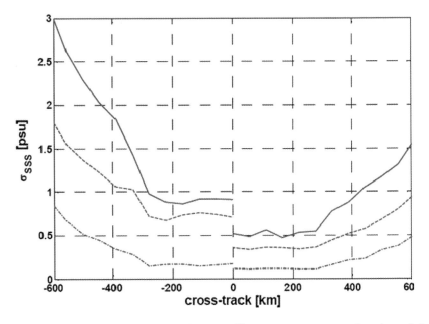

Fig. 3. SSS rms retrieval error without auxiliary parameters as a function of the pixel distance to the satellite ground-track for zero wind speed. The retrieval problem is formulated in terms of the first Stokes parameter I (left) or [T_h, T_v] (right) for SST = 5°C (solid), 15°C (dashed) and 25°C (dotted).

The part of the retrieval algorithm that performs the iterative comparison between model and data uses a cost function to be minimised that incorporates reference values and associated uncertainties (as weights) for the external geophysical parameters that provide information on the sea state conditions, and that will be themselves adjusted during the convergence process. For every pixel on the SMOS FOV, the comparison is made using all the available angular measurements acquired in consecutive satellite snapshots. The number of measurements of each pixel depends on the pixel's cross-track distance to the satellite ground-track. As this distance increases, the pixel is imaged fewer times; the angular variation is reduced (Figure 2d), and the instrument's noise increases, which translates into a degraded performance in terms of the quality of the retrieved parameters. At the sub satellite track the number of observations is 73, increasing up to 80 at ~220 km from the ground-track, and then rapidly decreases down to 11 at ~550 km due the irregular shape of the FOV.

Figure 3 shows an example of the retrieval performances that can be expected when equation (9) is applied to a single satellite overpass and there is no auxiliary information on roughness parameters or sea surface temperature. For cold waters, the rms salinity retrieval error is in the 0.5–1 range in the central part of the FOV, which means that in order to achieve the 0.1 accuracy, spatio-temporal averaging will be necessary. On the other hand, the use of auxiliary parameters may help to improve the retrieval accuracy, even though in some cases, when the auxiliary parameter is too far from the actual one it leads to erroneous solutions.

A simulation of space-time averaging, in a 10°×10° region of the NA using different auxiliary information on wind speed and sea surface temperature (Sabia *et al.* 2005), indicates that over 30 days and 1°×1° boxes, the retrieved rms SSS error reduces to 0.055–0.3, and in 2°×2° boxes to 0.032–0.29, depending on the set of auxiliary parameters used.

5. Conclusions

The development of the SMOS mission is an important step forward to a future operational measurement of Sea Surface Salinity from space. This chapter has reviewed the fundamentals of this new type of instrument and its sea surface salinity retrieval capabilities. Besides implementing a new technical approach, SMOS will test several aspects of real time operations that will allow improving the design for future follow-on missions. SMOS has characteristics very complementary to those of Aquarius-SAC/D, the US – Argentina salinity mission (Koblinsky *et al.* 2003), based on the real aperture radiometer concept (with higher radiometric performances but smaller spatial and temporal coverage) to be launched in 2009. A combination of the results of these two kinds of missions will also be a test for improved products and measurement strategies. ESA is now proposing to its member states the preparation of a SMOSops program that should launch a series of satellites after 2012 to ensure salinity operational measurement.

Acknowledgements

The authors thank the international science and technology research community that has made possible the development of the microwave salinity measurement concepts, as well as ESA, the SMOS Science Advisory Group, and all the colleagues that have contributed to the implementation of the mission. This chapter is a contribution to the MIRAS-4 project funded by the Spanish National Program on Space (ESP2005-06823-C05).

References

Acero-Schertzer CE, Hansen DV, Swenson MS (1997) Evaluation and diagnosis of surface currents in the National Centers for Environmental Prediction's ocean analyses. J Geophys Res 102: 21037–21048

Anterrieu E, Picard B, Martin-Neira M, Waldteufel P, Suess M, Vergely JL, Kerr Y, Roques S (2004) A strip adaptive processing approach for the SMOS space mission. In: IEEE Int Geosci Rem Sens Symp 2004 Proceed 3: 1922–1925

Ballabrera-Poy J, Murtugudde R, Busalacchi AJ (2002) On the potential impact of sea surface salinity observations on ENSO predictions. J Geophys Res 107: 8007–8017

Bará J, Camps A, Torres F, Corbella I (1998) Angular resolution of two-dimensional hexagonally sampled interferometric radiometers, Radio Sci 33: 1459–1473

Camps A (1996) Application of interferometric radiometry to Earth observation. Ph D Thesis, Univ. Politècnica de Catalunya, http://www.tdx.cesca.es/TDX-1020104–091741/

Camps A, Bará J, Corbella I, Torres F, (1997) The processing of hexagonally sampled signals with standard rectangular techniques: application to 2D large aperture synthesis interferometric radiometers. IEEE Trans Geosci Rem Sens 35: 183–190

Camps A, Corbella I, Vall-llossera M, Duffo N, Torres F, Villarino R, Enrique L, Julbe F, Font J, Julià A, Gabarró C, Etcheto J, Boutin J, Weill A, Rubio E, Caselles V, Wursteisen P, Martín-Neira M (2003) L-band sea surface emissivity: Preliminary results of the WISE-2000 campaign and its application to salinity retrieval in the SMOS mission. Radio Sci 38 (4): 8071–8079

Camps A, Font J, Vall-llossera M, Gabarró C, Corbella I, Duffo N, Torres F, Blanch S, Aguasca A, Villarino R, Enrique L, Miranda J, Arenas J, Julià A, Etcheto J, Caselles V, Weill A, Boutin J, Contardo S, Niclós R, Rivas R, Reising SC, Wursteisen P, Berger M, Martín-Neira M (2004a) The WISE 2000 and 2001 field experiments in support of the SMOS mission: sea surface L-Band brightness temperature observations and their application to sea surface salinity retrieval. IEEE Trans Geosci Rem Sens 42 (4): 804–823

Camps A, Vall-llossera M, Duffo N, Zapata M, Corbella I, Torres F, Barrena V (2004b) Sun effects in 2D aperture synthesis radiometry imaging and their cancellation. IEEE Trans Geosci Rem Sens 42: 1161–1167

Camps A, Vall-llossera M, Villarino R, Reul N, Chapron B, Corbella I, Duffo N, Torres F, Miranda JJ, Sabia R, Monerris A, Rodriguez (2005) The emissivity of foam-covered water surface at L-band: Theoretical modeling and experimental results from the FROG 2003 field experiment. IEEE Trans Geosci Rem Sens 43: 925–937

Camps A, Vall-llossera M, Corbella I, Duffo N, Torres F (2006) Improved image reconstruction algorithms for aperture synthesis radiometers. In: IEEE Int Geosci Rem Sens Symp 2006 Proceed: 1160–1163

Cooper NS (1988) The effect of salinity on tropical ocean models. J Phys Oceanogr 18: 697–707

Corbella I, Duffo N, Vall-llossera M, Camps A, Torres F (2004) The visibility function in interferometric aperture synthesis radiometry. IEEE T Geosci Rem Sens 42: 1677–1682

Dickson RR, Meincke J, Malmberg SA, Lee AJ (1988) The 'Great Salinity Anomaly' in the northern North Atlantic 1968–1982. Prog Oceanogr 20: 103–151

Font J, Lagerloef GSE, Le Vine DM, Camps A, Zanifé OZ (2004) The determination of surface salinity with the European SMOS space mission. IEEE Trans Geosci Rem Sens 42: 2196–2205

Font J, Boutin J, Reul N, Waldteufel P, Gabarró C, Zine S, Tenerelli J, Petitcolin F, Vergely JL (2006) An iterative convergence algorithm to retrieve sea surface salinity from SMOS L-band radiometric measurements. In: IEEE Int Geosci Rem Sens Symp 2006 Proceed: 1697–1701

Gabarró C, Font J, Camps A, Vall-llossera M, Julià A (2004) A new empirical model of sea surface microwave emissivity for salinity remote sensing. Geophys Res Lett 31: L01309

Gourdeau L, Verron J, Delcroix T, Busalacchi AJ, Murtugudde R (2000) Assimilation of TOPEX/Poseidon altimetric data in a primitive equation model of the tropical Pacific Ocean during the 1992–1996 El Niño-Southern Oscillation period. J Geophys Res 105: 8473–8488

Ji M, Reynolds RW, Behringer DW (2000) Use of TOPEX/Poseidon sea level data for ocean analyses and ENSO prediction: Some early results. J Climate 13: 216–231

Klein LA, Swift CT (1977) An improved model for the dielectric constant of sea water at microwave frequencies. IEEE J Ocean Eng 2: 104–111

Koblinsky CJ, Hildebrand P, Le Vine DM, Pellerano F, Chao Y, Wilson WJ, Yueh SH, Lagerloef GSE (2003) Sea surface salinity from space: Science goals and measurement approach. Radio Sci 38: 8064–8069

Lagerloef GSE (2000) Recent progress toward satellite measurements of the global sea surface salinity field. In: Halpern D (ed) Satellites, oceanography, and society. Elsevier Oceanography Series 63, Amsterdam, pp. 309–319

Latif M, Roeckner E, Mikolajewicz U, Voss R (2000) Tropical stabilization of the thermohaline circulation in a greenhouse warming simulation. J Climate 13: 1809–1813

Le Vine DM, Zaitzeff JB, D'Sa EJ, Miller JL, Swift C, Goodberlet M (2000) Sea surface salinity: toward an operational remote-sensing system. In: Halpern D (ed) Satellites, oceanography and society. Elsevier Oceanography Series 63, Amsterdam, pp. 321–335

Lukas R, Lindstrom E (1991) The mixed layer of the western equatorial Pacific-Ocean. J Geophys Res 96: 3343–3357

Maes C (1999) A note on the vertical scales of temperature and salinity and their signature in dynamic height in the western Pacific Ocean: Implications for data assimilation. J Geophys Res 104: 11037–11048

Maes C, Picaut J, Belamari S (2002) Salinity barrier layer and onset of El Niño in a Pacific coupled model. Geophys Res Lett 29: 2206–2216

Martín-Neira M (2004) In-orbit external calibration and validation. ESA-ESTEC TEC-ETP/2004.103/MMN

Martín-Neira M, Goutoule JM (1997) A two-dimensional aperture-synthesis radiometer for soil moisture and ocean salinity observations. ESA Bull 92: 95–104

Mignot J, Frankignoul C (2003) On the interannual variability of surface salinity in the Atlantic. Clim Dynam 20: 555–565

Miller J, Goodberlet MA, Zaitzeff J (1996) Airborne salinity mapper makes debut in coastal zone. EOS Trans AGU 79: 173–177

Murtugudde R, Busalacchi AJ (1998) Salinity effects in a tropical ocean model. J Geophys Res 103: 3283–3300

Reul N, Chapron B (2003) A model of sea-foam thickness distribution for passive microwave remote sensing applications. J Geophys Res 108: 3321–3331

Ruf CS, Swift CT, Tanner AB Le Vine DM (1988) Interferometric synthetic aperture microwave radiometry for the remote sensing of the Earth. IEEE Trans Geosci Rem Sens 26: 597–611

Sabia R, Camps A, Reul N, Vall-llossera M, Miranda J (2005) Synergetic aspects and auxiliary data concepts for Sea Surface Salinity measurements from space. Final Report WP1400 - Towards best SSS Level 2 products, ESA ESTEC ITT 1-4505/03/NL/Cb

Silvestrin P, Berger M, Kerr Y, Font J (2001) ESA's second Earth Explorer opportunity mission: The Soil Moisture and Ocean Salinity mission - SMOS. IEEE Geosci Remote S Newsl 118: 11–14

Skou N, Hoffman-Bang D (2005) L-band radiometers measuring salinity from space: atmospheric propagation effects. IEEE Trans Geosci Rem Sens 43: 2210–2217

Swift CT, McIntosh RE (1983) Considerations for microwave remote sensing of ocean-surface salinity. IEEE T Geosci Elect 21: 480–491

Troccoli A, Balmaseda MA, Segschneider J, Vialard J, Anderson DLT, Haines K, Stockdale T, Vitart F, Fox AD (2002) Salinity adjustments in the presence of temperature data assimilation. Mon Weather Rev 130: 89–102

Vossepoel FC, Burgers G, van Leeuwen PJ (2002) Effects of correcting salinity with altimeter measurements in an equatorial Pacific Ocean model. J Geophys Res 107: 8001–8010

Yueh SH, West R, Wilson WJ, Li FK, Njoku EG, Rahmat-Samii Y (2001) Error sources and feasibility for microwave remote sensing of ocean surface salinity. IEEE Trans Geosci Rem Sens 39: 1049–106

Sea Ice Parameters from Microwave Radiometry

Stefan Kern[1], Lars Kaleschke[1], Gunnar Spreen[1], Robert Ezraty[2],
Fanny Girard-Ardhuin[2], Georg Heygster[3], Søren Andersen[4], and
Rasmus Tonboe[4]

[1] Institute of Oceanography, University of Hamburg, Germany
[2] DOPS/LOS, IFREMER, Plouzane, France
[3] Institute of Environmental Physics, University of Bremen, Germany
[4] Danish Meteorological Institute, Copenhagen, Denmark

Abstract. Microwave (MW) radiometry has been playing a key role for
the observation of sea ice parameters at global scale for more than three
decades now. Among these parameters are sea ice concentration, drift, and
type. Recent advances in satellite technology and algorithm development
enable to further expand the parameter range, to refine the spatial resolution, and to apply MW radiometry also at regional scale. The present paper
informs about some key physical properties of sea ice. It informs briefly
about data acquisition and about the most common retrieval techniques
and gives examples of their application to MW data.

1. Introduction

Sea ice effectively reduces the ocean-atmosphere energy exchange. It has a
high short-wave albedo and thus reflects a large part of the incident short-
wave radiation energy. Sea ice formation affects water mass characteristics
and the thermo-haline circulation by brine rejection. The freshwater input
during sea ice melt stabilizes the top few ten meters of the ocean. Sea ice
drift can cause sea ice to melt far away from its origin. The annual transport through, *e.g.* Fram Strait amounts about 10% of the total Arctic sea
ice mass. Sea ice protects the coasts from erosion and, finally, can hamper
shipping and off-shore industries. The observed decrease of the Arctic sea
ice extent (Figure 1) and other changes have led to the conclusion, that sea
ice cover changes can be taken as an indicator of amplified climate warming in polar regions. One backbone of these observations are satellite passive
microwave (PM) data.

V. Barale, M. Gade (eds.), *Remote Sensing of the European Seas.*

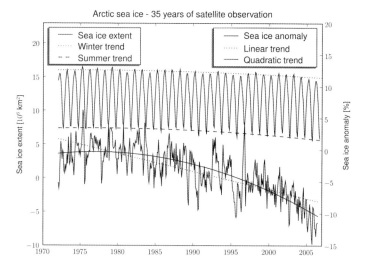

Fig. 1. Observed mean monthly Arctic sea ice extent (top) and its anomaly (bottom) together with trends for summer and winter (extent) and the entire year (anomaly) as based on the 30-year record by Cavalieri *et al.* (2003), and the sea ice index (Fetterer and Knowles 2002, updated 2007). The different records are merged by fitting the mean and standard deviations of the overlapping periods.

2. Theory and data

2.1 Physical properties and microwave signature of sea ice

Passive microwave (PM) sensors measure the intensity of the emitted MW radiation, usually expressed as brightness temperature, T_B, which is the MW emissivity times the temperature according to the Raileigh-Jeans approximation for MW wavelength. The typically used frequencies (wavelengths) are 7–90 GHz (40–3 mm). Physical sea ice parameters such as salinity, crystal structure, porosity, and ice-surface parameters wetness and roughness determine the MW emissivity. This emissivity is thus in turn a function of the penetration depth of MW radiation into sea ice and the partitioning between surface and volume scattering of MW radiation. Open water has a MW emissivity of about 0.5, see Figure 2, *e.g.* it is radiometrically "cold", at incidence angles commonly used by satellite PM sensors (50°). Young ice (less than 30 cm thick) is transparent to MW radiation below a certain thickness threshold (a few millimetres to some centimetres). Its emissivity is thus similar to that of open water; above this thickness threshold it increases quickly to values around 0.9 (Carsey 1992).

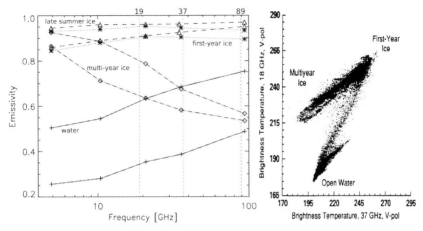

Fig. 2. Left: Observations of vertically (V) and horizontally (H) polarized emissivity of sea ice and sea water at an incident angle of 50° (Svendsen *et al.* 1983; Onstott *et al.* 1987). Right: Scanning Multichannel Microwave Radiometer brightness temperature observations for the Arctic: 18 GHz, V-polarized vs. 37 GHz, V-polarized (Carsey 1992).

Microwave emissivities of first-year (FY) ice take values around 0.9 (see Figure 2, left), because near-surface salinities are generally lower, and therefore the penetration depth of the MW radiation can increase to up to a few centimetres. Together young and FY ice are regarded as seasonal ice. Summer melt causes a desalination of typically the upper 5–50 cm of the sea ice, leaving lots of air bubbles in this upper part. These cause volume scattering, which, together with the low salinity, leads to a smaller emissivity (and thus smaller T_B values) for this so-called multi-year (MY) ice compared to FY ice (Figure 2).

Sea ice is often covered with snow. Dry snow usually is transparent at the mentioned frequency range, however, it may change the ice surface (ice snow interface) properties, *e.g.* by insulating the ice against the atmosphere and thereby causing a ice-snow interface temperature increase, which affects the brine volume close to the ice surface. The increase in conductivity by an increasing liquid water content during summer causes snow to become the dominating source of the emitted MW radiation (Carsey 1992). Summer melt causes melt ponds: ponds of (usually fresh) melt water on top of the ice. Typically, they have a size of 10 m² and can cover regularly up to 40% of the summer sea ice area (Perovich *et al.* 2002). Note, that their size and coverage varies with melt progress and ice type. As these ponds appear as open water, they decrease the MW emissivity of the sea ice.

Baltic Sea ice is brackish, seasonal ice. Its salinity and density is comparable to that of MY ice. Hence, the PM signature of Baltic sea ice depends on ice-surface roughness and wetness and the snow properties rather than the sea ice salinity. Reviews of the physical properties and the MW signature of sea ice are given by Carsey (1992) and Lubin and Massom (2006).

2.2   Satellite specifications

Satellite PM sensors typically receive MW radiation at multiple frequencies and two polarizations (Table 1). Frequencies are chosen such that the atmospheric contribution to the observed MW radiation (atmospheric effect) is usually quite small. Currently operating polar-orbiting PM sensors are the Special Sensor Microwave/Imager (SSM/I) (the successor of the Scanning Multichannel Microwave Radiometer, SMMR) and the Special Sensor Microwave/Image Sounder aboard spacecraft of the Defense Meteorological Satellite Program, and the Advanced Scanning Microwave Radiometer (AMSR-E) aboard the Earth Observations Satellite (EOS) Aqua.

Table 1. Specifications of SSM/I (left) and AMSR-E (right). Field-of-View (FOV) dimensions are given across- times along-track. Polarization is horizontal (H) and vertical (V), except for SSM/I, 22 GHz (only V). Sampling distances are 25 km (SSM/I, except 85 GHz: 12.5 km) and 10 km (AMSR-E, except 89 GHz: 5 km).

SSM/I		AMSR-E	
Frequency (GHz)	FOV (km × km)	Frequency (GHz)	FOV (km × km)
		6.925	41 × 71
		10.65	25 × 46
19.35	43 × 69	18.7	15 × 25
22.234	40 × 50	23.8	18 × 31
37.0	29 × 37	36.5	8 × 14
85.5	13 × 15	89.0	4 × 6

These sensors scan conically (constant surface incident angle) and provide T_B values at the frequencies given in Table 1 for about 1400 km wide swaths. PM data acquired by SSM/I and AMSR-E cover the both polar regions entirely twice daily. The large feld-of-view (FOV) size, however, does not enable small-scale analyses of sea ice. Here, *e.g.* active microwave (AM) sensors with synthetic aperture are more suitable.

3.   Methods and applications

3.1   Sea ice concentration

The percentage cover of an area with sea ice is termed sea ice concentration, C. Algorithms for the retrieval of C are often based on an empirical relationship between the MW emissivity and C, and so-called tie points: emissivity or T_B values that are typical of open water (OW), FY ice, and MY ice (Figure 2, left). Figure 2, right, shows a scatterplot of T_B values observed at two different frequencies over the Arctic Ocean. Clustering is evident at typical T_B values of the given surface types, yielding estimates for the required tie points. The accuracy with which a single value of C is retrieved depends on the spatial resolution achieved with the sensor (Table 1), its measurement noise, and on the used tie points.

The T_B observed over consolidated sea ice tends to form a linear cluster while that over OW tends to cluster around a single point (Figure 2, right). This sort of scheme can be used to calculate C, as is done, e.g. in the Comiso-Bootstrap algorithm (CBA) (Comiso 1986; Comiso et al. 2003) The Bristol algorithm (Smith 1996) uses a coordinate-transformed 3-dimensional version of this relationship. The NORSEX algorithm (Svendsen et al. 1983) includes radiative transfer (RT) model calculations in order to mitigate the inherent atmospheric effect. With RT model runs, the net surface T_B is obtained from satellite T_B data using standard atmospheric quantities. This value is used together with in-situ emissivity measurements and surface temperature data to obtain C. The NASA-Team algorithm (NTA) and NASA-Team 2 algorithm (NT2) (Cavalieri et al. 1984; Markus and Cavalieri 2000) use the T_B polarization difference (PD) at one frequency, e.g. 19 GHz, divided by the sum of horizontally and vertically polarized T_B (polarization ratio) and the gradient ratio (difference of two vertically polarized T_B at two different frequencies, e.g. 37 and 19 GHz, divided by their sum) to obtain partial (FY, MY) and total C values.

Smaller FOV size and sampling distance of 85/89 GHz data (Table 1) enables to obtain C at a finer spatial resolution (Svendsen et al. 1987) despite a considerably larger atmospheric effect, particularly over open water. Strategies to minimize this effect are, e.g. to apply RT model runs (Kern 2004), or to use this data only over consolidated ice (Pedersen 1998). One example of the latter one is the ARTIST (Arctic Radiation and Turbulence Interaction STudy) Sea Ice (ASI) algorithm (Kaleschke et al. 2001; Spreen et al. 2007). By using PD values typically observed for 0% and 100% ice cover by satellite at 85/89 GHz as tie points (i.e. including common high-latitude atmospheric conditions) together with a third-order polynomial fitted through these points, C is retrieved.

Figure 3 shows C as obtained with the ASI algorithm from AMSR-E data on a 6.25 km grid around time of the minimum sea ice area in 2006 ($4.5 \cdot 10^6$ km^2). For comparison the mean September 50% isolines of C of two five-year periods: 2002–2006 (ASI algorithm, AMSR-E) and 1979–1983 (CBA, SMMR), are superimposed. Albeit based on data of different satellite PM sensors with different spatial and temporal resolution, Figure 3 suggests, that after summer considerably less sea ice is found in the East Siberian Sea today compared to the early 1980's.

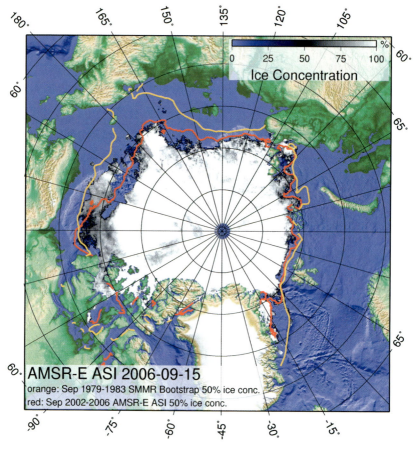

Fig. 3. C map of the Arctic as obtained with the ASI algorithm from AMSR-E 89 GHz data for Sep. 15, 2006. The red contour marks 50% ice concentration in the average C distribution for Sep. 2002–2006 (ASI algorithm applied to AMSR-E data). The orange contour shows the same for Sep. 1979–1983 (CBA applied to SMMR data). Missing data at the pole is flagged blue.

Only a few studies in literature deal with PM Baltic Sea ice properties at MW frequencies above 18 GHz (*e.g.* Kurvonen *et al.* 1996; Mäkynen and Hallikainen 2005). Standard methods to obtain sea ice parameters from satellite PM data are not directly applicable, because the low salinity influences the physical and MW sea ice properties and therefore cause different tie points. Moreover, PM sensors cannot compete with the high spatial resolution achieved with Synthetic Aperture Radar (SAR). The large FOV size (Table 1) in combination with the relatively small size of the Baltic Sea, and the high number of islands and fjords causes many mixed pixels (*i.e.* land and sea surface in one FOV). Sea ice parameter (and other parameter) retrieval for these pixels is substantially hampered (Bennartz 1999). In order to reduce the land influence the ASI algorithm (with tie points adapted to the Baltic Sea) can be applied to 85/89 GHz PM data to obtain C with fine spatial resolution (Table 1).

Figure 4 shows an example in comparison to an Aqua/MODIS image acquired at a wave-length of 670 nm in the Gulf of Riga. The C map based on PM data reveals the main ice-cover details as identified in the MODIS image. A synergy of satellite AM and PM data (*e.g.* Kouraev *et al.* 2004) can be advantageous, particularly along the coasts.

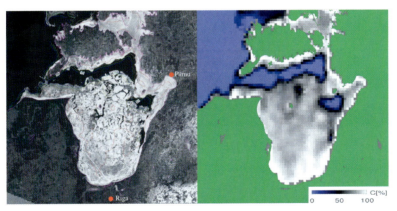

Fig. 4. Sea ice cover in the Gulf of Riga as observed simultaneously with MODIS at 670 nm, grid spacing: 250 m (left), and AMSR-E at 89 GHz, grid spacing: 3.125 km, using the ASI algorithm (right), Mar. 15, 2003. Grey pixels along the coast belong to the land mask.

3.2 Sea ice motion

Since 1979, a network of about 20 drifting buoys deployed on Arctic ice is maintained in the framework of the International Arctic Buoy Program

(IABP). These buoys provide continuous local measurements but with sparse coverage (Colony and Thorndike 1984). In order to obtain sea ice motion fields satellite data, *e.g.* of SAR (Kwok *et al.* 1990), or PM sensors (Kwok *et al.* 1998) can be used. PM data enables to obtain such fields globally and daily, which is an advantage compared to SAR (gaps due to limited data coverage), however, at limited spatial resolution (Table 1).

Existing methods are often based on tracking common features on pairs of sequential satellite maps and yield ice-motion fields at a scale of one to a few ten kilometres. One of these methods is the Maximum Cross Corre-lation (MCC) between successive and lagged maps (*e.g.* Kwok *et al.* 1998). The MCC method does not enable detection of rotation (Ezraty *et al.* 2007a). Another method is the wavelet analysis (Liu and Cavalieri 1998). In order to obtain independent ice motion vectors from PM data, at-tention has to be paid to the overlap between adjacent grid cells and the time-lag between sequential images. From AMSR-E 89 GHz data, drift vectors can be obtained every 31.25 km with a 2-day time-lag (SSM/I: 62.5 km, 3-day time-lag) (Ezraty *et al.* 2007b).

Figure 5 shows an AMSR-E drift map for Feb. 2–5, 2005. Gaps in this map are caused by a poor T_B contrast, hampering proper feature tracking.

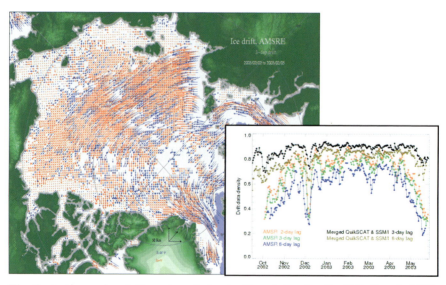

Fig. 5. Arctic sea ice drift vectors derived with a 3-day lag for Feb. 2 to 5, 2005, from AMSR-E 89 GHz data (both polarizations). Grid resolution is 31.25 km. Drift vectors of less than one pixel are marked with a dot. In red: identical drift for both polarizations, in blue: selection of one polarization. Insert: Times series of drift data density for winter 2002/03 for AMSR-E and merged SSM/I-QuikSCAT at lags of 2, 3 and 6 days.

This can be circumvented by combining PM data with AM data, *e.g.* with SeaWinds/QuikSCAT scatterometer data. Using the same MCC technique for both data sets, Ezraty *et al.* (2007a) produce a "merged" ice-motion product. In this product, the number of valid drift vectors is increased by 15%, particularly during early fall/spring (see Figure 5, insert). The "merged" product also contains fewer outliers than the individual products.

3.3   Sea ice type

The multi-frequency, dual-polarization capability of current and past satellite PM sensors (*e.g.* SMMR and SSM/I) enables to obtain the partial coverage of different ice types from satellite PM data, *e.g.* of MY ice (NORSEX Group 1983; Cavalieri *et al.* 1984) and of thin ice (TI) (Cavalieri 1994). Another method to identify and monitor TI is given by the Polynya Signature Simulation Method (PSSM) by Markus and Burns (1995). This method uses an iterative classification of resolution-enhanced SSM/I 37 and 85 GHz data to identify thick ice, TI and OW. Maps of the distribution of particularly OW and TI are used to study, *e.g.* polynya dynamics in relation to ice export (Kern *et al.* 2005).

4.   Summary and future perspectives

The physical and MW properties of sea ice enable to retrieve a number of sea ice parameters (concentration, drift, type) using data acquired by the satellite PM sensors SSM/I and AMSR-E. These sensors offer data coverage of both polar regions twice daily with a spatial resolution between about 5–15 and 50 km. The PM signature of sea ice can be characterized by its MW emissivity, which depends primarily on the salinity, and further on porosity, crystal structure, and ice-surface wetness and roughness (secondary parameters); for the Baltic Sea and other regions with brackish or fresh water the MW emissivity depends on the secondary parameters rather than on the salinity.

The sea ice concentration, C, is commonly retrieved from satellite PM data using an empirical relationship between C and the observed brightness temperature, T_B, together with tie points: typical T_B values of open water and 100% sea ice. The sea ice motion field can be derived using Maximum Cross Correlation to track common features on pairs of successive and lagged satellite PM maps; this is the most common technique. The present paper briefly describes a few methods and gives examples of the application of advanced C and ice-motion retrieval methods using SSM/I and AMSR-E data for the Arctic and the Baltic Sea. Note that the addressed algorithms and the shown examples give just a very limited view upon the various possibilities that exist using satellite PM data for sea ice parameter retrieval (Lubin and Massom 2006).

The present paper does not discuss the accuracy of the obtained products (see (Carsey 1992; Lubin and Massom 2006) for further reading). The validation of the products derived from satellite PM data requires quite a number of evenly distributed, reliable in-situ reference measurements. The associated large field-of-view size requires to involve a hierarchy of scales (a few meters to a few ten kilometres) into the validation of the product. Finally, the intra-annual variability of PM sea ice signature demands year-round validation. Consequently, most retrieval algorithms are validated with different, temporally and/or spatially limited data sets. A recent inter-comparison of different C retrieval algorithms using satellite PM data with independent data reveals, *e.g.* that almost all algorithms over-estimate the OW area compared to SAR (Andersen *et al.* 2007).

The synergy of PM and active microwave (AM) data would help to mitigate some of the shortcomings caused by the solely usage of satellite PM data for sea ice parameter retrieval (*e.g.* Kouraev *et al.* 2004; Ezraty *et al.* 2007b). Note, that albeit such a synergy would potentially improve the accuracy of current algorithms to obtain, *e.g. C*, it is not a substitute for a standardized evaluation of sea ice parameter retrieval algorithms (Andersen *et al.* 2007). Such an evaluation would help to more accurately quantify the effect of sea ice import, formation and melt on the northern European Seas and the European climate. Definitely required is such a synergy in regions like the Baltic Sea, where a large number of fjords and islands hampers sea ice parameter retrieval with satellite PM data substantially due to the mixed pixel effect (Bennartz 1999). The most promising C retrieval algorithm in this context using satellite PM observations is probably the ASI algorithm, because it uses 85/89 GHz data, which are provided at the finest spatial resolution possible with today's satellite PM sensors, and which are most sensitive for a discrimination of ice (a few millimetres thickness are sufficient) and water.

References

Andersen S, Tonboe RT, Kaleschke L, Heygster G, Toudal Pedersen L (2007) Inter-comparison of passive microwave sea ice concentration retrievals over the high concentration Arctic sea ice. J Geophys Res 112: co8004, doi:10.1029/2006JC003543

Bennartz A (1999) On the use of SSM/I measurements in coastal regions. Journal of Atmospheric and Oceanic Technology 16: 417– 431

Carsey FD (1992) Microwave Remote Sensing of Sea Ice. Geographical Monograph 68, American Geophysical Union, Washington, DC

Cavalieri DJ (1994) A microwave technique for mapping thin sea ice. J Geophys Res 99 (C6): 12561–12572

Cavalieri DJ, Gloersen P, Campbell WJ (1984) Determination of sea ice parameters with the NIMBUS-7 SMMR. J Geophys Res 89: 5355–5369

Cavalieri DJ, Parkinson CL, Vinnikov KY (2003) 30-year satellite record reveals contrasting Arctic and Antarctic decadal sea ice variability. Geophys Res Lett 30 (18): 1970, doi:10.1029/2003GL018031

Colony R, Thorndike S (1984) An estimate of the mean field of Arctic sea ice motion. J Geophys Res 89 (C6): 10623–10629

Comiso JC (1986) Characteristics of Arctic winter sea ice from satellite multispectral microwave observations. J Geophys Res 91: 975–994

Comiso JC, Cavalieri DJ, Markus T (2003) Sea ice concentration, ice temperature, and snow depth using AMSR-E data, IEEE Trans Geosci Rem Sens 41 (2): 243–252

Ezraty R, Girard-Ardhuin F, Poille JF (2007a) Sea ice drift in the central Arctic combining QuikSCAT and SSM/I sea ice drift data. User's manual, version 2.0, 2006. Available at IFREMER/CERSAT: *http://www.ifremer.fr/cersat/*

Ezraty R, Girard-Ardhuin F, Croize-Fillon D (2007b) Sea ice drift in the central Arctic using the 89 GHz brightness temperatures of the Advanced Microwave Scanning Radiometer. User's manual, version 2.0, April 2006. Available at IFREMER/CERSAT: *http://www.ifremer.fr/cersat/*

Fetterer F, Knowles K (2002, updated 2006) Sea ice index. Boulder, CO: National Snow and Ice Data Center. Digital media.

Kaleschke L, Heygster G, Lüpkes C, Bochert A, Hartmann J, Haarpaintner J, Vihma T (2001) SSM/I sea ice remote sensing for mesoscale ocean-atmosphere interaction analysis. Canadian J Rem Sens 27 (5): 526–537

Kern S (2004) A new method for medium-resolution sea ice analysis using weather-influence corrected Special Sensor Microwave/Imager 85 GHz data. Int J Rem Sens 25 (21): 4555–4582

Kern S, Harms I, Bakan S, Chen Y (2005) A Comprehensive View of Kara Sea Polynya Dynamics, Sea ice Compactness and Export from Model and Remote Sensing Data. Geophys Res Lett 32 (15): L15501, doi: 10.1029/2005GL023532

Kouraev AV, Papa F, Mognard NM, Buharizin PI, Cazenave A, Cretaux J-F, Dozortseva J, Remy F (2004) Synergy of active and passive satellite microwave data for the study of first-year sea ice in the Caspian and Aral Seas. IEEE Trans Geosci Rem Sens 42 (10): 2170–2176

Kurvonen L, Hallikainen M (1996) Classification of Baltic Sea Ice types by airborne multifrequency microwave radiometer. IEEE Trans Geosci Rem Sens 34: 1292–1299

Kwok R, Curlander JC, McConnell R, Pang SS (1990) An ice motion tracking system at the Alaska SAR Facility. IEEE J Oceanic Eng 15 (1): 44–54

Kwok R, Schweiger A, Rothrock DA, Pang SS, Kottmeier C (1998) Sea ice motion from satellite passive microwave imagery assessed with ERS SAR and buoy motions. J Geophys Res 103 (C4): 8191–8214

Liu AK, Cavalieri DJ (1998) On sea ice drift from the wavelet analysis of the Defense Meteorological Satellite Program (DMSP) Special Sensor Microwave Imager (SSM/I) data. Int J Rem Sens 19 (7): 1415–1423

Lubin D, Massom R (2006) Polar Remote Sensing – Volume I: Atmosphere and Oceans. Praxis Publishing Ltd, Chichester, UK

Mäkynen M, Hallikainen M (2005) Passive microwave signature observations of the Baltic Sea ice. Int J Rem Sens 26 (10): 2081–2106

Markus T, Burns BA (1995) A method to estimate subpixel-scale coastal polynyas with satellite microwave data. J Geophys Res 100 (C3): 4473–4487

Markus T, Cavalieri DJ (2000) An enhancement of the NASA Team sea ice algorithm. IEEE Trans Geosci Rem Sens 38 (3): 1387–398

Onstott RG, Grenfell TC, Mätzler C, Luther CA, Svendsen EA (1987) Evolution of microwave sea ice signatures during early summer and midsummer in the marginal ice zone. J Geophys Res 92 (C7): 6825– 6835

Pedersen LT (1998) Development of new satellite ice data products. In: Sandven S., et al., (Eds) IMSI Report No. 8, NERSC Technical Report 145, Nansen Environmental and Remote Sensing Center, Bergen, Norway

Perovich DK, Tucker III WB, Ligett KA (2002) Aerial observations of the evolution of ice surface conditions during summer. J Geophys Res 107 (C10): 8048, doi:10.1029/2000JC000449

Smith DM (1996) Extraction of winter sea ice concentration in the Greenland and Barents Seas from SSM/I data. Int J Rem Sens 17 (13): 2625–2646

Spreen G, Kaleschke L, Heygster G (2007) Sea Ice Remote Sensing Using AMSR-E 89 GHz Channels. J Geophys Res (in press)

Svendsen EA, Kloster K, Farrelly B, Johannessen OM, Johannessen JA, Cmpbell WJ, Gloersen P, Cavalieri DJ, Mätzler C (1983) Norwegian Remote Sensing Experiment: Evaluation of the Nimbus 7 scanning multichannel microwave radiometer for sea ice research. J Geophys Res 88: 2781–2791

Svendsen EA, Mätzler C, Grenfell TC (1987) A model for retrieving total sea ice concentration from a spaceborne dual-polarized passive microwave instrument operating near 90 GHz. Int J Rem Sens 8 (19): 1479–1487

Introduction to Microwave Active Techniques and Backscatter Properties

Peter Hoogeboom[1,2] and Ludvik Lidicky[2]

[1] Delft University of Technology, International Research Centre for Tele-
communications and Radar, Delft, The Netherlands
[2] TNO Defense, Security and Safety, The Hague, The Netherlands

Abstract. The present article introduces active microwave techniques that are used for remote sensing of the European seas, focusing on RAdio Detection And Ranging (RADAR) and Synthetic Aperture Radar (SAR). In section 1 the ranging principle and associated topics are introduced. As an example of this technique the radar altimeter is described. Radar altimeters employ the ranging technique for one-dimensional sea surface measurements from space. In section 2, two-dimensional imaging systems will be discussed. The addition of resolving power in a second dimension perpendicular to radar ranging leads to imaging devices that exist in many varieties. The major imaging systems that are in use today for remote sensing of the sea will be described, *e.g.* rotating beam radars, Side looking airborne radar, Synthetic Aperture Radar and polarimetric SAR. Section 3 deals with matters concerning radar backscatter: the basic concepts and definitions, speckle, and backscatter properties of the ocean. Furthermore, Bragg scattering, angular, wind and wave dependences of the backscatter, leading to the main oceanographic applications, are discussed.

1. Radar: range positioning

This section introduces the radar as an instrument used to measure distances. First, some theoretical background and basic definitions will be established. Then practical applications will be described.

1.1 Radar ranging principle

Imagine the situation in Figure 1: a transmitter TX sends a pulse in a direction of interest. The illuminated area is limited by the directive properties of the antenna. The pulse is reflected by a distant object. The antenna captures the

V. Barale, M. Gade (eds.), *Remote Sensing of the European Seas.*
© Springer Science+Business Media B.V. 2008

signal and passes it to the receiver RX, which filters the signal and samples the echo. The time t that has elapsed since transmission of the pulse can be determined from the output samples. The pulses travel at the speed of light c. The range r to the target can be found from t:

$$r = \frac{ct}{2},$$

(1)

where c is the speed of light. The radar pulse has to travel twice the range, from radar to object and back, hence the factor 2 in the formula. In its simples form, this radar is known as a non-coherent pulse radar. It measures amplitude and time-delay. The phase of the received signal is ignored.

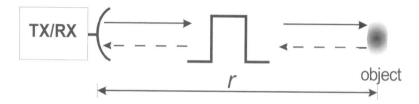

Fig. 1. Principle of the range measurement with a pulsed radar system.

1.2 Range ambiguities

To monitor changes in the object or to determine its speed by measuring Doppler shift, a coherent pulse-Doppler radar is required. Modern SAR systems, scatterometers and altimeters belong to this category. Such radars measure received amplitude and phase. Successive phase measurements from successive pulses reveal the Doppler frequency shift due to motion of the reflecting object.

The coherent pulse-Doppler radar transmits pulses at a rate called *Pulse Repetition Frequency* (PRF). Pulse repetition is also used to increase the signal-to-noise ratio (SNR) of the radar measurement by integration. The higher the PRF, the more SNR is achieved, and the higher is the maximum Doppler shift that can be measured. But a high PRF is a potential source of range ambiguities. An ambiguity occurs when reflections from two different distances appear at the same position on the radar timeline. Consider Figure 2. The radar timeline is indicated, with several transmitted pulses. In the top drawing, echoes A and B are found in response to the first transmitted pulse. They are received before the transmission of the second pulse.

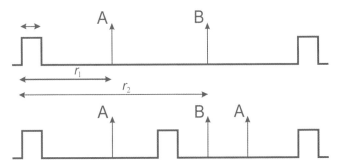

Fig. 2. Unambiguous (above) and ambiguous (below) range sampling with radar pulses.

In the lower part of Figure 2, the PRF is increased. Echo B is still from the first pulse, but the radar could erroneously interpret this echo as coming from the second pulse, hence at a shorter distance. In fact, if an echo is present at this shorter distance, it would interfere with the long distance echo B, causing a range ambiguity. To avoid this problem, one has to ensure that all significant targets are at a distance closer than the so-called *unambiguous range*

$$r = \frac{c}{2PRF} \tag{2}$$

For satellite systems, this is impossible. The distance between the satellite and the Earth is simply too large. Pulses cannot be sent at a PRF meeting the unambiguous range, because of other PRF requirements, *i.e.* a PRF that is high enough to sample all Doppler shifts due to the motion of the satellite unambiguously. However, by clever interpretation the unambiguous range can be compensated for the distance between the satellite and the first echoes received from the ground. As a result, the unambiguous range requirement in (2) can be relaxed. The range difference between the near and far end of the ground area that can be imaged unambiguously by the satellite, is limited by the required PRF and can be derived from (2).

1.3 Pulse compression and range resolution

To obtain high range resolution in radar, pulses with a wide spectrum must be transmitted. This can be achieved in two ways: The first way is easy to understand. One could transmit very narrow, in theory rectangular, pulses, which represent a wide frequency spectrum. This method is applied with brute force non-coherent radar systems. Figure 3 illustrates how two equal point size objects that are close together can be separated if the pulse length is in the order of the distance between these objects. The dip in the received signal is just large enough to separate the two objects. This determines the radar resolution. A practical way of defining the radar resolution is through the pulse width τ of the transmitted pulse, measured at the half-power points, as shown in Figure 3 right-hand side. The pulse width τ in seconds hence determines the resolution h in meters:

$$h = \frac{c\tau}{2} \tag{3}$$

For example, a pulse width of 10 nanoseconds corresponds to 1,5 meter resolution. The frequency bandwidth B of this pulse is approximately equal to $1/\tau$, or 100 MHz in this example. Note that this definition of resolution, which is commonly used in radar, differs from the resolution definition for optical systems. The latter is often based on the visibility of line pairs at a threshold contrast level. A difference up to a factor of 2 may occur.

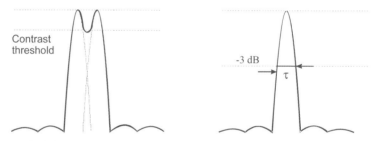

Fig. 3. Optical line pair resolution definition (left) applied to radar pulses and range resolution definition for a radar system (right).

The second way to achieve high resolution is to transmit longer pulses haped in such a way that they will have a broad bandwidth. The involved manipulation of the carrier wave implies the use of a coherent pulse-Doppler radar. An example of a common pulse shape used in radar is a *chirp*. Roughly speaking, as long as the bandwidth of the narrow pulse in the first example and the long, chirped pulse are the same, transmitted information (and thus the system resolution) will be the same.

To retrieve the resolution from the long, chirped pulse responses, an operation called *matched filtering* is needed. A matched filter performs a correlation of a received signal with a replica of the transmitted waveform, hence the term "matching". The matched filter output shows narrow sharp peaks at centre positions of the original pulses, see Figure 4. Since a matched filter transforms long pulses into short ones, this operation is also called *pulse compression*. Pulse compression is often used in coherent radar systems for power budget reasons. The average output power of the radar is important in achieving sufficient SNR. If a radar transmits shorter pulses to obtain finer geometrical resolution, the peak output power must be increased to maintain the same SNR. Due to technical limitations, a longer pulse of lower peak power is often preferred.

Fig. 4. Pulse compression with a matched filter (h); a long chirp (left) is converted to a narrow, high energy pulse (right).

A theory dealing with design of chirped pulses is called *waveform design* (Levanon 2004). Also, the pulse compression procedure influences system resolution. Mostly, the *Maximum Likelihood Estimation* (MLE) is used to perform the pulse compression. The resolution h obtained is given by equation 17.3, where again the pulse length τ is 1/B, the bandwidth of the pulse. As an alternative, *Minimum Mean Square Error Estimation* (MMSEE) can be used to enhance the resolution. However, MLE is easier to perform and it is more robust in the presence of noise (Haykin 2006).

1.4   Applications

A typical application of a ranging device in the remote sensing of seas is the altimeter. Its basic setup is provided in Figure 5. It is a down-looking coherent pulse-Doppler radar mounted on a satellite platform. The radar has a pencil-beam antenna which is pointed perpendicular to the Earth surface. The use of a narrow pencil beam enables application of the altimeter over oceans and seas, even close to the coast. This is of importance for applications in the European seas.

The radar transmits chirped pulses to obtain a high range resolution and sufficient SNR. Since sea water constitutes a very good reflector, the sea surface behaves as a mirror reflecting radar pulses back to the satellite. The incoming pulses are compressed as shown in Figure 4. Each range measurement provides an actual height of the instrument above the sea surface. If the position and distance of the satellite to the centre of the Earth are known, the measurement represents a sample of the water level at the Earth's surface. Calibration and correction for atmospheric delays are needed to bring the results to a level that is of interest to the user community. Nowadays, the accuracy of the measurements after correction and calibration is at millimetre level.

Fig. 5. Space borne radar altimeter measurement concept.

As the satellite orbits around the Earth, a complete 3-dimensional picture of the water level is created. In addition to the general height measurement, the altimeter provides average waveheight and windspeed. Waveheight is deduced from the shape of the reflected pulse, whereas the wind speed can be found from the backscatter level. Although more information can be retrieved, most often, altimeters are used to determine wave height and wind speed in addition to the primary height measurement.

2. Azimuth positioning, synthetic aperture radars

This section will take radar instruments from one dimensional ranging to the next level: two-dimensional radar sensing and signal processing in range and azimuth will be introduced. Here, azimuth means a direction perpendicular to the range direction, parallel to the antenna motion direction. Historically, the non-coherent radar systems used for range and azimuth positioning evolved from one-dimensional systems described above. By moving or rotating the antenna, adjacent azimuth locations are imaged in addition to the ranging. Finally, the requirement for higher resolution has led to the development of Synthetic Aperture Radar (SAR). A more complex coherent pulse-Doppler radar system is required for this application and also extensive digital processing capability is needed.

2.1 Rotating beam radars

Rotating beam radars are traditionally non-coherent radars that use a rotating antenna with a narrow beam width in the direction of rotation (azimuth). The image is formed by plotting the range measurements for each azimuth angle at the correct position in a 2-dimensional display. If the systems are used at long ranges, the azimuth resolution, which degrades with range, becomes noticeably crude. For a long time, rotating beam radars have been mainly used for (maritime) traffic applications, *e.g.* as ship radars to monitor the presence and movements of other ships. They are also used as coastal radars and as harbour radars and at airports to monitor the traffic situation. Nowadays rotating beam radars are also in use for remote sensing of the seas. Wave fields can be imaged by these radars and they can visualise surface slicks from oil spills etc. The radar itself being stationary, delivers an image at every rotation of the antenna. The resulting pile of 2D wave images can be analysed in a computer to provide wave spectra, water current and water depth information.

2.2 Side-Looking Airborne Radar (SLAR)

From a technical point of view SLARs strongly resemble the rotating beam radar. The antenna rotation is replaced by an antenna translation, by mounting a non-coherent radar on an aircraft in such a way that it looks to the side(s) of the aircraft. Again, a 2D image is created. SLAR systems are in use throughout Europe for monitoring coastal areas and seas. They are capable of detecting the positions of ships and can detect oil spills. In these applications the crude azimuth resolution at longer ranges is not a major drawback. The relatively low system cost and ease of data processing, when compared to SAR systems, make them a favourable choice.

2.3 Principle of SAR

The major drawback of SLAR systems, *i.e.* the crude azimuth resolution, is overcome by the SAR principle. In radar systems, the azimuth resolution is determined by the antenna footprint, which increases with range. To maintain sufficient resolution at long range, one could increase the antenna size. The larger the antenna, the smaller the antenna footprint, the finer the resolution. However, a fine resolution at a long range will place unrealistic requirements on the antenna diameter. Consider a satellite carrying a radar. Many satellites orbit at heights of about 800 km. At such a radar range, the antenna footprint will cover in azimuth several kilometres. To image ocean features, the azimuth resolution of such a system is unacceptable.

In order to circumvent this problem, the idea to *synthesise* a radar antenna aperture was suggested; hence the term *synthetic aperture radar*. The principle of a SAR system is the following: a moving, coherent pulse-Doppler radar takes range measurements along the path of its motion. These measurements are coherently processed to obtain a 2D image in range and azimuth direction. Since the azimuth measurements along the path are coherently combined, the same azimuth resolution is obtained as if a large antenna was used.

Another way of understanding the principle of SAR is to consider the change in Doppler frequency as the antenna moves along the objects to be imaged. For each object that appears in the beam, a positive Doppler shift is recorded as the radar approaches the object. In the middle of the beam the Doppler shift is zero, as the radar passes by. Finally, the Doppler shift is negative as the radar moves away from the object, until it is no longer within the scope of the radar beam. The radar has thus recorded a Doppler frequency chirp for each object. Through the use of the pulse compression technique previously described, this long chirp can be compressed to a narrow pulse.

Consider Figure 6. Suppose the radar moves in the azimuth direction and looks in the range direction at four point targets depicted on the right. The range positions of incoming pulses for each azimuth position are depicted on the left. The response for each object extends in range due to the long chirped pulses that are transmitted and in azimuth due to the long history that is recorded by the antenna. The goal of SAR processing is to obtain the focused points on the right from the data on the left. This task is achieved by means of 2D matched filtering. Observe that the shape of the signatures is range dependent. This complicates the processing. Various methods have been proposed (Carrara 1995, Soumekh 1999).

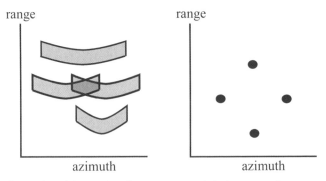

Fig. 6. Two-dimensional SAR signal responses and their processing.

The width of the signatures in azimuth depends on the length of the antenna: the longer the antenna, the narrower the beam, and the narrower the signature. This means fewer points to process, but also a reduced resolution. The maximum azimuth resolution that can be obtained with a side looking SAR system is approximately half the antenna length in azimuth direction. Apparently, with SAR there is no dependence of resolution on range or wavelength, which makes it a powerful imaging technique.

2.4   Azimuth ambiguities and moving objects in SAR

Azimuth ambiguities occur if the scatter in an image pixel is not uniquely related to one azimuth position. They are caused by strong reflections entering through the side-lobes of an antenna. Furthermore, moving objects are displaced from their actual position in the SAR image, which also can be considered as an ambiguity.

The SAR system is designed in such a way that stationary targets illuminated by its antenna's main lobe will be imaged unambiguously. Since PRF plays the role of a sampling frequency in the azimuth direction, one makes sure that this spatial sampling frequency is in a proper relation to the spatial extent of an illuminated area (radar footprint), *i.e.* that the Nyquist sampling criterion is satisfied. If a strong reflection enters through a side-lobe, it falls outside the intended spatial extent. This violates the Nyquist criterion and leads to aliasing. Normally, squinted targets are too weak to manifest themselves in the processed image because they are suppressed by the gain difference between the main lobe and side-lobes. Strong scatterers, however, are sometimes visible. They appear as misplaced objects or 'ghosts'.

Objects having a speed component in the direction of the radar are Doppler shifted. This Doppler shift leads to a squinted projection of the object in the SAR image. The squint angle is proportional to the object's speed. If its speed is low, the object is properly sampled and it mostly appears misplaced in the azimuth direction. Typical examples are ships separated from their wakes (see Figure 7). This example is taken from the first ESA ERS-1 image ever released, which was acquired on 27 July 1991. It shows an area of the North Sea, north of the Frisian island Terschelling in The Netherlands, with several ships and their wakes. The ships are slightly displaced from the wakes due to their speed. Note that in calm conditions the ship wake can be seen over distances of 20 km and more.

If the speed of an object is high, it will not be properly sampled by the radar. This is comparable to the previously discussed situation of the squinted object that entered through a side-lobe. In this case, the object will be smeared and is no longer visible in the SAR image.

Fig. 7. First (false colour) ERS-1 image of ship wakes in the North Sea near Terschelling, The Netherlands. © ESA 1991.

2.5 Polarimetric SAR

The single quantity measured by a radar is the strength of the backscatter. In remote sensing this quantity is used in many ways to obtain quantitative information from the sea. Sometimes more information about scattering mechanisms, shape and structure can be obtained by combining measurements from various polarisations. This can be achieved by a polarimetric SAR. Such a radar has two receiving antennas, which receive the vertical and the horizontal polarisation, respectively.

The transmitter alternatively sends out vertically and horizontally polarised waves. Consequently, all four polarisations (HH, HV, VH, VV) and their phase differences are measured and recorded in a scattering matrix. This matrix is the starting point for polarimetric analysis. Note that the information content of a polarimetric radar is four times the amount of a normal radar. Hence, the data stream is a challenge.

The polarimetric technology is nowadays also available on some satellite radar systems. The measurement of the cross polarised components is a challenging task, as the signal level is approximately a factor 10 smaller than the co-polarised. For this reason and also to limit the amount of data, some polarimetric modes are limited to a subset of the scattering matrix, for example HH, VV and their phase difference, or one cross polarisation (in theory HV equals VH) and one co-polarisation.

3. Radar backscatter

Microwave reflections are different from optical reflections, which the human eye is experienced in interpreting. Radar backscatter is often presented as an image and may at first glance resemble an optical picture, but it is not. If the interpreter does not realise this, the interpretation may be erroneous. In understanding microwave reflections, the wavelength should be taken into account. The relative size of the scatter cells (shapes and structures, relative to the wavelength) on an object determines the reflection behaviour, apart from conductive and dielectric properties. A scratched mirror is useless for reflecting an optical image. However, radio waves with a wavelength 1000 times longer will not be influenced by the scratches and will be reflected as if the mirror was perfect. Figure 8 illustrates several surface scattering situations, from forward scattering, occurring on smooth surfaces, to omni-directional scattering on rough surfaces.

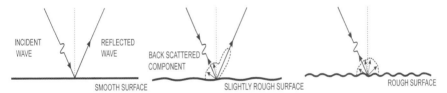

Fig. 8. Surface scattering mechanisms for smooth, slightly rough and rough surfaces.

3.1 Scattering definitions

Microwave scattering from objects may occur in various forms. The properties of the object determine what type of reflection occurs. A distinction can be made between single point target reflection, multiple-points reflection, also known as extended or surface scattering, and volume scattering. In discussing sea surfaces, the dominant scattering type is surface scattering, as the electromagnetic wave does not penetrate into the water. Ships, buoys and other objects in the sea will contribute target reflections consisting of one or more point targets. Depending on wavelength, sea ice can be penetrated by microwaves. In this case, not only surface scattering is important, but also the volume scattering from the volume that is illuminated by the radar radiation.

3.1.1 Radar cross-section

Consider the well known radar equation

$$P_{rec} = P_{tr} \frac{G^2 \lambda^2}{(4\pi)^3 R^4} \sigma \qquad (3)$$

where P_{rec} is received power, P_{tr} transmitted power, G the antenna gain, and λ the radar wavelength. σ stands for the radar cross-section. It is measured in square meters. The radar cross-section of an object can be thought of as the area of a perfectly conducting sphere that would give rise to the same level of reflection as observed from the object. Radars for remote sensing applications are usually calibrated in such a way that their outputs are radar cross-section numbers. For reasons of large dynamic range, the numbers are often given in decibels. This logarithmic compression reduces the dynamic range and enables easy calculations.

3.1.2 Sigma naught for surface scattering

A reflecting surface can be thought of as a collection of point targets.

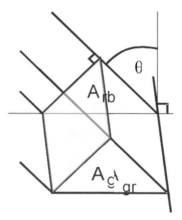

Fig. 9. Definition for ground referenced and radar beam referenced radar back-scatter coefficients.

The radar backscatter per unit area of this surface is known as sigma naught or σ_0. It is a dimensionless quantity, indicating the radar cross-section in square meters per square meter earth surface. This results in

$$\sigma_0 = \frac{\sigma}{A_{gr}} \tag{4}$$

A_{gr} is the resolution cell projected on the ground (Figure 9). σ_0 is mostly used with satellite systems. For sea skimming systems like ship radars another definition is sometimes used, based on the cross section of the radar beam A_{rb}, rather than the illuminated ground area A_{gr}. The backscatter coefficient γ can be easily related to σ_0:

$$\gamma = \frac{\sigma_0}{\cos \theta} \tag{5}$$

3.2 Speckle

Objects that have a larger size than the resolution of the radar will deliver several smaller reflections to the radar, which are then combined in the radar system into one reflection. Such reflections will possess special properties. The individual reflections are combined in amplitude and phase, known as coherent addition.

Two reflections with equal amplitudes, but opposite phase, will cancel each other out. If the reflections are in phase, the amplitude will double. Phase changes are largely due to small distance variations between the scattering object and the radar antenna. A moving sea surface gives rise to many phase changes. The radar reflection will show a continuously changing behaviour in amplitude and phase. This effect is known as speckle or clutter and should be separated from the mean radar backscatter value.

3.3 Sea backscatter

Several mechanisms contribute to the radar backscatter of the sea. The main effect is Bragg reflection. The backscatter level is influenced by wind, waves, currents and surface slicks, like oil pollutions or natural slicks. Other important effects are the angular dependence of radar backscatter, the dependence on wind speed and wind direction, and on polarisation. Finally, the moving sea surface decorrelates rapidly. Any coherent radar measurement process like SAR image formation will be hampered by the decorrelation. Nevertheless, many interesting features can be observed in the seas, especially in the European seas. These seas possess many delta areas with strong tidal currents and areas with layered structures of saline and fresh waters or water masses of different temperatures.

3.3.1 Bragg scattering

Bragg scattering is a resonant scattering mechanism, originally discovered and described by Bragg in 1913 for X-ray reflections from molecular grates in solid materials. The same mechanism is responsible for radar reflections on wind-generated short gravity-capillary waves. These waves can have wavelengths in the mm to dm range and increase to longer waves as a function of the fetch of the wind over the surface. The reflections are combined in the radar antenna. At this point, a coherent (amplitude and phase) addition of the reflections is performed. Those reflections that are perfectly in phase will lead to a high amplitude. All other reflections are cancelled out Figure 10 depicts the geometry.

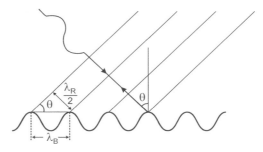

Fig. 10. Bragg resonant scattering on sea waves.

Electromagnetic waves with a wavelength λ_R fall onto the surface at an incidence angle θ. Capillary waves with a wavelength λ_B fulfil the resonance condition if

$$\frac{\lambda_R}{2} = \lambda_B \sin \theta. \tag{6}$$

Note that the radar wave will travel to the water surface and back, doubling the travelled distance, hence the factor 2. Wind is needed to image the ocean surface, because without the short gravity-capillary waves the surface will not be visible in the radar image.

3.3.2 Wind dependence

The radar backscatter level depends on the wind speed, because the spectral density for the Bragg resonant wave is related to the wind speed and the radar backscatter in turn is related to this spectral density. Consider a case where the radar is looking upwind and at a constant incidence angle. In this situation, a nearly perfect linear dependence is found between the radar reflectivity in dB (logarithmic scale) and the logarithm of the wind speed. The absolute value of the reflectivity depends on radar parameters, like polarisation and radar frequency, and on the viewing geometry.

3.3.3 Incidence angle dependence

The generalized radar backscatter dependence on incidence angle for moderate wind speeds is shown in Figure 11. The theoretical Bragg curves are shown. For extreme incidence angles (near nadir and near grazing), the Bragg resonant scattering mechanism is dominated by other effects.

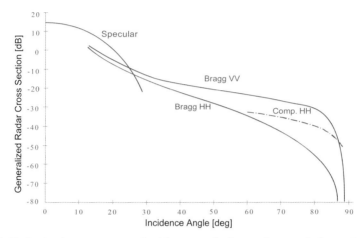

Fig. 11. Radar backscatter versus incidence angle for moderate wind speeds.

Specular reflections occur from $0°$ (nadir reflection) to $30°$. The upper value is in line with the maximum wave steepness of $60°$. This stems from the fact that normal sea waves, fulfilling Stokes laws, exhibit a minimum crest angle of $120°$. If this value is exceeded, the wave will break, which leads to strong spikes in the radar signal at large incidence angles, *i.e.* $80°$ to $90°$ (near grazing). In this angular region, multipath effects and shadowing occur as well. The average backscatter coefficient strongly reduces when the incidence angle approaches $90°$ or grazing.

3.3.4 Polarisation dependence

The conductive and dielectric properties of seawater lead to a distinct behaviour of the radar backscatter for various polarisations. The polarisation of an electromagnetic wave (Horizontal or Vertical) indicates the plane of the electric field of the wave. In the specular region ($0°–30°$), there is no noticeable difference between horizontal and vertical radar wave polarisation. In the Bragg dominated region ($30°–70°$), the vertical polarisation yields a stronger return signal. However, sea spikes and breaking waves generate strong returns at horizontal polarisation. At larger angles, above $60°$, these effects dominate the Bragg scattering as indicated by the dashed line in Figure 12. This line is based on composite models. At extreme angles near grazing, a cross-over occurs and the horizontal polarised signal is stronger than the vertical polarised Bragg scatter.

3.3.5 Decorrelation

When the radar return signal from a single resolution cell is monitored over some time, one may observe random variations in amplitude and phase, leading to incoherency of the time series. This decorrelation effect is of particular importance for SAR systems, as they achieve a high resolution by combining measurements from various antenna positions along a straight track during the SAR aperture time. If this aperture time exceeds the decorrelation time, the SAR image may not achieve the expected resolution in azimuth (antenna flight) direction. In radar range direction, the resolution is not altered. As a consequence, waves are sometimes not well imaged. They can be slightly rotated, broken up or blurred. Typical decorrelation times for the sea surface are 10–100 millisecond, depending on sea state, radar frequency and incidence angle. SAR aperture times vary with systems, resolution and observation distance. Typical values are 50 milliseconds to 1 second for satellite SAR systems.

3.4 Wind vector determination

Radar can be used to determine the wind vector over open seas, *i.e.* away from the coast. There are various methods, based on SAR, radar altimetry and wind-scatterometry. The latter is the usually applied and most accurate method. It is based on the measurement of wind speed and wind direction using a multi-beam radar instrument.

The radar backscatter level not only depends on wind speed, but also on wind direction with respect to the radar look direction. Figure 12 shows the general azimuth dependence. At upwind (the radar looks into the wind), the highest backscatter occurs. Downwind values are also high, but slightly lower than the upwind value. At crosswind positions, minima occur in the curve. A 3 beam scatterometer, as used by ESA on their satellites, with a $45°$ azimuth angle separation between the beams, is a good solution to determine the wind vector, even though an $180°$ ambiguity may occur, *e.g.* between the 2 crosswind positions in Figure 12. Such ambiguities can be resolved from a regional wind field analysis.

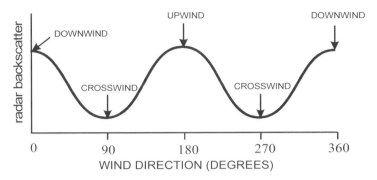

Fig. 12. Azimuth dependence of radar backscatter for the sea surface.

3.5 Modulation processes

The wind generated Bragg waves are easily influenced by external factors like sea waves, bottom topography, slicks and precipitation. This modulation leads to their visibility in SAR images. The ocean wave imaging and bottom topography visualisation will be discussed.

3.5.1 Long gravity waves

Long gravity sea waves introduce several modulation effects in the radar backscatter, leading to a generally good visibility of such waves in radar images. The main effects are slope modulation and hydro-dynamical modulation. Slope modulation is caused by the change of the local incidence angle of the radar wave due to the local ocean wave inclination. The effect depends on the average incidence angle. This modulation effect has a strength in the order of a few dB.

Hydro-dynamical modulation is caused by the motion of the surface due to the passing sea waves. This motion will lead to a stretching and straining of capillary wave fields, thereby modifying their amplitudes and thus the radar backscatter. Furthermore, this motion of the surface may lead to Doppler velocity shifts in the radar signal, which can lead to a non-linear mapping of the sea wave in the radar image. The effect can be compared to the displacements of ships from their wake due to their velocity.

3.5.2 Bottom features

Coastal seas with limited depth and delta areas can express bottom features under certain conditions in the radar image, even though the radar wave does not penetrate into the water. These effects were first discovered in

SLAR images over the North Sea by De Loor (De Loor and Brunsveld van Hulten 1978). The mechanism of bottom topography imaging is explained in Figure 13.

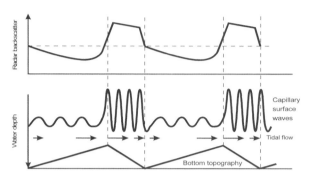

Fig. 13. Sea bottom topography imaging mechanism.

Consider a sea surface where the wind is generating capillary waves. A tidal water current moves a water body over the undulated shallow water (depth less than approx 25 meter). Water particles will accelerate over the underwater dunes and decelerate over the valleys to maintain a constant total water flow. On the surface, these changes in local water current will modulate the capillary waves. During accelerations they are stretched, lowering their amplitude. This causes a reduced radar backscatter. When decelerating, the capillary waves are compressed, thus intensifying the radar backscatter. As the disturbed capillary wave spectrum quickly decays to the equilibrium spectrum, the effect in the radar image also quickly disappears as indicated in the upper part of Figure13.

References & bibliography

Carrara WG, Goodman RS, Majewski RM (1995) Spotlight SAR: Signal processing algorithms. Artech House

De Loor GP, Brunsveld Van Hutten HW (1978) Microwave measurements over the North Sea. Bound-layer Met 13: 119–131

Levanon N, Mozeson E (2004) Radar Signals. Wiley & Sons

Haykin S (2006) Adaptive radar signal processing. Wiley & Sons

Henderson FM, Lewis AJ (1998) Principles and applications of imaging radar. Wiley & Sons

Skolnik MI (2001) Introduction to radar systems. McGraw-Hill

Soumekh M (1999) Synthetic aperture radar signal processing. Wiley & Sons

Ulaby FT, Moore RK, Fung AK (1981–1986) Microwave remote sensing. vol. I, vol II, Addison-Wesley, vol. III, Artech House

Scatterometer Applications in the European Seas

Ad Stoffelen

KNMI, AE de Bilt, the Netherlands

Abstract. The EUMETSAT Advanced Scatterometer ASCAT on MetOp-A was launched on 19 October 2006 as the third wind scatterometer currently in space joining up with the ESA ERS-2 and the NASA SeaWinds scatterometers. Scatterometers measure the radar backscatter from wind-generated cm-size gravity-capillary waves and provide high-resolution wind vector fields over the sea. Wind speed and wind direction are provided with high quality and uniquely define the mesoscale wind vector field at the sea surface. The all-weather ERS scatterometer observations have proven important for the forecasting of dynamical and severe weather. Oceanographic applications have been initiated using winds from Sea-Winds on QuikSCAT, since scatterometers provide unique forcing information on the ocean eddy scale. Together, ERS-2, ASCAT and SeaWinds provide good coverage over the oceans and are now used routinely in marine and weather forecasting. In the coming years, further progress in high resolution processing, closer to the coast, and with improved geophysical interpretation is expected.

1. Introduction

The all-weather capability of a scatterometer provides unique wind field products of the most intense and often cloud-covered wind phenomena, such as polar front disturbances and tropical cyclones (for example, see Figure 1). As such, it has been demonstrated that scatterometer winds are useful in the prediction of extra-tropical cyclones (Stoffelen and Beukering 1997) and tropical cyclones, *e.g.* Isaksen and Stoffelen (2000). Moreover, the high-resolution near-surface winds as provided by scatterometers are very relevant in driving the ocean circulation (*e.g.* Chelton *et al.* 2003; Liu and Xie 2006), which in turn plays a major role in the climate system and in water life (*e.g.* fishery).

V. Barale, M. Gade (eds.), *Remote Sensing of the European Seas.*
© Springer Science+Business Media B.V. 2008

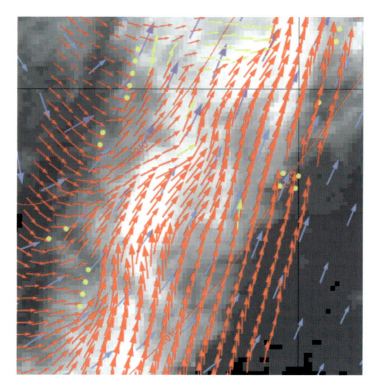

Fig. 1. ERS-2 scatterometer winds (red) on 28 August 2006 13:00 Z showing a train of atmospheric waves in the North Atlantic at 25W and 40N. Yellow arrows and dots are quality-flagged ERS-2 scatterometer cells. The blue and purple arrows depict simultaneous background Numerical Weather Prediction, NWP, model winds (KNMI HiRLAM) that generally do not resolve such weather phenolmena. The METEOSAT Infra-Red background image is consistent with the scatterometer surface winds. © EUMETSAT (from www.knmi.nl/scatterometer). The missed Rossby train resulted in a bust NWP forecast the next day in the Netherlands and England.

At the moment the ESA ERS-2 and the NASA SeaWinds scatterometer on QuikSCAT provide respectively a regional quasi real-time and a global near-real time data stream. EUMETSAT continues the global scatterometer mission with the ASCAT scatterometer on EPS/METOP launched 19 October 2006, and started a regional ASCAT dissemination at a 30-minute timeliness and a near-real time global service through KNMI. As such, continuity of both services is likely provided to the operational meteoro-logical community for another period of 15 years. In addition, high resolution surface wind fields will be maintained for the oceanographic community. Some main differences between these scatterometer systems are discussed in next section.

Specific challenges in scatterometer data interpretation are outlined thereafter and include:

- the non-linear relationship between backscatter and wind;
- lack of physical understanding and the consequent use of empirical modelling;
- rain contamination in case of K_u-band scatterometers such as SEASAT, NSCAT or SeaWinds;
- wind direction ambiguity.

Scatterometer wind processors have been developed that tackle all these problems with success. The characteristics of these wind products will be described before the scatterometer capabilities on sea ice detection are eluded to. Another capability of a scatterometer not described here concerns soil moisture (Wagner *et al*. 2003).

2. Scatterometer instruments

The principle of radar backscattering from a water surface is described elsewhere in the present volume. In Stoffelen (1998a) an overview is given of the different scatterometer systems, where both measurement geometry and radar wavelength and polarisation determine the wind information content. Besides these existing systems study reports exist of prognostic systems, such as a rotating fan beam scatterometer (Lin *et al*. 2003).

The current side-looking triple-fan-beam scatterometers exhibit a nadir gap, while rotating pencil-beam scatterometers do provide winds at the subsatellite track, albeit of reduced quality. Alternative fan beam concepts, such as a rotating fan beam scatterometer (Lin *et al*. 2003), can provide broad swath winds as well, but have not yet been build.

Current operational scatterometers use the C- and K_u-band radar wavelengths of respectively 5 and 2 cm. Both bands provide adequate wind vector sensitivity, but Ku band wavelengths provide substantial rain column attenuation (Portabella and Stoffelen 2001), while C band is basically immune to this. Less often, at very extreme and large-scale rain fall or at very low winds both Ku and C bands exhibit loss of the wind signal due to rain impact on the water surface. A third determining issue on scatterometer design represents the use of polarisation. ASCAT and ERS scatterometers use only vertical copolarisation (VV), while SeaWinds also uses the horizontal copolarisation (HH). The HH signal is weaker, but does not saturate for extreme winds (40 m/s), whereas VV does. To measure these hurricane-force winds, HH co-polarisation measurements thus seem essential.

3. Geophysical interpretation

Triple-fan-beam scatterometers such as ERS and ASCAT provide three backscatter measurements at each Wind Vector Cell, WVC, across the swath and which may be represented on a conical surface in 3D measurement space as shown in Figure 2 (Stoffelen and Anderson 1997a). The elonguated dimension of this surface is due to wind speed variations along the orbit causing quasi-linear backscatter variations, while the circular dimension represents wind direction sensitivity. The combination of a quasi-linear and circular dependency on respectively wind speed and direction, provides a rather smooth sensitivity to the WVC-mean wind vector, while other geometries such as a conical-scanning pencil beam have less favourable wind vector sensitivities, like, *e.g.*, on SeaWinds. The variation in the sum of the wind vector sensitivities at each radar beam, *i.e.*, the total wind vector sensitivity, has a direct effect on wind retrieval (inversion) capability as explained in detail by Stoffelen and Portabella (2006). Figure 3 (right panel) depicts relative peaks and throughs in the retrieved wind direction distribution due to instrument geometry.

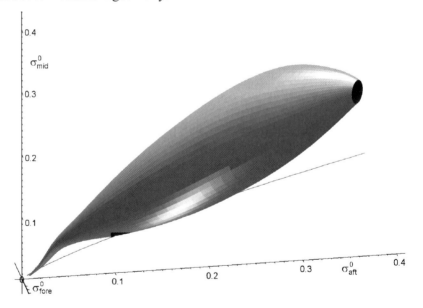

Fig. 2. ERS scatterometer mid swath (Wind Vector Cell 11) conical model surface spanning along wind speeds from 0–40 m/s, from left to right, and with wind directions along the radar mid beam, at the top and across the mid beam below. The visible outer surface is close to a hidden inner conical sheet, resulting in a 180 degrees wind ambiguity. The line is associated with sea ice backscatter measurement triplets which vary along the line for variable ice age.

ERS scatterometer backscatter measurements over the globe closely follow the conical surface in backscatter space (Figure 2) and the scatter of triplet points perpendicular to this cone generally represents only an equivalent vector RMS error of 0.5 m/s. Since two parameters are sufficient to describe a surface, it empirically follows that the amplitude and direction of the ocean forcing largely determine radar backscatter. This forcing is expressed as surface stress or friction velocity in atmospheric surface layer models. Besides from scatterometers, no other well-calibrated methods exist to measure the large-scale sea surface forcing. This poses a real problem in developing and validating a so-called Geophysical Model Function (GMF), relating backscatter to wind. Moreover, theoretical physical developments do not achieve the radiometric accuracy of the instrument of 0.2 dB (*e.g.* Janssen *et al.* 1998). A solution to this problem has been to develop an empirical radar backscatter GMF with real winds (Stoffelen and Anderson 1997b) or equivalent neutral winds as input.

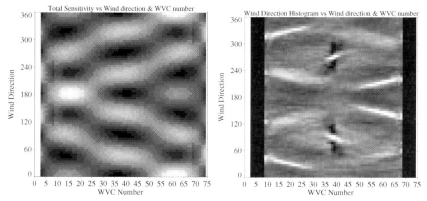

Fig. 3. Left: SeaWinds total wind vector sensitivity as a function of WVC number and wind direction for a wind speed of 8 m/s. The sensitivity values are plotted in grey scale, ranging from black (troughs) to white (peaks). Right: Retrieved relative wind direction distribution as a function of WVC number and ECMWF 12-hourly predicted wind direction, appearing in a unnatural pattern associated with total sensitivity (left). The retrieved SeaWinds distribution is divided by the ECMWF distribution to compute the relative distribution. Bins below values of 0.6 (min at 0.2) are black and white occurs above 2.0 (max at 5.6), with a logarithmic grey scale in between. Only WVCs 9–68 are processed (Stoffelen and Portabella 2006).

The GMF is inverted to obtain the wind vector from a set of radar cross section measurements in a particular WVC (Stoffelen and Portabella 2006). The GMF winds, such as from C band model CMOD5 (Hersbach *et al.* 2007), and long series of comprehensive buoy measurements provide a good basis to intercompare scatterometer, Numerical Weather Prediction (NWP), model wind and buoy wind or stress data in a triple collocation exercise (Stoffelen 1998b), a sophisticated methodology to determine random NWP model and observation errors and, at the same time, the systematic geophysical wind retrieval error (calibration) with respect to the *in situ* buoy winds. Figure 4 illustrates the relationship between CMOD5 winds, real winds, equivalent neutral winds, and friction velocity (Portabella and Stoffelen 2006) that has been found after triple collocation in both the tropical and extratropical regions.

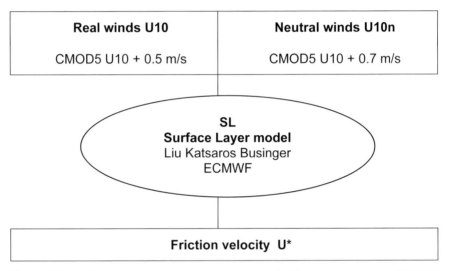

Fig. 4. Schematic of recommended scatterometer wind/stress conversion. CMOD5 winds at 10m height are used as basis. Either real or neutral 10m winds may be transformed to stress by either Liu-Katsaros-Businger, ECMWF or any similar surface layer model (Portabella and Stoffelen 2006).

In many applications it is the relative motion between ocean and atmosphere that is relevant, *e.g.*, for ship warnings and air-sea interaction. *In situ* data, however, measure the absolute motion with respect to an earth reference. Scatterometers do measure the relative motion (friction velocity) between atmosphere and ocean as depicted in Figure 5 (Kelly *et al.* 2001).

The spatial consistency and detail in scatterometer sea surface winds is unprecedented. For many applications, however, such as storm surge and wave prediction, marine warnings, ocean forcing, etc., NWP analysis winds are used as input, lacking mesoscale detail. For real-time marine and oceanographic research applications it is of interest to characterise the differences between the scatterometer and NWP products. Figure 6 shows the mean wind stress curl from SeaWinds and from the NCEP NWP model,

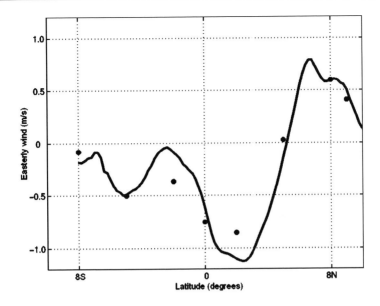

Fig. 5. Close correspondence between average ocean current measurements from ship cruises in 1999 (line) and difference between TAO buoy and QuikSCAT winds (dots). Reproduced from Kelly *et al.* 2001.

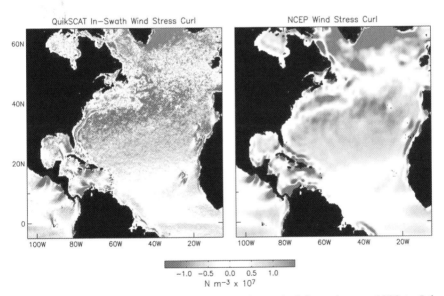

Fig. 6. 4-year average wind stress curl over the period from August 1999 to July 2003 as computed from SeaWinds scatterometer winds (left) and the NCEP operational NWP model (right). Adapted from Chelton *et al.* 2004.

illustrating the difference in scales. Note the sharp features in the scatterometer panel near islands, in the coastal region, *e.g.*, Gulf of Mexico, and near the Gulf Stream at the US east coast (Chelton *et al.* 2004; Zechetto and Cappa 2001). The wind stress curl depicts the main atmosphere-ocean interaction and the smaller scales indicate that the ocean eddies of typical 10-km size are forced much more effectively. In next section a further elaboration on the spatial wind spectrum is given.

4. Scatterometer wind products

EUMETSAT set up Satellite Application Facilities (SAFs) providing software and data products and services, with SAF scatterometer activities coordinated by KNMI. An overview of the scatterometer wind processing steps is given in Figure 7. Scatterometer sea surface wind research and development in these areas lies at the basis of wind product innovation:

- quality control, rain (mainly for K_u-band) and ice screening;
- spatial filtering to reduce noise and enhance wind quality;geophysical modeling of the radar signal and its noise properties;GMF non-linear inversion;
- determination of information content;
- ambiguity removal (spatial filter for a unique wind vector field);
- real-time and archive wind and stress processors; and
- active web monitoring and quality assurance.

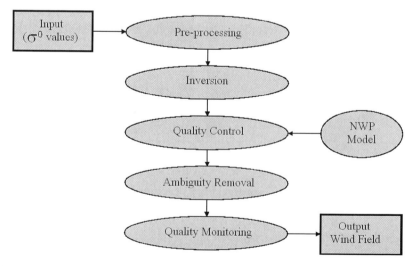

Fig. 7. Overview of scatterometer wind processing from the basic backscatter σ_0.

Wind product and service enhancement are the main goals of these activities. KNMI currently processes a global demonstration 25-km product, a global OSI SAF QuikSCAT 100-km product, an experimental SeaWinds 25-km product, and a North-Atlantic ERS-2 25-km product, and distributes it to the international meteorological community. Moreover, ASCAT and ERS-2 scatterometer winds are available in quasi real-time through the EUMETSAT ASCAT Early Advanced Re-transmission Service (EARS).

A visual presentation of these products is available[1], both in vector and flag presentation, together with product monitoring information. Global maps of wind speed are provided over the last 22 hours (as in Figure 8), segregated in ascending and descending orbit tracks and stored for 5 days. By regional selection on these maps more detailed regional plots become available (as in Figure 1 and Figure 9). The link also provides documentation, scientific publications, and software products.

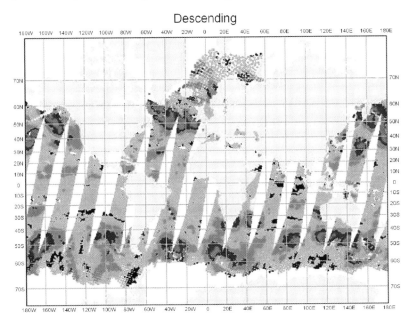

Fig. 8. Grey-coded SeaWinds 100-km wind speeds. (www.knmi.nl/scatterometer)

The standard KNMI 100-km QuikSCAT product has been developed for Numerical Weather Prediction (NWP) assimilation and it is verified to compare best with independent ECMWF NWP winds and is thus suitable for NWP assimilation. At higher resolutions more random noise is present

---

[1] www.knmi.nl/scatterometer

in SeaWinds data (Vogelzang 2006). SeaWinds noise reduction at 25 km resolution is achieved by implementing the so-called Multiple Solution Scheme (MSS; Portabella and Stoffelen 2003). Here the improvement is brought by using wind vector probability information, as computed in the inversion step, in combination with 2D-VAR spatial constraints on rotation and divergence of the resulting wind field, *i.e.* meteorological balance constraints. The improvements in MSS are mainly due to the reduction of erratic noise; while coherent mesoscale structures remain present and become more visible due to the noise reduction. In Figure 9 the MSS 25-km product is shown on the right. It is clear that important mesoscale details, potentially useful for short range weather forecasting and nowcasting, are added.

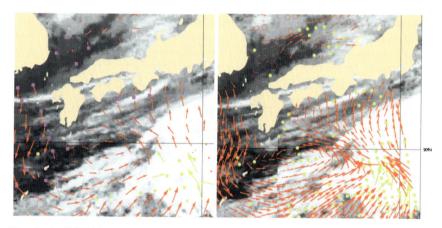

Fig. 9. QuikSCAT 100-km (left) and corresponding 25-km (right) wind product showing additional mesoscale detail. Yellow dots denote internal inconsistency, while yellow arrows fail a spatial consistency check. Collocated GOES IR cloud imagery is provided underneath in grey. Data around Japan 24.9.2006 0600 UTC.

A general quality of the different wind data sources lies in the represented spatial scales. NWP models lack scales below 100 km, scatterometers present scales down to 25 km and *in situ* measurements are local. On the other hand, in terms of application, a local wind is only locally useful, while a field of cell-averaged winds comprehensively represents the conditions over a large area. As such, these wind observations cannot be mutually compared without taking into account spatial representation and the common wind variability spectrum. Stoffelen (1998b) takes this into account in a triple collocation comparison method and provides wind observation error measures. Table 1 presents recent values obtained from triple collocation studies, showing the good quality of NWP data in the extratropics, but relatively poor quality in the tropics.

Buoy data appear rather uncertain on the scatterometer scale, since the error is computed for the mean wind in a 50-km scatterometer cell. As such the wind variability on scales smaller than 50 km is considered part of the buoy wind error.

Table 1. Vector error estimates of buoy, scatterometer and ECMWF model winds by triple collocation method for a data set collocated with tropical buoys, second column, and extra-tropical buoys, last column (Portabella and Stoffelen 2006; Hersbach *et al.* 2007).

Vector RMS error [m/s]	TAO/PIRATA buoys Tropical	NDBC/MEDS/UKMO buoys Extra-tropical
Buoy	1.5	1.5
Scatterometer	1.2	1.6
ECMWF model	2.0	2.1

5. Sea ice

The ERS scatterometer backscatter measurement space has proven to be extremely useful for the derivation of a Geophysical Model Function (GMF) over water surfaces and for tuning the GMF inversion algorithm to produce winds. Over other types of surface, a similar analysis methodology may be performed. In the measurement space sea ice points lie close to a line (grey line in Figure 2), indicating a single geophysical parameter dependency. The C-band sea ice GMF can thus be modeled by a dependency on incidence angle and one geophysical parameter, called ice age a (de Haan and Stoffelen 2000). The in the sea ice community commonly used $\sigma^0(\theta = 40°)$ is then equal to GMF(a, $\theta = 40°$); parameter a is plotted in Figure 10. Since wind and ice points can generally be well discriminated in measurement space (Figure 2), KNMI uses both the sea ice and wind GMF for a Baysian water/ice discrimination in its scatterometer wind products.

Fig. 10. Illustration of map of retrieved sea ice GMF parameter a (ice age) in grey around the North Pole in winter 1998 for subsequently overlaid orbits. While swath patterns are clearly visible over water surfaces, the ice age retrieval over sea ice is very consistent for all WVCs and no lines due to swath edges appear here.

6. Outlook

With the launch of ASCAT the series of C-band scatterometers is continued with now improved coverage of the ocean surface wind and ice in the current two decades. EUMETSAT provides timely and free user services in collaboration with KNMI for the SeaWinds, ERS-2 and ASCAT scatterometers. The ASCAT 25-km sampled wind product has been released in early 2007 as a demonstration data stream meeting the main wind user requirements. Research and development will continue within the EUMETSAT SAFs to enhance the wind processor capability towards a 12.5-km sampled product, nearer to the coast, and with improved geophysical interpretation. In the NWP SAF methods will be tested within 2D-VAR for high resolution scatterometer data assimilation.

Timely, highly accurate, spatially consistent and broad swath scatterometer winds provide mesoscale structures over water surfaces. Since these structures are relatively short-lived, sampling twice a day from one polar satellite is rather limited. Although concrete plans exist in China and India to launch a scatterometer, no guarantee yet exists for improved temporal sampling in near-real time. Reinforced international collaboration between space agencies is sollicited to strengthen the ocean surface wind mission.

Acknowledgements

The products described in this chapter are developed within the KNMI scatterometer team that is supported by the EUMETSAT SAFs. The NASA Ocean Wind Vector Science Team with NOAA participation and the ASCAT Science Advisory Group chaired by EUMETSAT/ESA support the international scatterometer community.

References

Chelton DB, Schlax MG, Freilich MH, Milliff RF (2004) Satellite measurements reveal persistent small-scale features in ocean winds. Science 303: 978–983

de Haan S, Stoffelen A (2000) Ice Discrimination Using ERS scatterometer, OSI SAF R&D report, http://www.knmi.nl/scatterometer/publications/pdf/SAFOSI _W_icescr-knmi.pdf

Hersbach H, Stoffelen A, de Haan S (2007) An improved C-band scatterometer ocean geophysical model function: CMOD5, J Geophys Res 112 (C3): doi:10.1029/2006JC003743

Isaksen L, Stoffelen A (2000) ERS-Scatterometer Wind Data Impact on ECMWF's Tropical Cyclone Forecasts. IEEE T Geosci Rem Sens (special issue on Emerging Scatterometer Applications) 38 (4): 1885–1892

Janssen PAEM, Wallbrink H, Calkoen CJ, van Halsema D, Oost WA, Snoeij P (1998) VIERS-1 scatterometer model. J Geophys Res 103 (C4): 7807–7831

Kelly KA, Dickinson S, McPhaden MJ, Johnson GC (2001) Ocean currents evident in satellite wind data," Geophys Res Lett 28 (12): 2469–2472

KNMI scatterometer site, http://www.knmi.nl/scatterometer

Lin CC, Stoffelen A, de Kloe J, Wismann V, Bartha S, Schulte HR (2003) Wind retrieval capability of rotating, range-gated, fanbeam spaceborne scatterometer. SPIE Proc 4881: 268–279

Liu WT, Xie X (2006) Measuring ocean surface wind from space. In: Gower J (ed) Remote Sensing of the Marine Environment, Manual of Remote Sensing,

3rd Edition, vol 6, Amer. Soc. for Photogrammetry and Remote Sensing, pp 149–178

Portabella M, Stoffelen A (2006) Development of a global scatterometer validation and monitoring. Visiting Scientist report for the EUMETSAT OSI SAF; http://www.osi-saf.org

Portabella M, Stoffelen A (2001) Rain detection and quality control of SeaWinds. J Atm and Ocean Techn 18 (7): 1171–1183

Portabella M, Stoffelen A (2003) A probabilistic approach for SeaWinds data assimilation. Quart J R Met Soc 130: 127–159

Stoffelen A, Portabella M (2006) On Baysian Scatterometer Wind Inversion. IEEE Trans Geosci Rem Sens 44: 1523–1533

Stoffelen A (1998a) Scatterometry", thesis RUU, 1998; http://pablo.ubu.ruu.nl/~ proefsch/01840669/ inhoud.htm

Stoffelen A (1998b) Error modeling and calibration: towards the true surface wind speed. J Geophys Res 103 (C4): 7755–7766

Stoffelen A, Anderson d (1997a) Scatterometer Data Interpretation: Measurement Space and inversion. J Atm and Ocean Techn 14 (6): 1298–1313

Stoffelen A, Anderson D (1997b) Scatterometer Data Interpretation: Transfer Function Estimation and Validation. J Geophys Res 102 (C3): 5767–5780

Stoffelen A, van Beukering P (1997) The impact of improved scatterometer winds on HIRLAM analyses and forecasts. BCRS study contract 1.1OP-04, HIRLAM technical report 31, published by IMET, Dublin, Ireland

Vogelzang J (2006) On the quality of high resolution wind fields. Report for the EUMETSAT NWP SAF, NWPSAF-KN-TR-002, Version 1.2; http://www.metoffice. gov.uk/research/interproj/nwpsaf

Wagner W, Scipal K, Pathe C, Gerten D, Lucht W, Rudolf B (2003) Evaluation of the agreement between the first global remotely sensed soil moisture data with model and precipitation data. J Geophys Res 108 (D19): 4611

Zecchetto S, Cappa C (2001) The spatial structure of the Mediterranean Sea winds reveale*d by ERS-1 scatterometer. Int J Rem Sens 22 (1): 45–70

Radar Altimetry: Introduction and Application to Air-Sea Interaction

David Kevin Woolf [1,2] and Christine Gommenginger [2]

[1] Environmental Research Institute, North Highland College, UHI Millennium Institute, Thurso, United Kingdom
[2] Laboratory for Satellite Oceanography, National Oceanography Centre, Southampton, United Kingdom

Abstract. Radar altimeters are among the more common satellite-borne Earth Observation instruments with a long history including continuous data since 1991. They are also exceptionally versatile providing information on sea level, ocean dynamics, wind speed and a number of wave parameters. Radar altimetry is a "point" rather than a "swath" instrument so that sampling resolution by a single altimeter is relatively poor in both space and time. Also the sizeable footprint restricts use of altimetry to greater than 10 km from any coast, with some restrictions at a greater distance. The sampling limitations do not negate the usefulness of altimetry, either in isolation or in combination with other instruments or modelling. Climatologies of sea level and wave parameters built from altimetry provide a unique perspective on both a regional and global scale. Measurements of individual storm events provide a test for both wave and storm-surge modelling and could be useful in Near Real Time applications. New interpretations of satellite altimeter waveforms and dual-frequency data, and new advanced altimeter concepts, continuously present new products to broaden the geophysical applications of radar altimetry.

1. Introduction

Radar altimeters are one of the most versatile satellite-borne instruments and altimeter data is extensive (due to many lengthy satellite altimeter missions) and is readily available. Satellite altimetry began in the 1970s with Skylab. GEOSAT provided the first multi-year data set in the late 1980s and there has always been at least one and up to five functioning satellite altimeters since the early 1990s. Satellite altimetry has been influential in the progress of oceanography (Fu and Cazenave 2001) though its application

V. Barale, M. Gade (eds.), *Remote Sensing of the European Seas.*

to near-shore studies has been less extensive. Here we describe the merits of satellite altimetry as a tool for the study of the European Seas.

Altimeters view the surface below at vertical incidence and measure the signal returned from a small footprint below. All current satellite-borne radar altimeters are a form of pulse-limited radar altimeter. An idealized version of the emitted pulse and the return waveform is shown schematically in Figure 1. The effective footprint diameter may be as little as 2 km (the theoretical limit for a flat sea) but increases to 5.5 km for a 3-metre swell. Standard "Geophysical Data Records (GDR)" contain one-second averages for each parameter, corresponding to a surface path of 6 km. If changes in wave heights occur on the scale of 1 km or less these will be undetected by satellite altimeter. Also, if land appears within a footprint the data is useless. Satellite altimeters also have a problem "locking in" to the sea surface immediately after the surface track leaves land with the result that data up to several 10 s of km from the shore is unreliable. These issues are less critical for more recent altimeters such as ENVISAT RA-2.

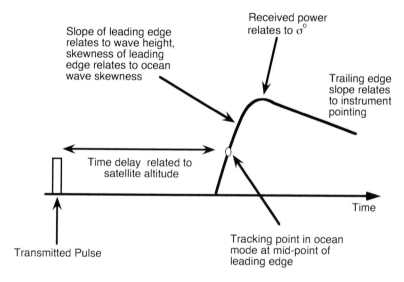

Fig. 1. Idealization of altimeter pulse and return waveform, labeled with features of the waveform and related geophysical variables.

A second set of problems arises when compiling a climatology, when, rather than just the reliability of individual measurements, the adequacy of sampling is a major issue. At any instance, an altimeter only measures within a single footprint, and thus the sampling is poor compared to a "swath" instrument. For offshore climatologies, the low sampling is usually combatted by "gridding" where all measurements within a given time period (typically one month) and within a defined area (typically one to two degrees of latitude and longitude) are pooled to form an adequate sample. Implicit in this procedure is the assumption that any measurement within these bounds is a random sample of an identical population irrespective of the precise time and place. This assumption is clearly unreasonable in the coastal regions where changes in the climate of waves will certainly take place on spatial scales less than the ≥ 100 km scales of a practical grid. Thus, coastal applications are inevitably limited to investigating variability on a scale of 10 km or greater (as imposed by the footprint size) and only then if a means of adequate sampling is achieved.

Standard geophysical values retrieved from altimetry are sea surface height from the time delay of the return (with careful corrections for "orbital", "atmospheric" and "sea state bias" effects; Fu and Cazenave 2001), significant wave height (from the slope of the leading edge of the waveform) and wind speed (from the backscatter coefficient, σ_0, alone or from both σ_0 and significant wave height). Here we will mainly focus on the utility of wave height and sea level data for coastal and air-sea interaction studies, but other geophysical retrievals will also be mentioned and we look forward to new developments in altimeter technology and new applications. More detailed studies of both the circulation and sea level of the Mediterranean follow later in this volume.

2. Sea level

The most widespread use of satellite altimetry data is for altimeter-derived sea surface height anomalies to study ocean currents and especially current variability associated with ocean eddy activity (Fu and Cazenave 2001). This application requires careful processing of the time delay data records (including tidal corrections) and is restricted to "geostrophic currents" which are strictly related to the spatial gradient of dynamic surface height. This application is less appropriate for shallow waters, partly due to the difficulty in correcting for tides with sufficient accuracy and the greater significance of non-geostrophic currents. Instead, sea level studies are a more obvious application. Woolf *et al.* (2003b) have shown that both the seasonal variability in sea level around Europe and the inter-annual variability are identified within altimeter data (gridded at $1° \times 1°$, monthly). In particular, sea level is expected to respond to the strongest mode of regional atmospheric variability, the North Atlantic Oscillation (NAO); and this is captured in the altimeter data (Figure 2). Tide gauge data provide very detailed site-specific data and validate the satellite altimeter data, while the satellites give a broad perspective on spatial features.

Figure 2 shows actual sea level, but where we wish to exclude the direct effect of atmospheric pressure (to calculate sea surface height anomalies and thence currents) we normally apply the standard GDR inverse barometer correction. Standard corrections are not always sufficient in coastal regions and improved corrections are advisable. Vignudelli *et al.* (2005) describe some improved procedures, by which more accurate sea level anomalies (2–3 cm rms compared to sea truth for seasonal and longer time scales) can be achieved in a challenging coastal environment.

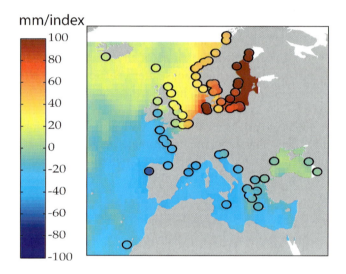

Fig. 2. The sensitivity (rate of change per index) of wintertime sea level to North Atlantic Oscillation. Values in circles are calculated from tide gauges at those locations. The remaining values (on an identical scale) are derived from a $1° × 1°$ climatology of sea level from Topex (Woolf *et al.* 2003b).

3.   Wave climate

Another climatological application of altimetry is the study of wave and wind climate. Altimetry provides near global coverage of wave height and was soon found to be useful, first as a test of wave model forecasts (*e.g.* Romeiser 1993) and also for assimilation into wave models (*e.g.* Young and Glowacki 1996). We will concentrate on significant wave height, but first note that several other wave parameters and wind speed can also be estimated from satellite altimetry. Wave period is a particularly important parameter and a new approach to estimating wave period has now been demonstrated (Gommenginger *et al.* 2003; Quilfen *et al.* 2004; Caires *et al.* 2005). Wave height and wave period are both important to shipping, offshore engineering and renewable energy. Additionally, wave parameters and wind speed are both important to air-sea interaction and in particular the air-sea exchange of gases (Woolf 2005), allowing the retrieval of gas transfer velocities from satellite altimetry (Fangohr and Woolf 2007).

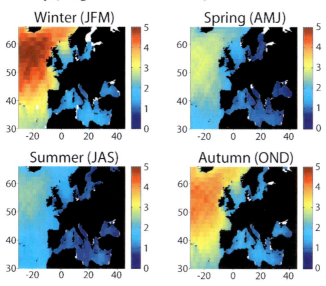

Fig. 3. The seasonality of wintertime significant wave height (in metres). Values are derived from a $1.5° \times 1.5°$ climatology of Topex for 1993–2004.

Gridded climatologies of significant wave height based on satellite altimeter data have been analyzed previously (Woolf *et al.* 2002 and 2003a), but here we analyze a new climatology based on Topex data from 1993–2004, concentrating on the European seas. Figure 3 presents the seasonal averages of significant wave height. Wave heights peak in the wintertime throughout the region. Wave heights are greatest in the north west of the region, due both to high local wind speeds and swell from the North Atlantic. Seas can also be rough in the wintertime in inland seas, notably in many parts of the Mediterranean. While the average seasonality is interesting, it does not tell the whole story. The region is subject to strong inter-annual variability. Woolf *et al.* (2002) identified wintertime variability in the North-Eastern Atlantic and Northern North Sea with two patterns of atmospheric variability, the NAO and the East Atlantic Pattern (EAP). The EAP strongly impacts the strength of westerly winds at 40–50°N and thereby affects wave heights at a similar latitude. However, these patterns do little to explain the substantial variability further east in the region. Recently, Grbec *et al.* (2003) have introduced the Mediterranean Oscillation Index (MOI) defined by the pressure difference between mid-northern Atlantic and South-Eastern Mediterranean. Here, we analyze the new Topex climatology of significant wave heights by a simple multiple linear regression on NAO, EAP and MOI. Anomalies of monthly data from December, January, February and March were calculated. The NAO and EAP indices published by the Climate Prediction Center[1] and MOI from Grbec *et al.* (2003) were used. As MOI was only available until mid 2002, the analysis is restricted to 39 months from January 1993 to March 2002.

Figure 4 shows results of this simple analysis are promising. In particular, MOI appears to explain a substantial fraction of the variability around Portugal, in the western Mediterranean and eastern Black Sea.

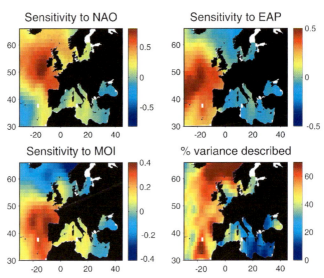

Fig. 4. The sensitivity of wintertime significant wave height to the North Atlantic Oscillation, East Atlantic Pattern and Mediterranean Oscillation Index (in metres/index) and the total fraction of interannual variability described by a linear relationship to all 3 indices. Values are derived from a 1.5° × 1.5° climatology of Topex for 1993–2002.

Wave climate changes in the last few decades have been very substantial (*e.g.* Bacon and Carter 1991, 1993; Gulev and Grigorieva 2004). The

---

[1] http://www.cpc.noaa.gov/ data/teledoc/telecontents.shtml

variability in wave parameters, wind speed and also sea level around Europe are of great practical significance. In particular, the vulnerability of the coastal zone and offshore activities (*e.g.* Tsimplis *et al.* 2005, Wolf and Woolf 2006) to future changes in the atmosphere is very important. Satellite altimetry can play an important role in establishing links between atmospheric variability and coastal vulnerability.

4.   Storms and storm surges

In addition to climatological studies, altimeter data can also provide data on individual events as shown in Figure 5. Warnings on storms and storm surges are based on numerical weather prediction and modelling of waves and surges. Currently, satellite altimetry is not fully integrated into this process, but there clearly is some significant potential for confronting model output with near-real-time altimeter data. Tsunamis can also appear in altimeter data (Allan 2005). There may be more attention to these capabilities in the future as operational oceanography matures.

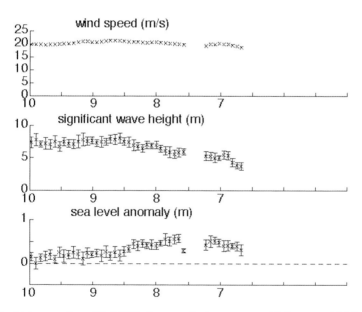

Fig. 5. Altimeter data (JASON) from early in a storm, 11 January 2005. All parameters are plotted against longitude (°W). The ground track travelled to the SE from the open Atlantic crossing between South Uist and Barra in the Outer Hebrides at 7.5°W reemerging in the Sea of the Hebrides. The combined surge and wave effects later caused serious damage on both islands.

5. New applications and techniques

Current retrievals are based on three parameters of the return waveform as described in the Introduction and Figure 1. In some cases (*e.g.* wave period and two-parameter wind-speed algorithms) two of the basic parameters are used in the estimate. Here, we discuss some more experimental applications that use additional information.

Some satellite missions, beginning with TOPEX and now continuing with JASON and ENVISAT RA-2, carry two altimeters operating at different electromagnetic frequencies. TOPEX and JASON carry K_u-band (13.6 GHz) and C-band (5.3 GHz) altimeters where the original function of the C-band altimeter was to provide a "ionospheric correction" to the sea surface height. However, alternative uses can be made for example of the two different values of the radar backscatter coefficient, σ_o^{Ku} and σ_o^C. Differences in the two values can result from both variation in atmospheric attenuation and the frequency-dependent interaction with the sea surface. The most established alternative application is rainfall estimation, which results from the much greater atmospheric attenuation by rain at Ku band than at C band (Quartly *et al.* 1996). A new and as yet un-validated application is to detect the suppression of surface ripples by surface films (Woolf and Ufermann 2005). This application is based on the principle that primarily it is surface waves shorter than 100 mm that are attenuated by surface films and this affects σ_o^{Ku} but not σ_o^C. True "slick conditions" are also apparent from very high values of backscatter (at either frequency) but the possible contribution of film damping cannot be separated from the role of wind speed using this information alone. A plot of $\sigma_o^{Ku} - \sigma_o^C$ against σ_o^C for a large sample of data is shown in Figure 6. Most of the points fall in a broad swath resulting from ordinary wind and wave effects. A qualitative identification of outliers as "rain", "film suppression" or "slick" is shown. This identification follows from an understanding of the phenomena and scattering theory, but requires verification.

Note that the detection of slick effects has a broad utility both for intrinsic interest and because their presence can degrade sea-surface-height data through "bloom effects" on the return waveform (Mitchum *et al.* 2004). Tournadre *et al.* (2006) have proposed a method of detecting this degradation. The methods proposed by Woolf and Ufermann identify both true slick conditions (whether or not this degrades sea-surface-height data) and the suppression of short surface waves by surface films in moderate winds. The latter is closely related to a reduction in gas transfer rates (*e.g.* Glover *et al.* 2007) and thus is an important phenomenon. The use of satellite altimetry for the study of slicks and of air-sea gas transfer is at an early stage, but is an important new avenue.

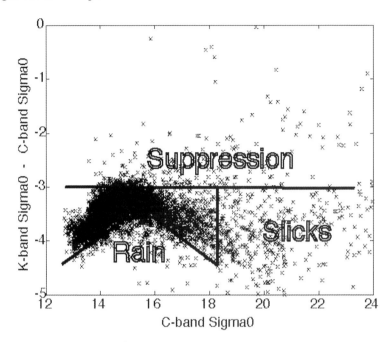

Fig. 6. A plot of $\sigma_0^{Ku} - \sigma_0^C$ against σ_0^C for a large sample of data with indications of the loci resulting from rain, suppression by surface films and slick conditions.

Further geophysical information can be extracted from altimeter return waveforms beyond the three basic parameters shown in Figure 1. The emergence of cleaner waveforms for recent satellite altimeters now also make it possible to retrieve ocean wave skewness by "re-tracking" wave-forms with an appropriate theoretical model (Gomez-Enri *et al.* 2006). Skewness is a measure of the ocean waves non-linearity ("peakier crests, flatter troughs") and is relevant to wave breaking and air-sea gas transfer. Figure 7 shows the kind of maps that can be obtained by retracking ENVISAT altimeter waveforms, thereby providing a unique new view of ocean wave non-linearity.

The sampling limitations of pulse-limited altimeters and increasing interest in coastal applications have stimulated a surge of new altimeter designs in recent years. The next generation of altimeters will consist of coherent systems with Delay-Doppler mapping capability and improved performance both in sampling and accuracy. Beam-forming on Cryosat's SIRAL altimeter (and on the replacement Cryosat-2 mission due in 2009) should yield resolution cells down to 300 m × 1 km instead of the ~10 km of pulse-limited systems. The high-rate coherent individual echoes currently collected by ENVISAT RA-2 point to a wealth of new applications based on complex signals at nadir, including the prospect of ocean wave encounter spectrum from altimeter phase fluctuations. Other innovative concepts such as the Doppler knife-beam altimeter (acquiring directional capability by using a rotating asymmetric knife-shaped beam: Karaev *et al.* 2005) promise ocean wave direction and orbital velocity products over wide swaths.

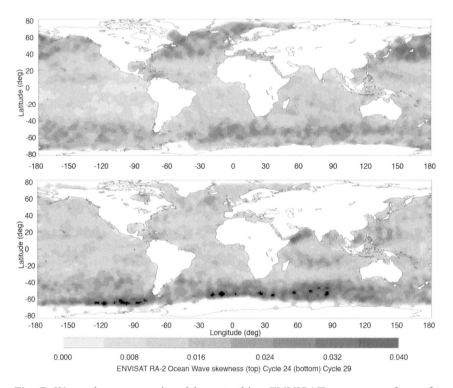

Fig. 7. Wave skewness retrieved by retracking ENVISAT ocean waveforms for (top) Feb/Mar 2004 and (bottom) July/Aug 2004 (Gómez-Enri *et al.* 2006).

6. Summary

A radar altimeter is a versatile instrument and numerous satellite missions have already provided an extensive data set, which should expand in the future. Its use in coastal zone is restricted by sampling issues and by quality issues very close to the coast. However, it has proved very useful for climatological studies (sea level and wave parameters) and has considerable potential as part of a Near Real Time monitoring system. Traditional altimeters may yield new products by fully exploiting the return waveform or by innovative use of dual-frequency data. Further products are foreseen, in particular for coastal applications, once advanced concepts such as knife-beam and Delay-Doppler altimeters reach fruition.

Acknowledgements

This research was mainly supported by Natural Environment Research Council through Core Strategic support of National Oceanography Centre, Southampton. Research on gas exchange and surface slicks is conducted as a contribution to the Centre for observation of Air-Sea Interactions and fluxes (CASIX) and also supported by EU INTAS (SIMP; 03-51-4987). We thank our colleagues in the Laboratory for Satellite Oceanography, NOC, for their assistance.

References

Allan T (2005) Detecting tsunamis: calling in the satellites. Marine Scientist 13: 12–14

Bacon S, Carter DJT (1991) Wave climate changes in the North Atlantic and North Sea. Int J Climatol 11: 545–558

Bacon S, Carter DJT (1993) A connection between mean wave height and atmospheric pressure gradient in the North Atlantic. Int J Climatol 13: 423–436

Caires S, Sterl A, Gommenginger CP (2005) Global ocean mean wave period data: Validation and description. J Geophys Res 110: C02003

Fangohr S, Woolf DK (2007) Application of new parameterizations of gas transfer velocity and their impact on regional and global marine CO_2 budgets. J Mar Sys 66(1–4): 195–203

Fu L-L, Cazenave A (2001) Satellite altimetry and earth sciences: a handbook of techniques and applications. Academic Press, San Diego

Glover DM, Frew NM, McCue SJ (2007) Air–sea gas transfer velocity estimates from the Jason-1 and TOPEX altimeters: Prospects for a long-term global time series. J Mar Sys 66(1–4): 173–181

Gómez-Enri J, Gommenginger C, Challenor P, Srokosz M, Drinkwater MR (2006) Envisat Radar Altimeter tracker bias. Marine Geodesy 29(1): 19–38

Gommenginger CP, Srokosz MA, Challenor PG, Cotton PD (2003) Measuring wave period with satellite altimeters: A simple empirical model. Geophys Res Lett 30(22): 2150–2153

Grbec B, Morovic M, Zore-Armanda M (2003) Mediterranean Oscillation Index and its relationship with salinity fluctuation in the Adriatic Sea. Acta Adriatica 44(1): 61–76

Gulev SK, Grigorieva V (2004) Last century changes in ocean wind wave height from global visual wave data. Geophys Res Lett 31: L24302, doi:10.1029/2004GL021040

Karaev VY, Kanevsky MB, Balandina GN, Challenor P, Gommenginger C, Srokosz M (2005) The concept of a microwave radar with an asymmetric knifelike beam for the remote sensing of ocean waves. J Atmos Ocean Tech 22: 1809–1820

Mitchum GT, Hancock DW, Hayne GS, Vandemark DC (2004) Blooms of sigma(0) in the TOPEX radar altimeter data. J Atmos Ocean Tech 21(8): 1232–1245

Quartly GD, Guymer TH, Srokosz MA (1996) The effect of rain on Topex radar altimeter data. J Atmos Ocean Tech 13: 1209–1229

Quilfen Y, Chapron B, Serre M (2004) Calibration/validation of an altimeter wave period model and application to TOPEX/Poseidon and Jason-1 altimeters. Marine Geodesy, 27: 535–549

Romeiser R (1993) Global validation of the wave model WAM over a one-year period using GEOSAT wave height data. J Geophys Res 98(C3): 4713–4726

Tournadre J, Chapron B, Reul N, Vandemark DC (2006) A satellite altimeter model for ocean slick detection J Geophys Res 111(C4): C04004, doi:10.1029/2005JC003109)

Tsimplis MN, Woolf DK, Osborn T, Wakelin S, Woodworth P, Wolf J, Flather R, Blackman D, Shaw AGP, Pert F, Challenor P, Yan Z (2005) Towards a vulnerability assessment of the UK and northern European coasts: the role of regional climate variability. Phil Trans R Soc A 363: 1329–1358, doi:10.1098/rsta.2005.1571

Vignudelli S, Cipollini P, Roblou L, Lyard F, Gasparini GP, Manzella G, Astraldi M (2005). Improved satellite altimetry in coastal systems: Case study of the Corsica Channel (Mediterranean Sea). Geophys Res Lett 32: L07608, doi:10.1029/2005GL0222602

Wolf J, Woolf DK (2006) Waves and climate change in the north-east Atlantic. Geophys Res Lett 33, L06604, doi:10.1029/2005GL025113

Woolf DK (2005) Parametrization of gas transfer velocities and sea-state-dependent wave breaking. Tellus, 57B, 87–94

Woolf DK, Ufermann, S (2005) Can dual-frequency altimetry be used to identify surface slicks? In: Proc. 2nd Workshop on Remote Sensing of the Coastal Zone, Porto, Portugal, pp 95

Woolf DK, Challenor PG, Cotton PD (2002) The variability and predictability of North Atlantic wave climate. J Geophys Res 107(C10): 3145, doi:10.1029/2001JC001124

Young IR, Glowacki TJ (1996) Assimilation of altimeter wave height data into a spectral wave model using statistical interpolation. Ocean Engng 23(8): 667–689

15 Years of Altimetry at Various Scales over the Mediterranean

Paolo Cipollini [1], Stefano Vignudelli [2], Florent Lyard [3], and Laurent Roblou [4]

[1] National Oceanography Centre, Southampton, UK
[2] Consiglio Nazionale delle Ricerche, Istituto di Biofisica, Pisa, Italy
[3] Laboratoire d'Etude en Géophysique et Océanographie Spatiale, Toulouse, France
[4] Noveltis, Ramonville-Saint-Agne, France

Abstract. The present article reviews the application of altimetry in the Mediterranean Sea, to study the circulation and sea surface height variability, both at basin scale and in specific regions. The improvements needed to fully exploit the 15-year record of data close to the coast are also discussed. These range from improved tidal models, to specialized atmospheric corrections, to *ad hoc* screening of data in proximity of the coast. Some of these improvements are already underway while others are the focus of forthcoming programs.

1.   Introduction and rationale

Understanding the oceanic variability of the Mediterranean Sea and its response to global climate change is crucial to predict its evolution, its feedback on regional climate, and its impact on socio-economic activities along its coasts. Other contributions to this book deal with changes in wave climate and new applications of altimetry and reconstructions of long-term sea level variations (described elsewhere in the present volume). Here we focus on the variability in Mediterranean circulation as inferred from altimetry, covering a wide range of spatial scales (coastal to basin) and temporal scales (weekly to decadal). First we review the main findings from basin-wide studies, then we look at results over specific regions and finally we discuss current and planned research aiming at improving the quality of altimetric data over the basin and recovering as much information as possible near the coast.

V. Barale, M. Gade (eds.), *Remote Sensing of the European Seas.*

2.   Basin-wide studies of mediterranean variability

Early studies, either with TOPEX/POSEIDON (T/P) data such as Larnicol *et al.* (1995) and Iudicone *et al.* (1998), or with ERS-1 data such as Vignudelli (1997a), demonstrated that altimetry is able to infer useful oceanographic information over the Mediterranean, but suffered from sampling limitations due to the use of a single mission, that aliased a significant part of the mesoscale circulation and of its variability.

A more successful approach has emerged with the merging of data from different missions, made possible both by the improvement in orbit determination and geophysical and instrumental corrections for the ERS satellites (Scharroo and Visser 1998; Le Traon and Ogor 1998) and by the development of techniques for the optimal intercalibration, merging and interpolation of multisatellite data. Fieguth *et al.* (1998) and Ayoub *et al.* (1998), for instance, improved considerably the resolution over the Mediterranean by merging T/P and ERS-1 data. The global mapping technique developed by Le Traon *et al.* (1998) and Ducet *et al.* (2000) and currently adopted for the generation of data by SSALTO/DUACS, has allowed basin-wide studies making the most of the improved resolution, including Larnicol *et al.* (2002) who were able to characterize the major changes in sea level variability for the 1993–1999 period and found important interannual signals in the Ionian basin, and Pujol and Larnicol (2005).

In the early years of this decade, the simultaneous availability of up to five altimeters (T/P, ERS-2, EnviSat RA, Jason-1 and GFO) has resulted into a further improvement of the capability of multimission altimeter datasets to capture the mesoscale. Pascual *et al.* (2007) have recently revisited this issue over the Mediterranean, and show that a four-altimeter configuration allows a mapping of sea level and velocity with a relative accuracy of 6% and 23%, respectively, and increase the average eddy kinetic energy over the basin by 15% with respect to a two altimeter (Jason-1 + ERS) configuration. Their conclusion is that at least three altimeters are needed to capture the mesoscale circulation.

Figure 1 represents the RMS variability over the period 1993–2006 of the SSALTO/DUACS regional multimission Sea Surface Height Anomaly (SSHA) for the Mediterranean (the DT-MSLA merged "Upd" product) distributed by AVISO, and highlights some areas of enhanced variability, most notably the Alboran Gyres, the Algerian Basin, the Ionan Basin, and the Ierapetra Gyre SE of Crete. The dataset can be analyzed in frequency space to illustrate the distribution of the different temporal scales of variability across the basin, as shown in Figure 2. In order to do this we have computed the Fast Fourier Transform of the SSHA time series in any location and then partitioned the signal into frequency bands.

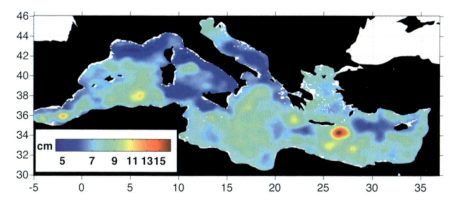

Fig. 1. RMS variability (cm) of SSALTO/DUACS multimission SSHA (DT-MSLA merged "Upd" product for the Mediterranean) over 1993–2006.

The low frequency variability map (Figure 2a) is dominated by a few regions that include the Ionian Sea, where a strong decreasing trend has been detected. The annual signal map in Figure 2b shows a striking peak (with an amplitude in excess of 15 centimetres, *i.e.* more than 30 cm peak-to-peak) in the Ierapetra Gyre, a narrow region which is also strongly energetic at periods longer than one year (also found to be virtually uncorrelated with the rest of the basin). Elsewhere the annual signal is almost uniform, with a small peak in the East Alboran Gyre, displaying significantly more variability at frequencies different from the annual one. The shorter time scales in Figures 2c, 2d and 2e highlight several areas of known mesoscale activity, most notably in the Alboran Sea and Algerian Basin which peak at 1.5–4.5 months, whereas the high frequency signal (T < 1.5 months) in Figure 2f is generally confined to a narrow coastal strip, a fact that underlines the need of a better temporal resolution of coastal dynamics, as we discuss below.

One important by-product of the merging of altimeter data with *in situ* data is the mean dynamic topography for the basin computed by Rio *et al.* (2007) and allowing a first estimation of absolute surface geostrophic currents. These clearly show many interesting patterns, such as the strong Mid-Mediterranean jet flowing along the coast of the Levantine basin, and several cyclonic/anticyclonic structures. This picture will further improve with the inclusion of gravimetric data, especially from the high-accuracy GOCE mission by ESA whose launch is scheduled for September 2007.

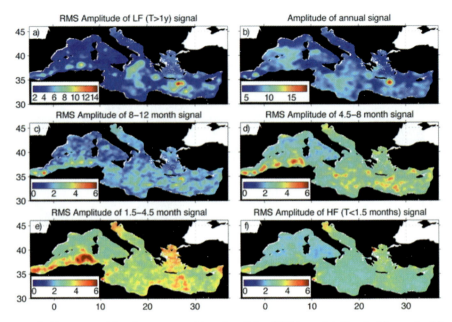

Fig. 2. Decomposition into spectral bands of the SSHA variability in Figure 1. All values are in cm. a) RMS amplitude of low-frequency signals (signals with period >1 year); b) amplitude of the single annual component; c) RMS amplitude of signals between 8 months and 1 year; d) RMS amplitude of signals between 4.5 and 8 months; e) RMS amplitude of signals between 1.5 and 4.5 months; f) RMS amplitude of high-frequency signals (signals with period <1.5 months)

3.   Applications of altimetry to specific regions

Several studies have applied altimetry to oceanographic problems over regional areas of the Mediterranean. Vignudelli (1997b) and Bouzinac *et al.* (1998) could detect the migration of anticyclonic eddies towards the interior of the Algerian Basin in T/P and ERS-1 data. Buongiorno Nardelli *et al.* (1999) combined altimeter and *in situ* observations in the Straits of Sicily to infer the 3-D structure of the circulation, in the first Mediterranean validation study of altimetry with hydrography, and found a strong barotropic component of the velocity field along the continental shelf. By including CTD casts from the MEDAR/MEDATLAS climatology and reconstructing the vertical structure through Coupled Pattern Reconstruction and multivariate EOFs, Buongiorno Nardelli *et al.* (2006) further clarify the important role of mesoscale processes in the area.

Vignudelli *et al.* (2000) studied the forcing driving the flow through the Corsica Channel by comparing water transport anomalies from a current-meter with altimeter-derived sea level differences between the two ends of the channel, whose time series are shown in Figure 3.

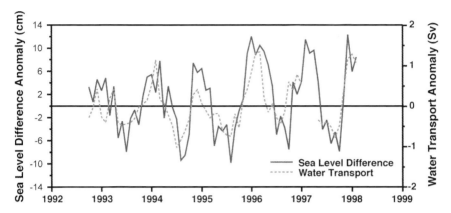

Fig. 3. Comparison of sea level difference between Tyrrhenian and Ligurian Seas, estimated along T/P track 044, and water transport from an *in situ* current-meter in the Corsica Channel (redrawn from Vignudelli *et al.* 2000).

The agreement between the seasonal current and the seasonal sea level difference indicates that the latter provides a significant component to the exchanges between the two basins, which in turn supports the findings by Astraldi and Gasparini (1992) that winter heat and water losses induced by air-sea interaction over the Liguro-Provencal Basin are important in driving the circulation of this part of the Mediterranean. Circulation in the Tyrrhenian and Ligurian seas was further investigated with a comparison between results from hydrographic (eXpendable Bathy-Thermograph, XBT) transects and altimeter data along the portion of T/P track 044 from Liguria to Sicily (Vignudelli *et al.* 2003). The comparison was done both in space (*i.e.* comparing near-simultaneous transects) and in time, and generally shows good agreement (within 2–4 cm RMS) for surface height changes associated with the time-varying part of the circulation. However this work did not include the segment of track 044 intersecting the channel (42.5–43.3 N), due to the lack of XBT data. A T/P-Jason ground track crossing the Channel (track 085) has been recently investigated with the data editing techniques developed for coastal altimetry (Vignudelli *et al.* 2005), see below.

Observation of eddies continues to be an important application of altimetry, see for instance Pascual *et al.* (2002) who used altimetry, sea surface temperature and CTD data to study the formation and evolution of an eddy in the Balearic Sea. A novel methodology for tracking of eddies in altimeter data, based on identifying closed contours of a parameter that measures the relative importance of deformation and rotation, has been developed by Isern-Fontanet *et al.* (2003) who illustrated it with an example in the Algerian Basin. Recently, Isern-Fontanet *et al.* (2006) have used this technique over the entire Mediterranean, detecting complex but well-defined patterns of eddy propagation.

4.   Improving altimetry from basin-scale to the coasts

Two fundamental challenges remain in altimetry over marginal seas: a) to improve the quality and availability of the corrections at regional scale and/or basin scale; and b) to recover a quality-controlled altimetric datum as close as possible to the coast. This would give access to 15 years of un-exploited data, certainly a very valuable asset to study coastal dynamics. An overview of these two issues was carried out by Anzenhofer *et al.* (1999), who described the generation of coastal altimeter data, discussed various retracking algorithms and their implementation and showed some examples based on the intermediate ERS waveform data. They concluded with recommendations on better (local) tidal modelling, careful screening of the data, improvement of the wet tropospheric correction and retracking.

The Mediterranean Sea, unique for its complex morphology of sub-basins bounded by an intricate shoreline and showing a huge variety of oceanographic features, is the perfect site to tackle the challenges described above. It is therefore not surprising that this basin has been the main focus of a few initiatives in the above direction.

An early attempt to custom-process coastal ERS-1 altimetric records was made by Manzella *et al.* (1997) around Corsica. They used a regional tidal model (Tsimplis *et al.* 1995), recomputed the wet tropospheric correction by recalibrating the model correction with the closest available radiometric estimate, and recovered measurements previously flagged as bad because of a high σ_0 value induced by sea surface flatness.

The ideas of customized tidal modelling and de-flagging were also followed by Vignudelli *et al.* (2000) over the Corsica Channel, using 1 Hz T/P data in combination with current meter and tide gauge data. Results were encouraging, showing that with simple improvements in the processing, the signal recovered at seasonal time scales was in good agreement with the *in situ* measurements and allowed useful oceanographic conclusions.

The concepts laid out by those pilot studies have been extended to the whole Northwest Mediterranean in the joint French-Italian ALBICOCCA

(ALtimeter-Based Investigations in COrsica, Capraia and Contiguous Areas) initiative for coastal altimetry. Here we describe three key elements of the ALBICOCCA strategy for the generation of the coastal products: better tidal modelling, improved modelling of high frequency atmospheric effects, and screening of the data to recover information previously flagged as bad. These, in principle, are valid for all coastal regions. The improvements in the ALBICOCCA area are quantified in Vignudelli *et al.* (2005).

4.1 Better tidal models

The earlier studies mentioned above have demonstrated that global tidal models, despite their impressive improvement during the past decade, are generally not accurate enough over the shallow water areas to properly remove tidal effects in the altimeter measurements. Because of their insufficient spatial resolution (that does not resolve rapid changes in tidal features) and of incorrect frictional dissipation, usual global model cannot represent tides over continental shelves below a decimetre error level. If the tides are not accurately removed, the residual signal will mask other oceanographic phenomena, being aliased at their typical frequencies by the sampling pattern of the altimeter. To avoid this, as an alternative to the FES2004 (Lyard *et al.* 2006) or GOT00 global tidal solution available in the T/P Geophysical Data Records (GDRs), an optimised tidal spectrum has been defined for the Mediterranean Sea, mainly based on MOG2D regional tidal solutions (Carrère and Lyard 2003) and combined with global model FES2004 for the diurnal component.

As an illustration of the improvement in tidal modelling, Table 1 summarizes the misfits for the major tidal constituents for both the regional MOG2D solution and the FES2004 global solution relative to a reference tide gauge data set in the Mediterranean Sea. We can notice that for the semi-diurnal and O1 constituents the MOG2D solutions are generally better (lower in bias and/or significantly less dispersed) than the FES2004. Instead, for the main diurnal constituent K1 in the optimal spectrum we choose the FES2004 solution, due to the much larger MOG2D phase bias.

4.2 Better correction of atmospheric effects

The atmospherically forced high frequency signal can be a source of large aliasing in the altimetric records at mid- and high latitudes. Over the open ocean the effects of atmospheric pressure loading are corrected by the inverted barometer approximation, but significant departures can be observed over continental shelves and marginal seas. In reality, the sea level variations depend both statically and dynamically on the meteorological forcing whereas the IB approximation formulates merely the hydrostatic equilibrium between the sea level and the applied atmospheric pressure gradients. Moreover, the IB assumption totally ignores wind-forced sea level variations, which can prevail particularly around the altimetric sampling periods. An improved modelling of the level response, following Carrère and Lyard (2003), has therefore been carried out in the Mediterranean Sea, using a regional mesh (another outputs of the MOG2D model).

Table 1 . Comparison of LEGOS MOG2D Mediterranean Sea tides and FES2004 solutions with a reference tide gauge data set from 45 stations in the Mediterranean Sea.

Tidal constituent		Amplitude difference (cm)	Phase lag (degrees)	Complex difference modulus (cm)
M2	FES2004	−0.1 +/−1.4	1.2 +/−17.7	0.6 +/−1.6
	MOG2D	0.0 +/−1.0	1.3 +/− 9.3	0.2 +/−1.3
S2	FES2004	−0.3 +/−0.8	−5.8 +/−20.5	0.4 +/−1.2
	MOG2D	0.3 +/−0.5	−1.6 +/−16.9	0.1 +/−0.7
N2	FES2004	−0.2 +/−0.4	4.5 +/−22.5	0.0 +/−0.4
	MOG2D	0.0 +/−0.2	−0.1 +/−12.0	0.0 +/−0.3
K1	FES2004	0.0 +/−1.3	3.0 +/−13.9	1.1 +/−1.0
	MOG2D	0.6 +/−0.7	15.5 +/−12.7	0.6 +/−1.0
O1	FES2004	0.2 +/−0.4	−8.8 +/−16.1	0.4 +/−0.4
	MOG2D	0.4 +/−0.4	10.0 +/−13.6	0.3 +/−0.5

Table 2 illustrates the gain in sea level variance reduction in the short period range (T < 20 days, where T is the period) when correcting sea level observations for 31 tide gauges with the global or regional MOG2D model instead of the IB parameterization.

Table 2. Residual sea level variance for T < 20 days when applying different atmospheric effect corrections to de-tided sea level at the 31 tide gauges around the Mediterranean for which data are available over the testing period (01/2002 to 06/2002).Values listed are the average of individual tide gauge variances.

	No correction	IB corrected	MOG2D-Glob correct.	Mog2D-Med correct.
Residual variance (cm²)	26,34	15,32	6,97	5,11
Variance reduction (%)		42	75	79

4.3   Better screening of data

In addition to the corrections described above, the ALBICOCCA processing adopted *ad hoc* filtering and screening techniques allowing to recover data that otherwise would be flagged as bad (Roblou and Lyard 2004). Figure 4 shows an example of data recovery along T/P track 222 for cycle 95 (22/04/1995). The original (GDR) sea level anomalies are affected by abnormal values in the wet tropospheric correction, and therefore flagged as bad. De-flagging and re-interpolation of the correction yields a reconstructed level profile. This exercise has been extended over the entire NW Mediterranean; the improved products allow a better characterization of the variability at seasonal scales (see Vignudelli *et al.* 2005). Moreover, GPS campaigns have been carried out in the ALBICOCCA framework, for the definition of a local geoid connecting the altimeter tracks to the tide gauge sites in Corsica and Capraia (Bonnefond *et al.* 2003).

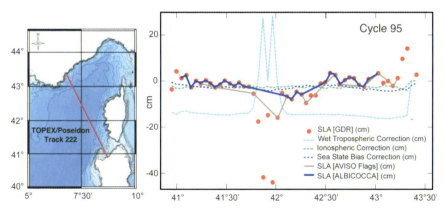

Fig. 4. Example of altimetric profile along track 222 for a given cycle (95) showing outliers in the wet troposphere correction. Red dots: uncorrected Sea Level Anomaly (SLA) and original corrections from the AVISO GDRs. Dashed lines: correction for atmospheric and surface effects. Solid brown line: SLA after application of the standard corrections from the GDR. Solid blue line: the new SLA profile after reconstructing the values with the ALBICOCCA technique described in Roblou and Lyard (2004).

4.4   Present and future of coastal altimetry

Several studies have dealt with the limitations of, and possible improvements to, coastal altimetry. The topic was the subject of a dedicated International Workshop in Beijing in July 2006, where Bouffard *et al.* (2006)

presented preliminary results from exploiting a multi-mission scenario and higher along-track data rates in the NW Mediterranean.

The growing demand for readily available, reprocessed coastal altimetry data is about to be met by some initiatives that build on the activities described above. One of these is the ALTICORE[1] (value-added ALTImetry for COastal REgions) project aiming at improving the monitoring capabilities of satellite altimeters over coastal areas. Access to distributed archives of data processed within ALTICORE will be via a Grid-compliant system, consisting of regional data centres, each having primary responsibility for regional archives, local corrections and quality control, and operating a set of web-services allowing access to the full functionality of data extraction (Vignudelli *et al.* 2006). This novel initiative focuses in particular on the European seas, including the Mediterranean and therefore reinforcing its role as the main laboratory for moving coastal altimetry towards operational status. It is complementary to other initiatives at European level, namely MERSEA[2] (Marine EnviRonment and Security for the European Area), which focuses on operational oceanography, and ECOOP[3] (European COastal sea OPerational observing and forecasting system), which aims at fine tuning MERSEA products to coastal applications.

Acknowledgments

The altimeter products were produced by SSALTO/DUACS and distributed by AVISO[4] with support from CNES. Coastal altimeter data were produced by CTOH[4].Current meter data were kindly provided by G.P. Gasparini and M. Astraldi of CNR/IOF La Spezia, Italy. ALTICORE is funded by INTAS (contract 05-1000008-7927)

References

Anzenhofer M, Shum CK, Rentsh M (1999) Coastal altimetry and applications, Tech Rep 464, Geodetic Science and Surveying, The Ohio State University Columbus, USA

---

[1] www.alticore.eu

[2] www.mersea.eu.org

[3] www.ecoop.eu

[4] www.aviso.oceanobs.com; www.legos.obs-mip.fr/observations/ctoh

Astraldi M, Gasparini GP (1992) The seasonal characteristics of the circulation in the north Mediterranean basin and their relationship with the atmospheric-climatic conditions. J Geophys Res 97: 9531–9540

Ayoub N, Le Traon PY, De Mey P (1998) A description of the Mediterranean surface variable circulation from combined ERS-1 and TOPEX/POSEIDON altimetric data. J Mar Sys 18 (1/3): 3–40

Bonnefond P, Exertier P, Laurain O, Menard Y, Orsoni A, Jan G, Jeansou E (2003) Absolute Calibration of Jason-1 and TOPEX/Poseidon Altimeters in Corsica. Mar Geodesy 26 (3/4): 261–284

Bouffard J, Vignudelli S, Lyard F, Birol F, Marsaleix P, Roblou L, Menard Y, Gasparini GP, Manzella G, Cipollini P (2006) Advances in Coastal Altimetry over the Northwestern Mediterranean. In: International Workshop on coast and land applications of satellite altimetry, Beijing, July 21–22, pp 26–27

Bouzinac C, Vazquez J, Font J (1998) Complex empirical orthogonal functions analysis of ERS-1 and TOPEX/POSEIDON combined altimetric data in the region of the Algerian current. J Geophys Res 103 (C4): 8059–8071

Buongiorno Nardelli B, Santoleri R, Zoffoli S, and Marullo S (1999) Altimetric signal and three-dimensional structure of the sea in the Channel of Sicily. J Geophys Res 104 (C9): 20585–20603

Buongiorno Nardelli B, Cavalieri O, Rio MH, Santoleri R (2006) Subsurface geostrophic velocities inference from altimeter data: Application to the Sicily Channel (Mediterranean Sea). J Geophys Res 111: C04007

Carrère L, Lyard F (2003) Modeling the barotropic response of the global ocean to atmospheric wind and pressure forcing: comparisons with observations. Geophys Res Lett 30 (6): 1275–1285

Ducet N, Le Traon PY, Reverdin G (2000) Global high-resolution mapping of ocean circulation from the combination of T/P and ERS-1/2. J Geophys Res 105 (C8): 19477–19498

Fieguth P, Menemenlis D, Ho T, Willsky A, Wunsch C (1998) Mapping Mediterranean Altimeter Data with a Multiresolution Optimal Interpolation Algorithm. J Atmos Oceanic Tech 15: 535–546

Isern-Fontanet J, García-Ladona E, Font J (2003) Identification of marine eddies from altimetry. J Atmos Oceanic Technol 20: 772–778

Isern-Fontanet J, García-Ladona E, Font J (2006) Vortices of the Mediterranean Sea: An Altimetric Perspective. J Phys Ocean 36: 87–103

Iudicone D, Santoleri R, Marullo S, Gerosa P (1998) Sea level variability and surface eddy statistics in the Mediterranean Sea from TOPEX/POSEIDON data. J Geophys Res 103 (C2): 2995–3012

Larnicol G, Ayoub N, Le Traon PY (2002) Major changes in Mediterranean Sea level variability from 7 years of TOPEX/POSEIDON and ERS-1/2 data. J Mar Sys 33/34: 63–89

Larnicol G, Le Traon PY, Ayoub N, De Mey P (1995) Mean sea level and surface circulation variability of the Mediterranean Sea from 2 years of TOPEX/ POSEIDON altimetry. J Geophys Res 100 (C12): 25163–25177

Le Traon PY, Nadal F, Ducet N (1998) An improved mapping method of multisatellite altimeter data. J Atmos Oceanic Tech 15: 522–533

Le Traon PY, Ogor F (1998) ERS-1/2 orbit improvement using TOPEX/ POSEIDON: The 2 cm challenge. J Geophys Res 103 (C4): 8045–8057

Lyard F, Lefevre F, Letellier T, Francis O (2006) Modelling the global ocean tides: modern insights from FES2004. Ocean Dynamics 56: 394–415

Manzella G, Borzelli GL, Cipollini P, Guymer TH, Snaith HM, Vignudelli S (1997) Potential use of satellite data to infer the circulation dynamics in a marginal area of the Mediterranean Sea. In: Proc 3rd ERS Symposium 3, ESA SP-414, pp 1461–466

Pascual A, Buongiorno Nardelli B, Larnicol G, Emelianov M, Gomis D (2002) A case of an intense anticyclonic eddy in the Balearic Sea (western Mediterranean). J Geophys Res 107 (C11): 3183–3193

Pascual A, Pujol MI, Larnicol G, Le Traon PY, Rio MH (2007) Mesoscale mapping capabilities of multisatellite altimeter missions: first results with real data in the Mediterranean Sea. J Mar Sys 65: 190–211

Pujol MI, Larnicol G (2005) Mediterranean sea eddy kinetic energy variability from 11 years of altimetric data. J Mar Sys 58: 121–142

Rio MH, Poulain PM, Pascual A, Mauri E, Larnicol G, Santoleri R (2007) A mean dynamic topography of the Mediterranean Sea computed from altimetric data, in-situ measurements and a general circulation model. J Mar Sys 65 (1–4): 484–508

Roblou L, Lyard F (2004) Retraitement des données altimétriques satellitaires pour des applications cotières en Mer Méditerranée. Tech Rep POC-TR-09 – 04, Pole d'Océanogr. Cotière, Toulouse, France

Scharroo R, Visser P (1998) Precise Orbit Determination and Gravity Field Improvement for the ERS Satellites. J Geophys Res 103: 8113–8127

Tsimplis MN, Proctor R, Flather RA (1995) A two-dimensional tidal model for the Mediterranean Sea. J Geophys Res 100: 16223–16239

Vignudelli S (1997a) Analysis of ERS-1 altimeter collinear passes in the Mediterranean Sea during 1992–1993. Int J Rem Sens 18 (3): 573–601

Vignudelli S (1997b) Potential use of ERS-1 and TOPEX/POSEIDON altimeters for resolving oceanographic patterns in the Algerian Basin. Geophys Res Lett 24 (14): 1787–1790

Vignudelli S, Cipollini P, Astraldi M, Gasparini GP, Manzella G (2000) Integrated use of altimeter and in situ data for understanding the water exchanges between the Tyrrhenian and Ligurian seas. J Geophys Res 105 (C8): 19649–19663

Vignudelli S, Cipollini P, Reseghetti F, Fusco G, Gasparini GP, Manzella G (2003) Comparison between XBT data and TOPEX/Poseidon satellite altimetry in the Ligurian-Tyrrhenian area. Annales Geophys 21: 123–135

Vignudelli S, Cipollini P, Roblou L, Lyard F, Gasparini GP, Manzella G, Astraldi M (2005) Improved satellite altimetry in coastal systems: Case study of the Corsica Channel (Mediterranean Sea). Geophys Res Lett 32: L07608

Vignudelli S, Snaith HM, Lyard F, Cipollini P, Birol F, Bouffard J, Roblou L (2006) Satellite radar altimetry from open ocean to coasts: challenges and perspectives. Proc SPIE 6406, 64060L: 1–12

Can we Reconstruct the 20th Century Sea Level Variability in the Mediterranean Sea on the Basis of Recent Altimetric Measurements?

Michael N. Tsimplis[1], Andrew G.P. Shaw [1], Ananda Pascual[2],
Marta Marcos[1], Mira Pasaric[3], and Luciana Fenoglio-Marc[4]

[1] National Oceanography Centre, Southampton, UK
[2] Institut Mediterrani d'Estuidis Avançats, IMEDEA (CSIC-UIB), Palma de Mallorca, Spain
[3] Andrija Mohorovicic Geophysical Institute, Faculty of Science, University of Zagreb, Zagreb, Croatia
[4] Institute of Physical Geodesy, Darmstadt University of Technology, Germany.

Abstract. The spatial sea level patterns, derived from Empirical Orthogonal Function (EOF) analysis of satellite altimetry data, are combined with the few available long tide-gauge records at the coasts of the basin, in order to reconstruct sea level variability in the Mediterranean Sea. Cross correlations in the basin are explored, in order to assess to what extent the period of altimetric measurements is typical of the longer term variability of the region. Areas of intense mesoscale activity are not well correlated with the rest of the basin and in particular with the coastal areas. Thus, reconstruction of sea level in such areas is not possible. The first EOF from altimetry and the first EOF derived from tide-gauges are well correlated and permit reconstructions to be developed. For the tide-gauges used, the reconstruction of the sea level for the past century, based on data from the last decade, is at least as successful in describing the variance at the long tide-gauges as reconstructions based on previous decades. The power spectrum of the tide-gauges shows energy at particular dominant frequencies albeit varying in time.

1. Introduction

Describing the Mediterranean sea level decadal variability and understanding its forcing is important in order to extract long term sea level trends and develop scenarios for the future of the coastal areas of the basin. Sea

V. Barale, M. Gade (eds.), *Remote Sensing of the European Seas.*
© Springer Science+Business Media B.V. 2008

level records from tide-gauges go back to the beginning of the 19th century, but are exclusively located at the northern coasts of the Mediterranean basin. By contrast, altimetry measurements provide spatially uniform observations in all but the closest to the coast 30–40 km, but span only the last couple of decades. As a result the spatial patterns derived from altimetry may not be representative of the long term mean conditions of the Mediterranean Sea, if one accepts that such mean conditions exist. Consequently a reconstruction of past sea level variability by the combination of the spatially sparse long time-series of tide-gauge data and the spatially complete but relatively short altimetric records depends on the degree by which the last 15 years differ from the last century.

In this paper we demonstrate a way of reconstructing the Mediterranean sea level variability by combining altimetry with tide-gauges. In order to explore the validity of the reconstruction we explore the extent of agreement in the basin wide variability as described by the altimetry and the available coastal tide-gauges and the agreement between the spatial sea level patterns in the 1990s and in previous decades.

2. Data and methodology

Altimetry data from two sources are used. The first altimetric dataset (TPJ) is based on TOPEX/POSEIDON and Jason-1 data. The along-track SSH anomalies are mapped to a 0.25 deg × 0.25 deg grid using 2D-planar Gaussian weighted averages with 150 km distance of half power (Fenoglio-Marc et al. 2006). The second dataset (Multi-sensor) is based on 13 years (January 1993-December 2005) of gridded Sea Level Anomaly (SLA) fields combining several altimeter missions (Le Traon et al. 2003 for details; Ssalto/Duacs system[1]). Merging multiple altimeter missions provides improved description of the mesoscale variability as well as sea level variations (Pascual et al. 2006, 2007).

Both altimetric datasets provide mean monthly values. The seasonal cycle and the trends were removed from both datasets. The two altimetric datasets are not independent as they both contain the TOPEX/POSEIDON and Jason-1 data. However, the second one is a multi-mission product with higher spatial resolution while the first one has lower spatial resolution but as it is based on one mission it avoids complications of merging data obtained from various platforms. It is part of this study to compare the use of these products within the scope of sea level reconstruction. Empirical Orthogonal Functions

---

[1] http://www.aviso.oceanobs.com/

(Toumazou *et al.* 2001) where used to extract the coherent variability of each dataset.

Monthly mean values of sea level, derived from tide-gauge measurements of the Permanent Service of Mean Sea Level (PSMSL), are also used (Woodworth and Player 2003). Only a few tide-gauge stations on the north coasts of the western Mediterranean and the Adriatic have data of good quality and long duration (Tsimplis and Spencer 1997). However, we use the data from the tide-gauges in Antalya and Alexandria and the Eastern Mediterranean Index (Tsimplis and Josey 2001) to describe the Eastern Mediterranean Region, and Ceuta, Cadiz and Tarifa to cover the proximity of the Strait of Gibraltar. A few years of the tide-gauge in Malta are also included for validation purposes (tide-gauge positions in Figure 1).

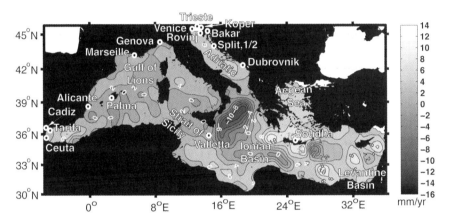

Fig. 1. Sea level trends for 1993–2005 estimated from the MultiSensor altimetry. The tide-gauges used are also shown.

Linear trends and the mean seasonal cycle, estimated as the mean of each month (full years only), were removed from each quality controlled tide-gauge record. Coherent sea level variability from tide-gauges was derived by using weighted EOFs in regions. The weight of each tide-gauge in each region was based on their spatial distribution (Table 1). For the Eastern Mediterranean a synthetic time series was also used (Tsimplis and Josey 2001). The regional indices were then combined into EOFs for the basin and were used for direct comparison with the EOFs derived from the two altimetry datasets.

The time variability of the EOFs from tide gauges were used in a linear regression against the corresponding EOFs of altimetry. The regression parameters were then used to extrapolate the altimetry EOFs back in time and then multiplied by the spatial patterns.

Table 1. The tide gauges used for the extraction of the regional indices.

	Lat °N	Lon °E	Period covered	Gaps %	Weight
Alboran Sea					
Cadiz	36.53	−6.28	1961–2001	2.4	1/3
Tarifa	36.00	−5.60	1943–2001	7.1	1/3
Ceuta	35.90	−5.32	1944–2002	5.9	1/3
Western Mediterranean					
Alicante	38.33	−0.48	1960–1997	3.1	1/2
Marseille	43.30	5.35	1885–2004	2.6	1/4
Genova	44.40	8.90	1884–1997	21.7	1/4
Adriatic Sea					
Venice	45.43	12.34	1909–2000	5.9	1/8
Trieste	45.65	13.75	1905–2004	6.0	1/8
Rovinj	45.08	13.63	1955–2002	0.5	1/8
Bakar	45.30	14.53	1930–2002	15.1	1/8
Split-1	43.50	16.38	1952–2002	2.3	1/8
Split-2	43.50	16.43	1954–2002	0	1/8
Dubrovnik	42.67	18.07	1956–2002	0.9	1/8
Koper	45.57	13.75	1962–1991	4.4	1/8
Eastern Mediterranean Index	N/A	N/A	1923–1997	0	1

A way by which one can check whether the last decade has been significantly different for the previous with respect to sea level variability is by deriving EOFs of smaller areas and checking whether the spatial patterns change or, put in another way, whether the reconstruction of the tide-gauges based on each decadal spatial pattern is always equally good. In order to do this we reconstructed, within each region, every tide-gauge record on the basis of EOFs derived from EOF analysis performed of parts of the tide-gauge records 10 yr long. Thus for the groups of stations (Table 1) we calculated EOFs for each decade and for the whole period covered by each subgroup of stations. Because the tide-gauge record length varied between regions the resulting EOFs in some regions cover only four decades, while in others cover almost a century.

Reconstructions of the tide-gauges in each region for the whole length of the tide-gauge were then made based on the spatial weights of the EOF of each decade as well as the spatial weights of the EOFs derived from the whole record. The assumption is that if all the reconstructions of one tide-gauge are equivalent, in terms of the variance explained by each reconstruction, then it does not matter which decade is used as the basis of extraction of the EOFs as all decades produce equivalent results. This would then be permitting the use of the spatial weights derived from altimetry as the basis of reconstruction of sea level variability, at least within these smaller regions. This, however, does not solve the problem of whether spatial patterns remain steady in time at larger scales.

3. Results

The spatial variability of sea level trends for the period January 1993 - December 2005 is shown in Figure 1. Smaller values can be found in the areas where the tide-gauges are located . Very large values both of positive and negative trends can be seen in some other areas for this period of observation. These are not long term trends in sea level but rather areas where decadal variability, mesoscale variability and the influence of the Eastern Mediterranean Transient has dominated the sea level signal (Fenoglio-Marc et al. 2002; Larnicol et al. 2002).

The possibility of using altimetry combined with the longest tide-gauges for the reconstruction of sea level variability depends on the correlation between the tide-gauge location and the rest of the Mediterranean Basin. In Figure 2 the correlation maps for several grid points of the MultiSensor product with the rest of the basin are shown. The locations used here are those of Genova, Trieste, Alicante, Malta, (which does not have a long term record but will be used below for validation) and Shoudas at the North of Crete which is used to explore the coherency of the Eastern basin because the Eastern Mediterranean Index used cannot be allocated to a specific position. It is not surprising that correlation is higher at the regions close to the location used as a reference point. However, there are areas, well known for mesoscale activity, in the southern parts of the eastern and western Mediterranean which appear less correlated or uncorrelated with the reference points. By contrast most of the northern Mediterranean coasts, the Gulf of Gabes and the Gulf of Sirte as well as most of the coastal regions in the Levantine basin appear to be correlated with all reference points at values higher than 0.6. This is an encouraging result indicating that the reconstruction for the coastal areas could well be successful even if away from the coast mesoscale variability affects the results. The TJP product gives comparable results (not shown) with the areas of agreement and disagreement larger due to the smoothing filter applied to it.

The mean correlation each grid point has with every other grid point is shown in Figure 2(f), which supports broadly the conclusions reached earlier, namely that there is significant correlation in all but the southern open sea areas of the eastern and western Mediterranean basins. This indicates that, any tide-gauge in the high mean correlation areas would be better than any tide-gauge in the low correlation areas. It also indicates that the longer tide-gauges of Trieste and Marseille are indeed in areas where the mean correlations are high and that the coastal areas in general appear well correlated even if the southern parts of the basin appear less correlated.

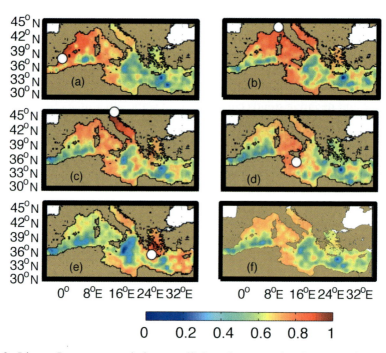

Fig. 2. Linear Pearson correlation coefficient between the deseasoned and detrended sea level time series at each grid point and at (a) Alicante, (b) Genova, (c) Trieste, (d) Malta, and (e) Soudha. The average correlation coefficients, where the value at grid point corresponds to the average of all the correlation coefficients between the sea level time series at that grid point and all other grid points (f).

The time variability of the EOFs from the two altimetric products and the tide-gauges are shown in Figure 3. The variance explained by the 1st EOF is 62%, 62% and 68% for TPJ, MultiSensor and tide-gauges respectively while the 2nd EOF explains 6%, 6%, and 17% respectively. The EOF of the tide-gauges stops in 1998 simply because many of the tide-gauge records have not been updated yet to 2006. This is a well known problem with tide-gauge data provision to the PSMSL database. For the common period the variability of the first EOF is very coherent for all three datasets. The correlation coefficient between the first EOFs of tide-gauge and altimetry is 0.82 and 0.81 for the TPJ and Multi-sensor product respectively. The second EOF of the tide-gauge is better correlated with the TPJ altimetry (0.57) than with the Multi-sensor altimetry (0.29). The two altimetric datasets also remain coherent in the remainder of the record (1999–2006) although some discrepancies are noticeable. The 2nd, 3rd and fourth EOFs of the tide-gauges describe different parts of the signal than those of the two altimetric products and they describe only 6% of the variance of the signal. The correlations between the 3rd and 4th tide-gauge based EOFs and altimeter are less than 0.30. However, the altimetric products are in reasonable agreement with each other. We can therefore suggest that the reconstruction of sea level within the Mediterranean from the time variability of long tide-gauges and altimetric records cannot go further than what is explained by the first and possibly the second EOF. The correlation between the EOFs of the altimetric records is 0.95, 0.75, 0.67, 0.75 for the first four EOFs respectively.

The spatial pattern of the first mode of the MultiSensor EOF (Figure 3) has indications of mesoscale signatures in the Ionian and Levantine Seas. However, it is more uniformly weighted with slightly higher weights in the north Adriatic. The other three modes indicate very strong regional patterns: the second in particular (Figure 3) determined by an oscillation between the Levantine basin and the rest of the Mediterranean, the third (not shown) an oscillation between the Alboran, Balearic Seas and the Ionian Seas and the fourth (not shown) an oscillation between the Adriatic and parts of the Levantine and Ionian. The interpretation of these features has already been discussed elsewhere (Larnicol *et al.* 2002). The modes from TPJ are in essence the same with lower spatial resolution caused by the filtering method (Figure 3). Because the first EOFs of both altimetric products are very well correlated with the tide-gauge EOFs, both can be used for the reconstruction of sea level backwards in time. However, the TPJ 2nd EOF appears better correlated with the tide-gauge EOFs than the Multi-sensor 2nd EOF, probably because the tide-gauges used are not close to the areas of mesoscale activity.

We use the tide-gauges in Valletta (Malta) and in Palma (Mallorca) which have not been used for the extraction of the tide-gauge EOFs as validation points. The tide-gauges are located in areas that have high average correlation (Figure 2) and which correlate well with the coastal areas of most of the Mediterranean and in particular with the location of the longest tide-gauges in Trieste and Genova. We correlate the temporal pattern of the first EOF with the tide gauge time-series. The correlation is 0.73 for Valletta and 0.53 for Palma. The reconstruction of the tide-gauges based on the first EOF is shown in Figure 4. The success of the reproduction is encouraging. Similar results are obtained for other tide-gauges. About 50% of the variability of the record of Valetta and around 22% of that of Palma is reconstructed by the first EOF (Figure 4). The 2nd EOF adds very little to variance explained in these stations and it is not shown.

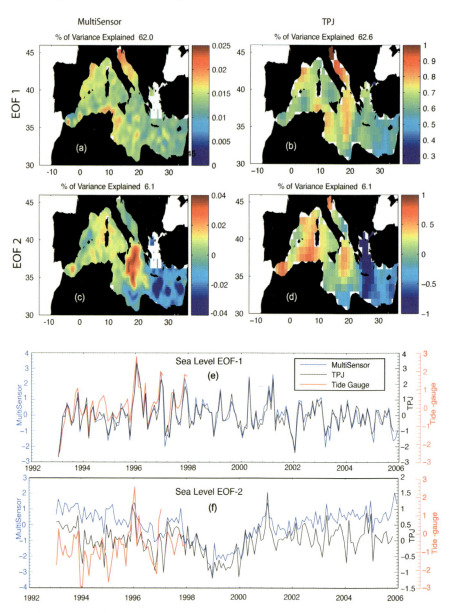

Fig. 3. The spatial patterns of the two first EOFs derived from the MultiSensor (a,c) and the TPJ altimetric product (b,d). The time variation of the first 2 EOFs estimated from the MultiSensor product (blue), the TPJ product (black) and the tide-gauge based indices (red).

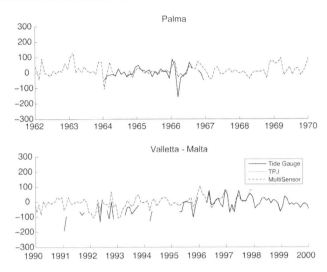

Fig. 4. Validation of the reconstruction in two tide-gauges, one (Valletta-Malta) during the era of altimetric mission. The other (Palma-Mallorca) for a period well before that era. Only the first EOF is used.

In order to reconstruct sea level fields for the past century we must ensure that the spatial and temporal scales of the variability are not dominated by transient signals. Thus there are two interrelated issues to be dealt with. First, to what extent the spatial patterns of sea level revealed by satellite altimetry are steady in time and second, whether the variability occurs at certain frequency bands rather than transient phenomena or noise. If there are frequency bands dominating the spectrum for the whole period of observational measurements then we may have the option of focusing on such frequencies, by filtering the data, and reconstructing the data on the characteristics of the dominant signal alone. However this is not what the tide-gauges reveal. Transient signals in the running power spectra are evident in many of the stations (Figure 5).

For example, in Marseille the power concentrated at the 2–10 years time scales indicates strong variability between 1935–1970. However, before that period there is no significant signal in this frequency band and even within this period the energy of the peak is not constant. A signal appears around 14 months at the early parts of the records. By contrast, Trieste shows more energy at the low frequency part of the spectrum at the beginning of the record with the signal disappearing after 1975. The signal at 14 months is in this station almost constant from 1960 onwards apart from the beginning of the record. Note that both the energy bands and the time intervals at which the peaks within these energy bands appear are variable within stations. In conclusion we cannot identify frequency bands which are permanently dominating the tide-gauge records. Thus we do not attempt to base the reconstruction on any particular frequency domain.

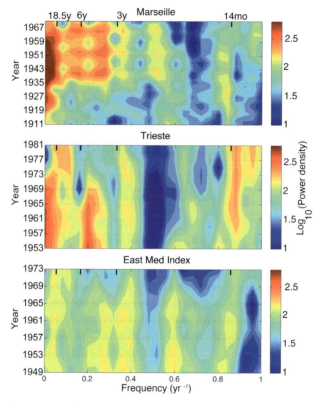

Fig. 5. Running spectra for Marseille, Trieste and the Eastern Mediterranean Index. Each spectrum was calculated by FFT on up to a 50 year long segment. The next spectrum in time was shifted by 2 yr. The resulting frequency resolution is 1/25 yr-1. The linear trend and the seasonal cycle were removed from the data.

We now turn into exploring the consistency of the observed variability during the last decade with the patterns observed in previous decades. This is done by using EOFs for the tide-gauges in the Adriatic and the western Mediterranean basin. In each region the EOFs were extracted for each decade. The spatial weights for each decade were then used to reconstruct each tide-gauge for the whole of the record. The percentage of the variability explained by each decadal reconstruction is used as the measure of the success of the reconstruction. Apart from the generally good ability of the first EOF to describe each tide-gauge, which is more than 70% for all but the tide-gauge in Genova, more important is that this ability is not restricted to a particular decadal reconstruction (not shown). This result implies that for the proximity of these groups of tide-gauges the spatial patterns have been steady during the previous century.

4. Conclusions

We have reconstructed sea level variability backwards into time by exploiting the correlation of tide-gauges and altimetry. The reconstruction is based on the assumption that the mesoscale variability has remained steady in time. A significant part of the detrended and deseasoned sea level variability (60%) in the Mediterranean is adequately described by use of EOFs whether these are based on the few tide-gauges located in the north coast of the western Mediterranean and the Adriatic or on the EOFs from the spatially complete altimetric datasets. As a result the coastal areas are well correlated everywhere apart from the coasts of the Alboran Sea.

This result provides confidence that coastal sea level can be reconstructed on the basis of any of the existing long tide-gauge records within the basin even if the spatial distribution of them is biased towards the European coasts. Such reconstruction is reliable in all the Mediterranean Sea with the exception of areas of strong mesoscale activity. As an example we have reconstructed the tide-gauge records of Valletta in Malta and Palma in Mallorca. The reconstructions based on the MultiSensor and the TPJ products are virtually equivalent in respect of the first EOF. The second EOF of the TPJ dataset is marginally better correlated with the 2nd EOF of the tide-gauge data than that of the Multi-sensor product. This is probably due to the higher spatial resolution of the latter. This results in the EOFs being affected by the mesoscale variability of the basin which is not present near the coast where the tide-gauges are located. However, the inclusion of the 2nd EOF adds very little to the variance explained.

In spite of the well known peculiarity of the 1990s, in the Mediterranean and globally, the reconstruction based on the spatial components of EOFs, derived from decadal segments of the tide-gauge records in the 1990s, is equally successful to reconstructions based on earlier decades. Thus, within each region, the spatial patterns can be used as the basis of extrapolation into the past. However, this work has not resolved whether this is true for the spatial patterns within the basin. Reconstructions of sea level based on the spatial altimetric patterns are reliable only to the extent that the regions of mesoscale activity have remained constant in time.

The power spectrum of the sea level variability at each tide-gauge appears to be dominated by particular frequencies. However, the time variability of the energy at these frequencies is neither steady in time at each station nor similar between stations excluding the possibility of reconstructing particular frequencies more successfully.

Acknowledgments

This paper was written during a visit of M. Tsimplis to IMEDEA and was completed during a visit by A. Pascual at NOC. The work is partly funded by the VANIMEDAT project (CTM2005-05694-C03/MAR) and is partly supported by the EC CIRCE project. M. Marcos acknowledges a post-doc fellowship funded by the Spanish Ministry of Education and Science.

References

Fenoglio-Marc L (2002) Long-term sea level change in the Mediterranean Sea from multi-satelitte altimetry and tide gauges. Physics and Chemistry of the Earth 27: 1419–1431

Fenoglio-Marc L, Kusche J, Becker M (2006) Estimation of mass variation and mean dynamic topography in the Mediterranean Sea from altimetry and GRACE/GOCE geoids. In: 3rd GOCE Users Symposium Proceedings

Larnicol G, Ayoub N, Le Traon PY (2002) Major changes in Mediterranean Sea level variability from 7 years of TOPES/Poseidon and ERS-1/2 data. J Mar Syst 33–34: 63–89

Le Traon PY, Faugère Y, Hernandez F, Dorandeu J, Mertz F, Ablain M (2003) Can we merge GEOSAT Follow-On with TOPEX/POSEIDON and ERS-2 for an improved description of the ocean circulation? J Atmos Oceanic Technol 20: 889–895

Pascual A, Faugère Y, Larnicol G, Le Traon PY (2006) Improved description of the ocean mesoscale variability by combining four satellite altimeters. Geophys Res Lett 33, L02611, doi:10.1029/2005GL024633

Pascual A, Pujol MI, Larnicol G, Le Traon PY, RioMH (2007) Mesoscale Mapping Capabilities of Multisatellite Altimeter Missions: First Results with Real Data in the Mediterranean Sea. J Mar Sys 65, 190–211, doi:10.1016/ j.jmarsys.2004.12.004

Toumazou V, Cretaux JF (2001) Using a Lanczos Eigensolver in the Computation of Empirical Orthogonal Functions. Month Weather Rev 129: 1243–1250

Tsimplis MN, Josey SA (2001) Forcing of the Mediterranean Sea by atmospheric oscillations over the North Atlantic. Geophys Res Lett 28 (5): 803–806

Tsimplis MN, Spencer NE (1997) Collection and analysis of monthly mean sea level data in the Mediterranean and the Black Sea. J Coast Res 13: 534–544

Woodworth PL, Player R (2003) The Permanent Service for Mean Sea Level: an update to the 21st century. J Coast Res 19: 287–295

Internal Waves Generated in the Straits of Gibraltar and Messina: Observations from Space

Werner Alpers[1], Peter Brandt[2], and Angelo Rubino[3]

[1] Institut für Meereskunde, Universität Hamburg, Hamburg, Germany
[2] Leibniz-Institut für Meereswissenschaften, IFM-GEOMAR, Kiel, Germany
[3] Dipartimento di Science Ambientali, Università Ca' Foscari, Venezia, Italy

Abstract. The Straits of Gibraltar and Messina are areas where strong internal solitary waves are generated by the interaction of tidal currents with shallow underwater ridges located within the straits. Remote sensing techniques and numerical simulations have been instrumental in studying the generation and propagation of internal solitary waves in these sea areas. It was a Synthetic Aperture Radar (SAR) image acquired by the American SEASAT satellite in 1978 that first revealed that long internal waves are generated in the Strait of Messina. Furthermore, SAR images acquired by the European Remote Sensing satellite ERS-1 and ERS-2 have revealed that trains of internal solitary waves generated in the Strait of Gibraltar propagate mostly eastward into the Mediterranean Sea, while westward propagating wave trains can only be supported by a seasonal thermocline.

1. Introduction

Internal waves are waves of the interior ocean. They can exist when the water body is stratified, *i.e.* when it consists of layers of different density. This difference in water density is mostly due to a difference in water temperature, but it can also be due to a difference in salinity as in the Strait of Gibraltar. Often the density structure of the ocean can be approximated by two layers. In this paper we will consider long internal waves propagating along the main pycnocline.

In the Straits of Gibraltar and Messina the tidal flow pushes the layered water body over the shallow ridges or sills located within the straits. A lee depression of the interface is thus generated which crosses the strait when the tide slackens. As demonstrated theoretically long time ago (Korteweg

V. Barale, M. Gade (eds.), *Remote Sensing of the European Seas.*

and de Vries 1895) such a travelling disturbance may disintegrate into trains of rank-ordered solitary waves. To first order, such internal waves do not produce an elevation of the sea surface as the familiar surface waves do, but they induce a variable horizontal surface current. This variable surface current gives rise to a change of the sea surface roughness, which, under favourable viewing conditions, can be detected by visible observations from ships, air- and spacecrafts. From space, internal waves can be detected very efficiently using a Synthetic Aperture Radar (SAR) which yields images of the sea surface with a spatial resolution of typically 25 meters. However, also optical, and, to a lesser extent, infrared and ultraviolet images acquired from satellites have been used to observe internal waves from space.

2. SAR imaging of internal waves

A linear internal wave propagating in a two-layer ocean causes a variable surface current which varies in magnitude and direction thus generating convergent and divergent surface flow regimes. The variable surface current interacts with the surface waves and modulates the sea surface roughness (Alpers 1985). According to Bragg scattering theory, the amplitude of small-scale sea surface waves, which obey the Bragg resonance condition (the so-called "Bragg waves") determines the backscattered radar power. Due to this hydrodynamic interaction of the Bragg waves with the variable surface current, their amplitude increases in convergent flow regions and decreases in divergent flow. As a consequence, the radar signatures of oceanic internal waves consist of alternating bright and dark bands.

However, sea surface manifestations of internal waves are not only visible on radar images. Under favourable viewing conditions they are also visible on images acquired in the visible, ultraviolet or infrared bands (see *e.g.* Artale *et al.* 1990; Mitnik *et al.* 2000). The reason is that the backscattered (or reflected) sunlight from the sea surface strongly depends on the sea surface roughness.

Tidally generated internal waves, which are in general nonlinear and dispersive, often evolve as trains of solitary waves. They are generated by the disintegration of long internal waves of tidal period. The resulting wave packets consist of several solitary waves. Since the first theory on solitary waves (soliton theory) was developed by Korteweg and de Vries (1895) hundreds of papers have been published dealing with this subject. Soliton theories applicable to the description of the generation and propagation of internal solitary waves predict that, if the depth of the upper layer is smaller than the depth of the lower layer, the resulting stationary internal solitary waves are waves of depression, *i.e.* the pycnocline is pushed down.

In Figure 1 a typical example a nonlinear internal wave packet consisting of three solitary waves of depression is depicted. The color coding denotes the water density and the arrows the velocity. This profile was measured north of the Strait of Messina on 25 October 1995 by a towed conductivity-temperature-depth (CTD) chain and by a vessel-mounted acoustic Doppler current profiler (ADCP).

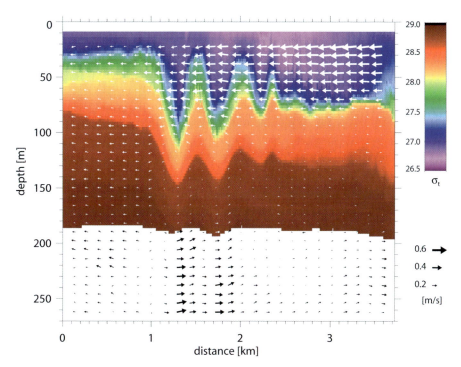

Fig. 1. Density distribution of the water column and distribution of the velocity north of the Strait of Messina measured by ship-borne sensors during the passage of a highly nonlinear internal wave packet on 25 October 1995.

The leading edge of a soliton of depression is always associated with a convergent surface flow region and the trailing edge with a divergent region. At the front of the internal soliton, the amplitude of thc Bragg waves is increased, while at the rear it is decreased. This is the reason why on SAR images the front section of an internal solitary wave of depression is bright and the rear section is dark (Alpers 1985).

However, when the wind speed is low and when surface slicks are present, the radar signatures of internal solitary waves may deviate from this scheme. When the wind speed is below threshold for Bragg wave generation, strong internal solitary waves often manifest themselves on radar images as bright bands which, most likely, are caused by breaking short-scale surface waves even in the absence of wind. When surface slicks are present, the internal waves often manifest themselves as dark lines caused by the accumulation of surfactants in the convergent surface flow regions.

Fig. 2. Astronaut photograph of the Strait of Gibraltar and adjacent sea areas taken from the Space Shuttle on 11 October 1984, at 12:22 UTC, showing sea surface manifestations of two internal wave packets propagating out of the Strait of Gibraltar into the Mediterranean Sea[1]. The land area in the upper right-hand section of the image is Spain, the thorn-shaped peninsula at the eastern entrance of the Strait attached to Spain is the British Crown Colony Gibraltar and the land area in the lower left is Morocco.

3.   Internal solitary waves in the Strait of Gibraltar

The Strait of Gibraltar connects the Atlantic Ocean with the Mediterranean Sea. The water body in the Strait of Gibraltar and its approaches consists of a deep layer of salty Mediterranean water and an upper layer of less salty Atlantic water. The mean depth of the interface between these two

---

[1] http://www.lpi.usra.edu/publications/slidesets/oceans/oceanviews/slide_13.html

layers slopes down from about 80 m at the Mediterranean side of the Strait to about 800 m at the Atlantic side. The relative change of density across this interface, which is mainly determined by the salinity difference, is about 0.002 (Lacombe and Richez 1982). The Strait of Gibraltar has a complex bottom topography including several ridges. The shallowest section in the Strait of Gibraltar is at the Camarinal Sill where the maximum water depth is 290 m. The interaction of the predominantly semidiurnal tidal flow with the sills inside the Strait, in particular with the Camarinal Sill, causes periodic deformations of the halocline in the sill regions (Armi and Farmer 1985; Farmer and Armi 1988) which then give birth to internal solitary waves with amplitudes as high as 80 m.

Unfortunately, the first SAR missions (SEASAT 1978, Shuttle Imaging Radar missions SIR-A 1981, SIR-B 1984, and SIR-C/X-SAR 1994) did not acquire images over the Strait of Gibraltar. However, during the Space Shuttle flight STS 41-G (the SIR-B mission) the US oceanographer-astronaut Paul Scully-Power took photos of the Strait of Gibraltar with a hand-held camera, which showed on 11 October 1984, at 12:22 UTC, impressive sea surface signatures of two internal wave packets (see Figure 2). Only after the launch of the first European Remote Sensing Satellite (ERS-1), in 1991, a large number of SAR images of the Strait became available. Many of them, in particular those acquired near spring tide, show sea surface manifestations of internal solitary waves (Brandt *et al.* 1996).

In Figure 3 a typical ERS-1 SAR image of the Strait of Gibraltar is depicted showing sea surface manifestations (or radar signatures) of an internal solitary wave packet consisting of more than 10 long internal waves. Note that the distance between the solitary waves in the packet decreases from front to rear and that the strength of the image intensity modulation of the solitary waves also decreases from front to rear indicating a successive decrease in amplitude of the solitary waves.

4.   Internal solitary waves in the Strait of Messina

The Strait of Messina separates the island of Sicily from the Italian peninsula and connects the Tyrrhenian Sea, in the north, with the Ionian Sea, in the south. In spite of the small tidal displacements encountered in the Mediterranean Sea, large gradients of tidal displacements are present along the Strait of Messina, because the semidiurnal tides in the Tyrrhenian and Ionian Seas are approximately in phase opposition. These gradients, acting on the water body constrained by the strait topography, force intense tidal currents, which can be as large as 3 m/s in the sill region (Vercelli 1925).

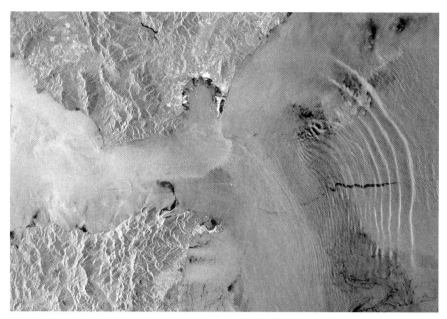

Fig. 3. ERS-1 SAR image acquired on 1 January 1993 at 22:39 UTC (orbit: 7661) showing sea surface manifestations of a packet of internal solitary waves generated in the Strait of Gibraltar and propagating eastward into the Mediterranean Sea. The dark line intersecting the packet results from an oil spill, probably released from a ship. Imaged area: 100 km × 50 km. © ESA

The hydrological peculiarities of the Strait of Messina attracted the attention of many ancient writers and philosophers. Homer (800 B.C.) makes two monsters, Scylla and Charybdis, responsible for the violent currents in the strait (Homer, Odyssey, 12th song, line 80–114). Aristotle (384–322 B.C.) argues that hollows in the sea floor and the interaction of two opposing wind-generated currents could produce such intense currents (Aristotle, Problema Physica, chap. 23) and in the poetry of ancient times, allegories alluding to the danger of sailing in the Strait of Messina can often be found ('Incidis in Scillam cupiens vitare Charybdim', Ovid, Metamorphosis).

As the Strait of Messina represents a barrier to the free water exchange between the Tyrrhenian and the Ionian Seas, significant horizontal and vertical density gradients are encountered in this region. According to the climatological density distribution, at all depths the water south of the strait is throughout the year denser than north of it. The knowledge of the presence of horizontal density gradients along the Strait of Messina enabled Defant (1940) to draw a picture of the tidally induced dynamics of this area: during northward tidal flow, the Ionian water overflowing the sill spreads under the Tyrrhenian water into the Tyrrhenian Sea. During southward tidal flow, the Tyrrhenian water, overflowing the sill, forms a surface jet that spreads into the Ionian Sea.

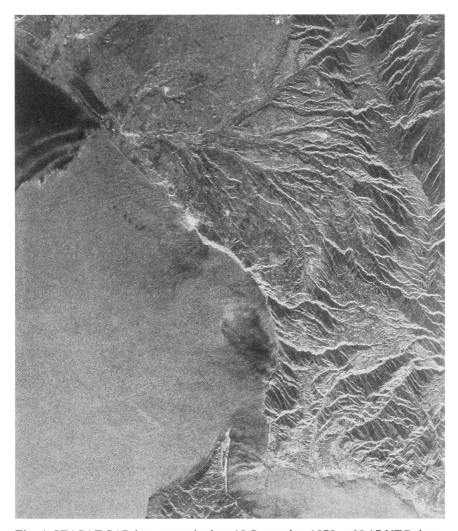

Fig. 4. SEASAT SAR image acquired on 18 September 1978 at 08:17 UTC showing sea surface manifestations of northward propagating internal solitary waves generated in the Strait of Messina (lower section). © NASA

The shallowest section in the sill region, in the centre of the Strait, has a depth of 90 m. While in the southern part the Strait bottom slopes down very steeply, to a depth of more than 800 m about 15 km south of the sill, the northern part has a gentler slope. Here the 400 m isobath is located about 15 km north of the sill. Throughout the year, two different layers of water are encountered in the Strait of Messina: the Tyrrhenian Surface Water and the colder and saltier Levantine Intermediate Water. In the vicinity of the Strait, these water masses are separated at a depth of approximately 150 m (Vercelli 1925). During most of the year, a seasonal thermocline is also present in the Strait which overlies this weak stratification.

Fig. 5. ERS-1 SAR image acquired on 11 July 1993 at 9:41 UTC (orbit: 10387) showing northward as well as southward propagating internal wave packets. Imaged area: 65 km × 65 km. @ ESA

The fact that (1) strong tidally induced currents are encountered in the Strait, (2) the water body is stratified, and (3) there is a shallow sill in the center of the Strait which disturbs the tidal flow, suggests that internal waves should be generated in the Strait of Messina. But it was not before 1978 that internal waves were detected in this strait. The first hint came from a synthetic aperture radar image which was acquired by the American SEASAT satellite on September 15, 1978 (see Figure 4).

The three rings visible on the SEASAT SAR image of the Tyrrhenian Sea north of the Strait were interpreted as sea surface manifestations of a train of internal solitary waves propagating northwards (Alpers and Salusti 1983). In the following years internal waves propagating northward as well as southward have been detected during several oceanographic campaigns by in-situ measurements (Sapia and Salusti 1987; Nicolò and Salusti 1991). Nonlinear internal waves propagating in a southward direction out of the Strait of Messina have also been detected on a Landsat 5 thematic mapper image acquired on 20 July 1984 at 09:30 LT (Artale *et al.* 1990).

A large number of spaceborne SAR images of the Strait of Messina became available after the launch of the First European Remote Sensing Satellite ERS-1 in 1991 and the Second European Remote Sensing satellite ERS-2 in 1994 (Alpers *et al.* 1996). A typical ERS-1 SAR image showing internal wave packets propagating northward as well as southward out of the strait is depicted in Figure 5. It was acquired on 11 July 1993 at 09:41 UTC, which was 20 min after the maximum northward tidal flow at Punta Pezzo (located at the Calabrian coast at the northern exit of the Strait).

The analysis of a large number of ERS-1 and ERS-2 SAR images acquired between 1991 and 1995 (Brandt *et al.* 1997) has shown that sea surface manifestations (or radar signatures) of internal waves are observed more frequently during periods when a strong seasonal thermocline is known to be present, *i.e.* during summer. Furthermore, sea surface manifestations of northward propagating internal solitary waves are quite uncommon (Brandt *et al.* 1997). Since the strong roughness bands are associated with large-amplitude internal solitary waves, the presence of such strong radar signatures on SAR images could be used as an indicator for an anomalous density distribution along the Strait of Messina that has its origin in fluctuations of larger-scale circulation patterns. Indeed, on one occasion, as an anomalous horizontal density distribution of the water masses was present in this region, the presence of strong internal waves north of the Strait was confirmed by in-situ measurements (Brandt *et al.* 1999). This was on 24 and 25 October 1995, when oceanographic measurements were carried out north and south of the Strait of Messina from the research vessel Alliance of the NATO Undersea Research Centre (Figure 1).

As mentioned before, internal waves generated in the Strait of Messina have also been detected on optical images. A very impressive example of such an optical image is depicted in Figure 6. It was acquired by the Advanced Spaceborne Thermal Emission and Reflection Radiometer (ASTER) on

NASA's Terra satellite on 11 August, 2003, when the sun was located just at right angle to illuminate the internal wave packets. ASTER has in the visible band a resolution of 15 m. On this image two wave patterns are visible: a strong one resulting from internal waves generated in the Strait of Messina and a weak one resulting from diffraction of the primary internal soliton.

Fig. 6. ASTER image acquired on 11 August, 2003, showing a strong southward propagating internal wave packet and a weak quasi- semicircular internal wave packet resulting from diffraction of the primary internal soliton propagating southward along the channel[2].

---

[2] http://earthobservatory.nasa.gov/Newsroom/NewImages/images.php3?img_id=17 628

5.   Discussion and conclusion

The advent of remote sensing of the ocean revealed the fascinating presence, in the straits of Gibraltar and Messina, of coherent, large patterns in the sea surface roughness, which have been recognized as surface manifestations of long internal waves. It was a SEASAT SAR image acquired in 1978 that first revealed that internal waves are generated in the Strait of Messina (Alpers and Salusti, 1983). Since then, a large amount of investigations has been devoted to the understanding of the nature, the generation, propagation, and dissipation of tidally induced long internal solitary waves. For instance, shore based marine radar observations from the rock of Gibraltar as well as airborne synthetic aperture radar observations contributed to clarify the generation and propagation characteristics of the internal solitary waves in the Strait of Gibraltar (Watson and Robinson, 1990; Richez 1994). Parallel, theoretical investigations addressed the hydrodynamics and the imaging mechanisms of such oceanic features (Brandt et al., 1996, 1997) thus contributing to deepen our knowledge on the complexity of the baroclinic dynamics of these sea straits.

Acknowledgements

We thank ESA who provided us with a large number of ERS and Envisat SAR images acquired over the Straits of Gibraltar and Messina.

References

Alpers W (1985) Theory of radar imaging of internal waves, Nature, 314, 245–247
Alpers W, Salusti E (1983) Scylla and Charybdis observed from space. J Geophys Res 88: 1800–1808247
Alpers W, La Violette PE (1993) Tide-generated nonlinear internal wave packets in the Strait of Gibraltar observed by the synthetic aperture radar aboard the ERS-1 satellite. Proc First ERS-1 Symposium - Space at the Service of our Environment, Cannes, 4-6/11/1992. ESA, Paris, ESA SP-359, pp 753–758
Alpers W, Brandt P, Rubino A, Backhaus JO (1996) Recent contributions of remote sensing to the study of internal waves in the Straits of Gibraltar and Messina. In: Briand F (ed) Dynamics of Mediterranean Straits and Channels, CIESM Science Series no.2, Bulletin de l'Institut Ocèanographique, Monaco, no. special 17: 21–40
Armi L, Farmer DM (1985) The internal hydraulics of the Strait of Gibraltar and associated sills and narrows. Oceanologica Acta 8: 37–46

Artale V, Levi D, Marullo S, Santoleri R (1990) Analysis of nonlinear internal waves observed by Landsat thematic mapper. J Geophys Res 95: 16065–16073

Brandt P, Alpers W, Backhaus JO (1996) Study of the generation and propagation of internal waves in the Strait of Gibraltar using a numerical model and synthetic aperture radar images of the European ERS 1 satellite. J Geophys Res 101: 14237–14252

Brandt P, Rubino A, Alpers W, Backhaus JO (1997) Internal waves in the Strait of Messina studied by a numerical model and synthetic aperture radar images from the ERS 1/2 satellites. J Phys Oceanogr 27: 648–663

Brandt P, Rubino A, Quadfasel D, Alpers W, Sellschopp J, Fiekas H (1999) Evidence for the influence of Atlantic-Ionian stream fluctuations on the tidally induced internal dynamics in the Strait of Messina. J Phys Oceanogr 29: 1071–1080

Defant A (1940) Scylla und Charybdis und die Gezeitenstroemungen in der Strasse von Messina. Ann Hydr Marit Meteor 5: 145–157

Defant A (1961) Physical Oceanography, vol 1 & 2. Pergamon Press, New York

Korteweg, DJ, de Vries G (1895) On the change of long waves advancing in a rectangular canal and a new type of long stationary waves. Phil Mag 5: 422

Lacombe H, Richez C (1982) The regime of the Strait of Gibraltar. In: Nihoul JCJ (ed) Hydrodynamics of Semi-Enclosed Seas. Elsevier, Amsterdam, pp 13–73

Mitnik L, Alpers W, Chen KS, Chen AJ (2000) Manifestation of internal solitary waves on ERS SAR and SPOT images: similarities and differences. Proc 2000 Int Geoscience and Remote Sensing Symposium (IGARSS'00) Hawaii, USA, 24–28 July 2000, vol 5: 1857–1859

Nicolò L, Salusti E (1991) Field and satellite observations of large amplitude internal tidal wave trains south of the Strait of Messina, Mediterranean Sea. Ann Geophys 9: 534–539

Richez C (1994) Airborne synthetic aperture radar tracking of internal waves in the Strait of Gibraltar. Prog Oceanogr 33: 93–159

Sapia A, Salusti E (1987) Observation of non-linear internal solitary wave trains at the northern and southern mouths of the Strait of Messina. Deep-Sea Res 34: 1081–1092

Vercelli F (1925) Il regime delle correnti e delle maree nello stretto di Messina. Comm. Int. del Mediterraneo, Venice, Italy

Watson G, Robinson IS (1990) A study of internal wave propagation in the Strait of Gibraltar using shore-based marine radar images. J Phys Oceanogr 20: 374–395

High Resolution Wind Field Retrieval from Synthetic Aperture Radar: North Sea Examples

Jochen Horstmann and Wolfgang Koch

GKSS Research Center, Institute for Coastal Research, Germany

Abstract. A methodology for retrieving high resolution ocean surface wind fields from satellite borne Synthetic Aperture Radar (SAR) data is introduced and validated. The algorithm is applicable to SAR data acquired at C-band at moderate incidence angles. Wind directions are extracted from wind induced streaks that are visible in SAR images and that are very well aligned with the mean surface wind direction. To extract the orientation of these streaks an algorithm based on the derivation of local gradients is utilized. Ocean surface wind speeds are derived from the Normalized Radar Cross Section (NRCS) using a geophysical model function that describes the dependency of the NRCS on the wind and imaging geometry. To validate the algorithm and demonstrate its applicability, SAR retrieved wind fields of the North Sea are compared to numerical atmospheric model results of the German Weather Service.

1. Introduction

Nowadays, several scatterometers (SCAT) are in orbit, which enable to measure wind fields with a resolution of up to 25 km on a global and operational basis independent of daylight and cloudiness. Originally SCAT were not designed to measure high resolution wind fields and therefore make it difficult to measure the highly spatially variable winds, which are especially important in coastal areas. However, satellite-borne Synthetic Aperture Radars (SAR) offer the unique opportunity to image the ocean surface with a very high resolution, typically below 100 m.

Since the launch of the European remote-sensing satellites ERS-l, ERS-2, and Environmental Satellite ENVISAT, as well as the Canadian Radar Satellite RADARSAT-1, SAR images have been acquired over the oceans on a continuous basis over the last 15 years. Their high resolution, together with their large spatial coverage, make them a valuable tool for measuring oceanographic parameters such as ocean surface winds, waves, and sea ice.

V. Barale, M. Gade (eds.), *Remote Sensing of the European Seas.*
© Springer Science+Business Media B.V. 2008

All the above mentioned SAR operate at C-band (5.3 GHz) with either vertical (V) or horizontal (H) polarization in transmission and reception and at moderate incidence angles between 15° and 55°. For this electromagnetic wavelength (~5 cm) and range of incidence angles the backscatter of the ocean surface is primarily caused by the small-scale ocean surface roughness on horizontal scales of 5 to 10 cm. This dominant scattering mechanism is called resonant Bragg scattering.

The ability of space borne microwave radars to measure the wind vector near the surface of the ocean relies on the fact that the near-surface wind field generates the small scale surface roughness, which increases with wind speed. For moderate incidence angles and wind speeds below ~30 ms^{-1}, the Normalized Radar Cross Section (NRCS) is typically largest when the wind blows directly toward the radar and decreases to a minimum when the wind direction is orthogonal to the radar look direction. Another smaller maximum in NRCS occurs when the wind blows directly away from the radar. The relation between the near-surface wind vector and NRCS can be described by an equation of the form

$$\sigma_0 = a(\theta) u^{\gamma(\theta)} \left(1 + b(u,\theta)\cos\Phi + c(u,\theta)\cos 2\Phi\right) \qquad (1)$$

where σ_0 is the NRCS, u the wind speed, Φ the relative angle between the radar look and wind direction, and θ is the nadir incidence angle. The quantities $a(\theta)$, $\gamma(\theta)$, $b(u,\theta)$ and $c(u,\theta)$ are empirical parameters that are functions of θ and sometimes u. Equation (1) captures the nature of the wind vector to NRCS relationship; specifically that NRCS is an exponential function of wind speed and a harmonic function of its direction (relative to the radar look). It is important to note that though the model function can relate wind speed and direction to NRCS, the reverse is not true. A specific NRCS can be associated with a large number of wind speed and direction pairs. This significantly complicates the inversion from NRCS to wind speed. If wind direction is known, then it is possible to use Equation (1) in conjunction with NRCS to estimate wind speed. An approach convenient for operational use is to use predicted wind directions from operational meteorological models (Monaldo et al. 2001) or to estimate the wind direction directly from linear structures in the SAR image (Vachon and Dobson 1996; Lehner et al. 1998; Horstmann et al. 2002). All these approaches have proven successful for moderate wind speed regimes. For an overview of the different methods and applications of SAR wind retrieval refer to Monaldo et al. (2004a).

In this paper, WiSAR, an algorithm to retrieve wind-fields from satellite borne SARs operating at the C-band is introduced. The algorithm is applied to retrieve wind fields from the Advanced SAR (ASAR) aboard the

European satellite ENVISAT. To demonstrate the applicability of WiSAR, ASAR-retrieved wind fields are compared to results of the operational numerical atmospheric model of the German Weather Service (DWD).

2.   Utilized data

For the investigations, ASAR data acquired by the European remote sensing satellite ENVISAT were used. The ENVISAT satellite operates in a sun-synchronous polar orbit at a height of 800 km. It has an orbital period of 101 min and is operating in a 35-day repeat cycle. The ASAR system is a right-looking system, which means that the imagery is acquired on the right-hand side with respect to the satellite flight direction (azimuth) perpendicular to the flight direction. ENVISAT ASAR acquires images at the C band (5.34 GHz) and can be operated at different polarization combinations. For this study, 61 ASAR images were acquired at either HH or VV polarization in transmission and reception. All ENVISAT ASAR imagery utilized were acquired in the ScanSAR wide-swath mode, which offers the largest coverage in the across-track direction (range). ScanSAR images are generated by scanning over the incidence angles and sequentially synthesizing images for different subswaths at incidence angles between 15° and 45°. In ScanSAR wide-swath mode, ENVISAT ASAR can image a swath width of up to 450 km with a spatial resolution of ~100 m.

3.   Wind retrieval from SAR

Ocean surface wind retrieval from SAR is a two-step process; in the first step, wind directions are retrieved, which are a necessary input in the second step. Wind directions are extracted from wind-induced streaks visible in the SAR image at different scales; typically above 200 m. Wind speeds are retrieved from the backscattered NRCS of the ocean surface utilizing a Geophysical Model Function (GMF), which describes the dependence of the NRCS on the wind and radar imaging geometry.

3.1   Wind direction retrieval

The most popular methodology for SAR wind-direction retrieval is based on the imaging of linear features at scales above 200 m. Most of these features are associated to wind-induced streaks and Marine Atmospheric Boundary Layer (MABL) rolls. Investigations of tower based Real Aperture Radar

(RAR) imagery by Dankert *et al.* (2003) have shown that wind-induced streaks can be seen at various scales between approximately 50 and 500 m. Their results encourage to focus on the smallest possible scales that can be utilized from space-borne SARs. In case of the SAR data used within this study the smallest scales are ~200 m (limited by the spatial resolution of the SAR system). Results of SAR wind-direction retrieval based on larger scale features (<3 km) often depict MABL rolls, which are more likely to significantly differ from the mean surface wind direction (Etling and Brown 1993). The orientation of the linear features imaged by the SAR can be retrieved by two methods, the Local-Gradient (LG) method (Horstmann *et al.* 2002; Koch 2004), which is applied in the spatial domain, and the Fast Fourier Transformation (FFT) method (Gerling 1986; Vachon and Dobson 1996, Lehner *et al.* 1998), which is applied in the spectral domain.

In case of the LG method the SAR image is smoothed and reduced to resolutions of 100, 200, and 400 m. This results in three SAR images representing spatial scales above 200, 400, and 800 m. From each of these images, local directions, defined by the normal to the LG, are computed leaving a 180° ambiguity. All pixels that are affected by non wind-induced features, *e.g.* land, surface slicks, and sea ice, are masked and excluded from further analysis, considering land masks and SAR image filters. The image filters are extracted from the SAR image itself considering parameters, such as, the mean and standard deviation of the image intensity as well as the retrieved LGs (for details refer to Koch 2004). Finally, from all of the resulting directions, only the most frequent directions in a predefined grid cell are selected. The 180° directional ambiguities can be removed if wind shadowing is present, which is often visible in the lee of coastlines, or other sources, *e.g.* weather charts, atmospheric models or *in situ* measurements, have to be taken into account.

3.2   Wind speed retrieval

With the wind direction in hand, we can use the NRCS measured by the SAR to retrieve the wind speed using GMFs of the form given in Equation (1). For C-band NRCSs acquired at VV polarization there are a number of GMFs available. These were determined empirically by evaluation of SCAT data acquired by ERS-1 and ERS-2 to co-located winds from the numerical model of the European Center for Medium-Range Weather Forecast (ECMWF). The most common C-band GMFs are the Cmod4 (Stoffelen and Anderson 1997), CmodIfr2 (Quilfen *et al.* 1998) and the recently developed Cmod5 (Hersbach *et al.* 2007).

Each of these GMFs is directly applicable for wind speed retrieval from C-band VV polarized SAR images (*e.g.* Vachon and Dobson 1996; Lehner *et al.* 1998; Horstmann *et al.* 2003). For wind speed retrieval from C-band SAR images acquired at HH-polarization (*i.e.* the SAR aboard RADARSAT-1), no similar well-developed model exists. To meet this deficiency a hybrid model function has to be applied that consists of one of the prior mentioned empirical models and a C-band polarization ratio (Horstmann *et al.* 2000; Thompson and Beal 2000; Vachon and Dobson 2000). The Polarization Ratio (PR) is defined as the ratio of HH-polarization NRCS to VV-polarization NRCS. The nature of the PR is still an active area of research and several different PR's have been proposed in literature (Thompson *et al.* 1998; Mouche *et al.* 2005). The PR proposed by Thompson *et al.* 1998 neglects wind speed and wind direction dependence and is given by

$$PR = \frac{\left(1 + \alpha \tan^2 \theta\right)^2}{\left(1 + 2 \tan^2 \theta\right)^2} \tag{2}$$

where α is a constant set to 0.6 to yield consistency with the measurements of Unal *et al.* (1991). Several different values for α have been suggested in literature. These vary between 0.4 and 1.2 (Horstmann *et al.* 2000; Vachon and Dobson 2000; Monaldo *et al.* 2001, Horstmann and Koch 2005). Comparisons of RADARSAT-1 SAR imagery produced from different SAR processing facilities showed that the different estimates of α may be due to the different radiometric calibrations of RADARSAT-1 SAR data. Mouche *et al.* (2005), proposed a C-band PR that is also dependent on wind direction. Their model was constructed using airborne RAR data acquired at C-band with both VV- and HH-polarization for moderate incidence angles and a wide range of wind speeds and wind directions.

Comparisons of C-band SAR retrieved wind speeds, using the Cmod4 at low to moderate wind speeds (up to ~20 ms^{-1}) resulted in errors of ~2 ms^{-1} (Monaldo *et al.* 2001; Horstmann *et al.* 2003; Monaldo *et al.* 2004b; Horstmann and Koch 2005). It is also well known that both, the Cmod4 and CmodIfr2 model, underestimate the wind speed at high winds (>20 ms^{-1}) when applied to SCAT and SAR data (Donnelly *et al.* 1999; Horstmann *et al.* 2005). The Cmod5 model was specifically designed to better estimate NRCS at higher wind speeds. It was constructed primarily using co-locations between ERS-2 SCAT backscatter triplets and ECMWF first-guess model winds. For extremely high wind conditions, results from aircraft campaigns were included (Donnelly *et al.* 1999). Differences between Cmod4 and Cmod5 for low to moderate wind speeds are relatively minor. At high wind speeds (≥ 25 ms^{-1}) the differences become significant. In particular, the NRCS from Cmod4 increases monotonically with wind speed for all incident angles, while that predicted by Cmod5 increases much more slowly. For lower incident angles ($\leq 25°$ to $30°$ depending on wind speed) the NRCS can even reach a maximum value and start to decrease with further increase in wind speed. Furthermore, dependence of the NRCS on wind direction becomes much smaller in the high wind regime than that of Cmod4.

4.   Validation of SAR wind field retrieval

In the following, wind fields are retrieved from ENVISAT ASAR images using the methodology introduced in Section 3. The ASAR-retrieved wind fields are compared to the results of the German Weather Service (DWD) model, which permits us to draw some conclusions on the accuracy of SAR wind retrieval.

4.1   Example of an ASAR wind field

Figure 1 shows an ENVISAT ASAR image of the Southern North Sea acquired in the ScanSAR mode with HH polarization. The ASAR wind fields were retrieved from the area corresponding to the grid cell in the DWD-model output, which corresponds to an average grid-cell size of approximately 45 km × 75 km. Wind directions were retrieved using the LG method and the wind speeds using the Cmod4 model with the polarization ratio according to Equation (2) with an $\alpha = 0.6$. The wind direction ambiguities could be removed due to the wind shadowing, which is especially visible at the west coast of Germany (lower right of Figure 1). The DWD model results are only available on a 6-h basis so that the model wind fields were interpolated to the exact ASAR acquisition times. In most parts of the image ASAR retrieved winds agree very well to the DWD model results both in magnitude and direction.

To demonstrate the high-resolution capability of ASAR wind field retrieval, an ASAR image is shown in Figure 2 (left hand side). The image shows the Southeastern part of the North Sea as well as parts of the Western Baltic Sea. In this example wind directions were retrieved on a 10 km grid and interpolated to the resolution of the wind speed grid (~500 m). On the right hand side of Figure 2 the mask resulting from the filtering of non wind induced phenomena is superimposed to the ASAR image. It can be seen that most of the Wadden Sea area at the coast of Germany is masked, as the shallow water together with the strong tidal currents leads to significant modulation of the NRCS. Also some clusters in the Southern North Sea are masked; theses areas are most likely affected by heavy rain, which can modulate the NRCS.

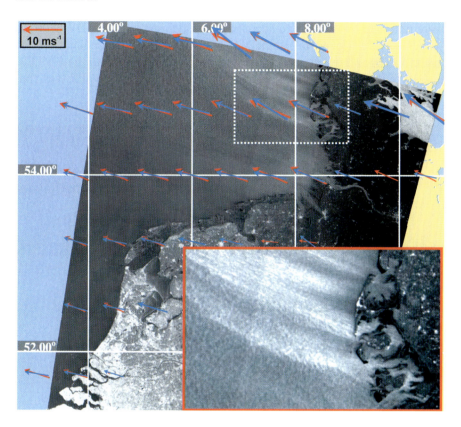

Fig. 1. ENVISAT ASAR image of the southern North Sea acquired at 09:53 UTC on 14. December, 2002. Superimposed to the images are the wind vectors resulting from the A SAR data (blue vectors) and from the model of the German weather service (red vectors). The contrast enhanced cutout shows the wind streaks as well as the wind shadowing due to the coast.

4.2   Comparison of the SAR wind-retrieval methods

To validate the wind retrieval algorithm WiSAR, a set of 61 ENVISAT ASAR ScanSAR data were utilized of which 9 were acquired at HH polarization. All ScanSAR retrieved wind fields were compared to the results of the numerical model of the DWD. Therefore the ASAR wind fields were retrieved from the area corresponding to the grid cell in the DWD-model output, resulting in an average grid-cell size of approximately 45 km × 75 km. The wind fields from the DWD model represent 6-h analysis and were therefore interpolated to the ASAR acquisition times. ASAR wind directions were retrieved using the LG method and the directional ambiguities were removed by taking into account the model results of the DWD. Wind speeds were retrieved by taking the ASAR-retrieved wind direction, the mean NRCS and incidence angle of each grid cell as input to the CMOD4 model. In the case of images acquired at HH polarization, the PR was considered according to Equation (2) with an $\alpha = 0.6$.

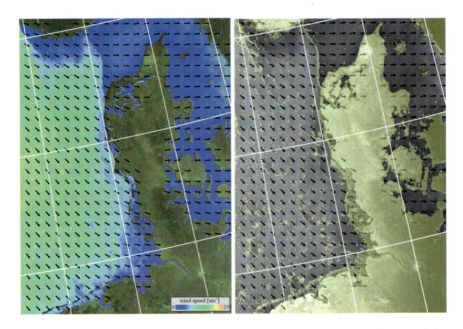

Fig. 2. ENVISAT ASAR image of the southeastern North Sea and parts of the western Baltic Sea. The image was acquired in VV polarization at 09:49 UTC on 19. December, 2005. Superimposed to the images are the wind directions resulting from the A SAR data (black arrows). The left hand side gives wind speeds on a resolution of ~500 m (color coded) and on the right hand side the masked areas, which can be excluded in he wind retrieval are highlighted.

The scatter plots resulting from the comparison of DWD model results to ENVISAT ASAR results are shown in Figure 3. If all grid cells are included the comparison of wind directions from ASAR to the wind directions of the DWD model result in a correlation coefficient of 0.92 and a root mean square (rms) error of 20.1° with a bias of −1.7°. The comparison to wind speeds resulted in a correlation coefficient of 0.75 and an rms error of 3.0 ms^{-1} with a bias of 1.1 ms^{-1}. If grid cells are excluded, that are partially covered by land and that are imaged by the ASAR by less than 20% (blue dots in Figure 3), the error decreases significantly. This comparison results in a correlation coefficient of 0.96 and a rms error of 17.2° with a bias of −2.4° for wind direction and a correlation coefficient of 0.85 and a rms error of 2.3 ms^{-1} with a bias of 0.6 ms^{-1} for wind speed. The later result is not unexpected as the coastal areas are affected by a signi-ficantly higher wind variability, which cannot be resolved by the numerical model due to its too-coarse resolution.

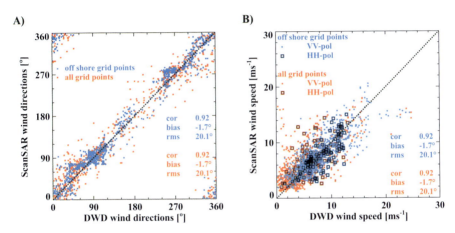

Fig. 3. Scatter plots of comparison between results of the German weather service (DWD) model and ENVISAT ASAR retrieved wind directions (A) and wind speeds (B). Red points represent all data, while blue points represent only grid cells that contain no land and that are imaged to more than 20% by the ASAR.

Concerning the comparison of the C band models, the largest differ-ences occur at wind speeds over 10 ms^{-1}. At moderate wind speeds, they agree fairly well. The models CmodIfr2 and Cmod5 are very similar to each other. Their main differences occur at wind speeds above 20 ms^{-1}, where Cmod5 estimates higher winds. Comparison of Cmod4-retrieved wind speeds to CmodIfr2-retrieved wind speeds shows that CmodIfr2 pre-dicts, on average, slightly higher wind speeds (0.4 ms^{-1}). For wind speeds below 4 ms^{-1} Cmod4 gives higher wind speeds. Comparison of CmodIfr2- and Cmod5-retrieved wind speeds shows that Cmod5 gives higher wind speeds at low 4 ms^{-1} and high 17 ms^{-1} wind speeds. Concerning the statis-tics resulting from the comparison of DWD to ASAR retrieved wind speeds Cmod4 gives the best results. However, especially at high wind speeds, Cmod4 and CmodIfr2 significantly underestimate the wind speeds (Horstmann *et al.* 2005).

Comparison with respect to the different PR models was performed using the CMOD4 model. The main differences between the PRs are seen in the biases, while the standard deviations of the PRs are very similar. It is significant that all PRs lead to an underestimation of the wind speeds. Taking the root-mean-square error as reference, the most suitable PR is according to Equation (3) using $\alpha = 0.6$. However, these results have to be validated considering a much larger set of ASAR data acquired in HH polarization.

5.   Conclusions and outlook

WiSAR a methodology to retrieve high resolution wind fields from C-band SAR images is introduced. The method retrieves the wind field in two steps. In the first wind directions are retrieved from wind-induced streaks by derivation of LGs at scales >200 m. Comparison of SAR-retrieved wind directions to DWD-model analysis resulted in a root mean square error of 17.2° with a bias of −2.4°. In the second step wind speeds are retrieved using the C-band models CmodIfr2, Cmod4, and Cmod5, which were developed for the VV-polarized C-band SCATs. As input to these models, the NRCS, incidence angle, and wind direction is needed, which all can be extracted from the SAR data. Comparison of ENVISAT ASAR-retrieved wind speeds to DWD-model analysis resulted in a root mean square error of 2.3 ms^{-1} with a bias of 0.6 ms^{-1} using the Cmod4 model. Both CmodIfr2 and Cmod5 show slightly larger errors.

The wind retrieval errors are significantly lower when the filters proposed by Koch (2004) are considered. Filtering is especially important when retrieving wind fields in the marginal ice zone as well as at low wind speeds, when non-wind-induced features are more likely to occur.

Comparison of the PR showed that all PRs lead to an underestimation of wind speeds retrieved from HH-polarized images. Due to the limited number of available HH-polarized ASAR data; a final conclusion on the best suited PR cannot be drawn. However, all PRs enable a good estimate of the surface wind speed with a similar accuracy to VV polarized SAR data. The good agreement of ASAR-retrieved wind directions and wind speeds to the DWD-model results shows the applicability of the LG method together with the Cmod4 model.

Future investigations will have to concentrate not only on the validity of the C-band models, especially concerning high wind speeds and polarization, but also effects due to fetch limitations. Concerning the wind-direction retrieval from wind-induced streaks, an investigation as to which scales are the most appropriate to infer the near-surface wind direction has to be carried out. Furthermore, the accuracy of the wind fields at different spatial resolutions has to be investigated.

Acknowledgements

The ENVISAT ASAR data were provided by the European Space Agency within the project BIGPASO. All numerical atmospheric-model results were kindly made available by the German Weather Service.

References

Dankert H, Horstmann J, Rosenthal W (2003) Ocean wind fields retrieved from radar-image sequences. J Geophys Res 108 (C11) doi:10.1029/2003JC002056
Donnelly WJ, Carswell JR, McIntosh RE, Chang PS, Wilkerson J (1999) Revised ocean backscatter models at C and Ku band under high wind conditions. J Geophys Res, 104 (C5): 11485–11498
Etling D, Brown R (1993) Roll vortices in the planetary boundary layer: a review. Bound-Layer Meteorol 18 (3): 215–248
Gerling T (1986) Structure of the surface wind field from SEASAT SAR. J Geophys Res 91 (C2): 2308–2320
Hersbach H, Stoffelen A, de Haan S (2007) An improved C-band scatterometer ocean geophysical model function: CMOD5. J Geophys Res 112 (C3): C03006, doi 10.1029/2006JC003743
Horstmann J, Koch W, Lehner S, Tonboe R (2000) Wind retrieval over the ocean using synthetic aperture radar with C-band HH polarization. IEEE Trans Geosci Rem Sens 38 (5): 2122–2131
Horstmann J, Koch W, Lehner S, Tonboe R (2002) Ocean winds from RADARSAT-1 ScanSAR. Can J Rem Sens 28 (3): 524–533
Horstmann J, Schiller H, Schulz-Stellenfleth J, Lehner S (2003) Global wind speed retrieval from SAR. IEEE Trans Geosci Rem Sens 41 (10): 2277–2286
Horstmann J, Koch W (2005) Measurement of Ocean Surface Winds Using Synthetic Aperture Radars. J Oceanic Eng 30 (3): 506–515
Horstmann J, Thompson DR, Monaldo F, Iris S, Graber HC (2005) Can synthetic aperture radars be used to estimate hurricane force winds? Geophys Res Lett 32: L22801, doi:10.1029/2005GL023992
Johannessen OM, Alexandrov VY, Frolov IY, Sandven S, Pettersson LH, Bobylov LP, Kloster K, Smirnov VG, Mironov YU and Babich NG (2006) Remote Sensing of Sea Ice in the Northern Sea Route - Studies and Applications, Nansen Centers Polar Series no.4. Springer Praxis.

Koch W (2004) Directional analysis of SAR images aiming at wind direction. IEEE Trans Geosci Rem Sens 42 (4): 702–710

Lehner S, Horstmann J, Koch W, Rosenthal W (1998) Mesoscale wind measurements using recalibrated ERS SAR images. J Geophys Res 103 (C4): 7847–7856

Monaldo F, Thompson D, Beal R, Pichel W, Clemente-Colon P (2001) Comparison of SAR-derived wind speed with model predictions and ocean buoy measurements. IEEE Trans Geosci Rem Sens 39 (12): 2587–2600

Monaldo F, Kerbaol V, Clemente-Colon P, Furevik B, Horstmann J, Johannessen J, Li X, Pichel W, Sikora T, Thompson D, Wackerman C (2004a) The SAR Measurement of Ocean Surface Winds: An Overview. Proc 2nd Workshop on Coastal and Marine Applications of SAR, ESA SP-Series SP-565, pp 15–32

Monaldo F, Thompson D, Pichel W, Clemente-Colon P (2004b) A systematic comparison of QuikSCAT and SAR ocean surface wind speeds. IEEE Trans Geosi Rem Sens 42 (2): 283–291

Mouche A, Hauser D, Daloze JF, Guerin C (2005) Dual polarization measurements at C-band over the ocean: results from airborne radar observations and comparison with ENVISAT ASAR data. IEEE Trans Geosci Rem Sens 43 (4): 753–769

Quilfen Y, Chapron B, Elfouhaily T, Katsaros K, Tournadre J (1998) Observation of tropical cyclones by high-resolution scatterometry. J Geophys Res 103 (C4): 7767–7786

Stoffelen A, Anderson D (1997) Scatterometer data interpretation: Estimation and validation of the transfer function CMOD4. J Geophys Res 102 (C3): 5767–5780

Thompson D, Elfouhaily T, Chapron B (1998) Polarization ratio for microwave backscattering from the ocean surface at low to moderate incidence angles. Proc Int Geoscience and Remote Sensing Symp, Seattle, pp 1671–1673

Thompson D, Beal R (2000) Mapping of mesoscale and submesoscale wind fields using synthetic aperture radar. John Hopkins APL Tech Dig 21 pp 58–67

Unal C, Snoeij P, Swart P (1991) The polarization-dependent relation between radar backscatter from the ocean surface and surface wind vectors at frequencies between 1 and 18 GHz. IEEE Trans Geosci Rem Sens 29 (4): 621–626

Vachon PW, Dobson F (1996) Validation of wind vector retrieval from ERS-1 SAR images over the ocean. Global Atmos Ocean Syst 5: 177–187

Vachon PW, Dobson F (2000) Wind retrieval from RADARSAT SAR images: Selection of a suitable C-band HH polarization wind retrieval model. Can J Rem Sens 26 (4): 306–313

Satellite Imaging for Maritime Surveillance of the European Seas

Harm Greidanus

Institute for the Protection and Security of the Citizen, Joint Research Centre, European Commission, Ispra, Italy

Abstract. Surveillance of ships at sea poses particular challenges for spaceborne sensors: small targets need to be detected, wide areas need to be surveyed, and both targets and background are anything but stationary. Incidental, snap-shot images find their application niches – among the existing technologies for ship monitoring – mostly in alerting to the presence of unknown targets and in surveying outlying areas. The main applications in the European seas cover fisheries control, pollution control and maritime border security, although operational use is still sporadic. The sensor of choice is SAR, because it allows ship detection over wide swaths and under many conditions. High-resolution optical sensors can provide additional information on ship classification, which is still difficult for SAR. Their narrow swaths, however, mostly limit their use to predetermined locations. Clutter from the sea surface hinders detection of the smallest ships, especially at high sea state, leading to false alarms and to less than 100% detection probability. For many applications these drawbacks can be kept within acceptable limits by proper choice of the swath width/resolution combination. Satellite sensors may also image ship wakes, from which information on ship speed and heading can be gleaned. Crucial to operational use is the ability for automatic analysis. This is relatively well developed for ship detection, and less well for classification and wake analysis, in SAR images, and quite immature for optical images.

1. Introduction

Ship detection has been perceived as a potential application of satellite remote sensing since its earliest days. Remote sensing images of the sea immediately reveal that ships can be seen from space with radar and optical sensors. A large collection of papers has been produced over the years, mainly on Synthetic Aperture Radar (SAR): a recent literature survey lists

343

V. Barale, M. Gade (eds.), *Remote Sensing of the European Seas.*
© Springer Science+Business Media B.V. 2008

474 references (Arnesen and Olsen 2004). Nevertheless, operational use of satellite remote sensing for ship detection has been slow to take off and it is not widespread. This is mainly because the maritime situation changes rapidly, so that in most cases constant monitoring is required. This is not possible with the present-day low Earth orbit satellites. Insofar as incidental sampling of limited areas suffices, for many of applications the results need to be available in real time. Only very recently have improvements in processing speed and communications links made this possible.

This paper aims to give a condensed overview of the state of the art of ship detection and classification from space (in section 2) and its applications (in section 3), with an emphasis on SAR. It is mainly based on the results of the DECLIMS project (DECLIMS 2007), the literature survey mentioned above and work at the Joint Research Centre (JRC).

2.   Technology

2.1   Ship imaging

In radar images, ships show up as bright dots against a dark background. The 'dot' may be unresolved or resolved, and in the latter case may or may not show structure, depending on target size, sensor resolution and sea state. The radar image of a ship is the result of the coherent summation of the radar echoes from all individual scattering elements within each resolution cell covering the ship. The scattering elements are metal structures on the ship and double bounce sea-ship configurations. The coherent summation dictates that the radar signature of a ship will be quite variable, depending on the exact angle under which the ship is viewed, and unpredictable. In addition, the SAR image of a target can become distorted in azimuth direction when the target is not stationary.

It is often said that radar is an all-weather sensor, but in the case of ship detection this is not quite accurate. Although radar is not limited by most atmospheric effects as such, its ability to detect ships is reduced by high wind and rough seas. The radar echoes of ships have to be detected against a background of sea clutter. The clutter level, both in average and standard deviation, increases with wind speed. In addition, high sea state causes the ships to pitch and roll which induces azimuth blurring in SAR images. Therefore, it is more difficult to detect ships at higher wind speeds.

In optical images, ships can reveal much more structure than in radar (Figure 1). This is due to the higher resolution of available optical satellite cameras and the absence of coherent backscatter effects or a synthetic aperture time. However, optical imaging is prevented by clouds or fog, hampered by haze and sun glint and impossible at night – all in contrast to radar imaging. Furthermore, those optical images that have a resolution high enough to give more information than radar cover only a small area.

Due to the above characteristics, SAR is the instrument of choice to survey more extended areas and find ships, whereas optical sensors are more suitable to focus on particular locations and to classify ships.

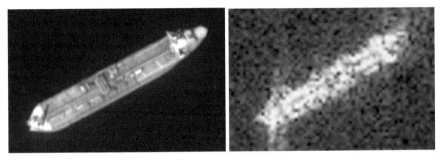

Fig. 1. The same, 350 m long, ship at anchor in the Bay of Gibraltar imaged by EROS-A (Panchromatic, 1.9 m resolution; left; © ImageSat International) and RADARSAT Fine (8 m resolution; right; © CSA/MDA).

2.2 Ship wakes

A ship creates a wake when moving through the water. The literature on SAR imaging describes four types of wakes, resulting from different hydrodynamic mechanisms (Lyden *et al.* 1988). Most frequently seen is the turbulent wake, created by a counter-rotating pair of vortex currents behind the ship. Where the vortex currents reach the surface they produce a current divergence or convergence (which of the two depends on the shape of the vessel's hull). Its centreline traces out the track of the ship (Figure 2). The track appears as an adjoining bright/dark line pair that can extend quite far behind the ship. Second is the Kelvin wake, which is characterised by a V-shape with the ship at its apex, plus a transverse wave pattern filling the V. The wavelength of the transverse waves is such that their phase speed equals the speed of the ship. In a SAR image, one or both arms of this V can be visible as bright line segments, and exceptionally the transverse wave pattern can be seen. The third type of wake is called narrow-V and is the result of direct creation of Bragg waves by the ship's hull. The opening angle of the narrow-V is related to the ship speed, as opposed to the opening angle of the Kelvin V which is a constant 39°. In a SAR image the narrow-V is bright. The fourth type of wake is the surface expression of an internal wave created by the moving ship, when it perturbs a shallow thermocline. The internal wave wake typically takes on the appearance of a stacked set of (possibly curved) V's behind the ship.

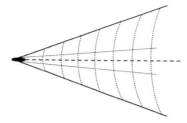

Fig. 2. Sketch of wake types. Dashed thick horizontal line: turbulent wake. Outer V: Kelvin wake envelope. Dotted curves inside outer V: transverse Kelvin wake. Inner thin V: narrow-V wake. The internal wave wake is not drawn.

The SAR imaging process peculiarly leads to a displacement in azimuth direction of a moving target. As a consequence, ships are often not located at the apex of their wake: the moving ship itself is displaced, but its wake is not (Figure 3, left). The amount of displacement is proportional to the radial component of the ship's speed.

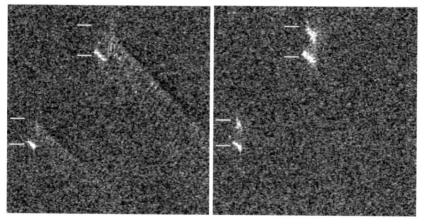

Fig. 3. Two large ships in the Baltic Sea with wakes, in HH (left) and HV (right) from ENVISAT ASAR AP, IS3 beam. Image size is 3.75 km, range direction is to the right. HH shows a turbulent wake for both ships, and exceptionally even a transverse Kelvin wake for the upper ship. HV shows enhanced backscatter originating from near the apex of the wake. The horizontal bar pairs point to the ship (lower) and the wake apex (upper); their vertical distance is the azimuth displacement. © ESA.

In addition to the four wakes types mentioned above, there is also the very turbulent white water around the ship, thrown up by the bow and by the propulsion at the stern. This is well visible as a bright area in an optical image, but seems to also create a marked signature in radar cross- (*i.e.* HV) polarisation. This is at least the author's interpretation of the bright region that is often seen in cross-pol (Figure 3, right) and that is located between a large ship and the apex of its wake as seen in co-pol – preferentially at steeper incidence angles (19°–6°).

Which of these wakes and which parts of them are actually visible in a given SAR image depends on many factors including ship speed, hull shape, radar properties, viewing angle and ambient sea conditions. The optimum sensor parameters for wakes are not the same as those for ships.

In optical images, ship wake signatures can resemble those in radar, especially when viewed in sun glint. The Kelvin wake is often visible, and in shallow water the turbulent wake may be expressed by turbidity effects.

2.3   Satellite sensors and image products

The first instrument to prove that SAR ship detection is possible from space was the 1978 L-band SEASAT. Since 1991, C-band space-based SAR images showing ships have been continuously available from the ERS-1, ERS-2, RADARSAT and ENVISAT satellites. ERS-1 re-started the developments for ship detection that were previously initiated with SEASAT, but it was not until RADARSAT data had been available for a few years that ship detection was seriously picked up as an application.

Whereas ERS-1 and -2 had no mode flexibility, RADARSAT and ENVISAT-ASAR offer several choices between wide swath/low resolution modes (up to 500 km/100 m) and narrow swath/high resolution modes (down to 50 km/8 m), and a range of incidence angles. ENVISAT has additional flexibility of polarisation. As the sea clutter is higher for steep incidence and VV polarisation, the HH polarisation and shallow incidence options make RADARSAT and ENVISAT more suitable for ship detection than ERS, whose fixed steep incidence and VV polarisation are in turn more sensitive to wake (and oil spill) imaging.

ENVISAT-ASAR is yielding a data corpus that is being assessed for the added value of dual channel polarimetry. Studies on these data and on airborne data (Yeremy et al. 2001) indicate that HH polarisation is optimal for detection of ships at higher incidence angles (> 35–45°) while cross-polarisation is optimal at steeper incidence. It can be beneficial to have HH and HV available at the same time; this reduces the possibility that a target is missed because it has low backscatter in one of the two channels. Their combination can help in discriminating ships from icebergs (Howell et al. 2004). It has been argued that the combination of HH and VV, or HV and VV under steeper incidence angles, is suitable for maritime surveillance because it is favourable for ship imaging in HH (or HV) and for wake (and oil spill) imaging in VV.

By far most experience with ship detection has been built up in C-band. SEASAT, some SAR missions on the Space Shuttle (in particular SIR-C/X-SAR) as well as airborne SAR measurements have shown that ship detection is also possible with L- and X-band SAR. New satellite SARs operate in L-band (ALOS-PALSAR launched January 2006) and X-band (TerraSAR-X and Cosmo-Skymed, both launched in June 2007), so more experience with ship detection at these radar frequencies is becoming available. In addition, these system are polarimetric, as is the C-band RADARSAT-2 (also due 2007), which will much extend the experience with polarimetric ship detection that has started to form with ENVISAT.

The full (4-channel) polarimetric capability of ALOS-PALSAR and RADARSAT-2 gives some interesting possibilities for ship detection and classification (Touzi *et al.* 2004; Yeremy *et al.* 2001). For detection, the synthesis of the send-receive polarisation pair produces maximum ship-sea contrast (such as circular); also use of the polarisation entropy or anisotropy has been proposed. For classification, polarimetric decomposition identifies scattering types (e.g., dihedral, cylinder) and their distribution over the ship. Full polarimetry should also be helpful in discriminating ships from false alarms. However, the decompositions are sensitive to errors in processing and calibration, and recent modelling studies indicate that very high resolutions is needed for meaningful polarimetry (to limit the number of scatterers in the resolution cell) (Margarit *et al.* 2004). The extra polarimetric channels, as well as high resolution, come at the expense of swath width, which many ship detection scenarios cannot afford.

There are many more optical than radar imagers already in orbit, or planned. Although their sub-meter resolution is attractive, most of them have swaths of no more than 15 km, restricting their use to surveillance of ports or specific coastal areas. SPOT-5 combines 2.5 m resolution with a 60 km swath and may be the most suitable one for maritime surveillance.

2.4 Analysis

CFAR detection in SAR images

The most widely used approach for automatic ship detection in radar images is Constant False Alarm Rate (CFAR) detection. Bright pixels are assumed to be ships; the threshold is set on the basis of an estimated Probability Density Function (PDF) of the clutter. Ships with radar backscatter below the threshold are not detected, while high clutter peaks will be detected as false alarms. Depending on the application, the threshold can be raised or lowered to adjust the detection and false alarm rates.

Normally, a parametric PDF is assumed. In that way, moments of the measured distribution of clutter pixel values can be used to define the PDF, enabling the fixing of the CFAR threshold. The simplest approximation is a Gaussian (two parameter) PDF. This is known to be an oversimplification, but it may be justified depending on the quality of the image, the number of looks, the targets to detect and the further steps of the entire detection algorithm. A better and also generally used approach is to assume a K distribution. This is a convolution of two Gamma distributions; one which describes the intrinsic backscatter variations from the sea surface due to spatially resolved processes such as long ocean waves, and the other which describes the speckle due to the coherent imaging. The K distribution PDF is characterised by three parameters.

Even a K distribution, however, does not always adequately describe the clutter PDF. Regularly the actual PDF has an extended tail on the high end side, leading to excess false alarms. More sophisticated PDF models are being considered (*e.g.* Delignon *et al.* 1997), but not widely used.

As the clutter tends to vary within the image as a consequence of local environmental variations, the PDF and the consequent detection threshold have to be determined locally. How wide an area to use for this is a compromise between statistical accuracy of the parameter estimates (larger area) and accurately following background variations (smaller area). Some authors advocate filtering before CFAR analysis. Interestingly, recommendations can be found for both high-pass filters and speckle reduction (low-pass) filters. Segmentation is a more sophisticated approach (Lombardo and Sciotti 2001) but may suffer from overlong processing times.

As ships in most cases show an extended signature, because they are resolved or due to azimuth smearing, a ship typically leads to a set of neighbouring pixels that exceed the detection threshold. These pixels have to be clustered into one single target. In the ideal case, this cluster reflects the shape of the ship. This gives the possibility of estimating length, width and heading of the ship – the latter with 180° ambiguity. The shape of the signature can indicate a false alarm due to ambient oceanic or atmospheric features. Morphological filters can be used in this context.

The entire above scheme has to be preceded by applying a land mask to avoid ship detections on land. For this, coast line vector files with global coverage are available. The overlaying of the coast line with the SAR image is done on the basis of the georeference information in the SAR image product header. This necessitates the availability of accurate satellite orbit and (for optical) attitude information in real time.

The final output of the detection process is a list of detected ship positions (latitude, longitude), possibly accompanied by attributes such as estimated length, width, heading, Radar Cross Section (RCS) and a reliability figure.

The latter should be defined on the basis of properties of the target cluster such as detection significance, resemblance of the signature to a ship-like shape, RCS, nature of the direct surroundings and the like.

Alternative approaches

There are a few alternatives to detection on the basis of pixels exceeding a threshold. One is using template matching (Kourti *et al.* 2001). Ship signatures are often characterised by a bright pixel surrounded by less bright pixels, with a preference for elongation in the azimuth direction due to azimuth blurring. Matching pixel clusters with such a pre-defined template can lead to good detection results and fewer false alarms. A drawback with this approach is that ship signatures, and therefore the required template shape and size, can be different for different vessel sizes and image modes.

Another approach is via the use of wavelets (Tello *et al.* 2005). Cross-correlation of the orthogonal components of a wavelet decomposition appears to be a good indicator of the presence of targets in a clutter background. This approach is also able to find targets that are not distinguished from the clutter background by brightness but only by their structure.

A quite different approach is on the basis of SAR sub-apertures. Normally, the sub-apertures are used for incoherent summation to attain speckle reduction. The sub-apertures correspond to images of the same scene taken from slightly different angles at slightly different times. It has been argued (Ouchi *et al.* 2004) that targets such as ships would not show a large variation over the sub-apertures, whereas sea clutter would, on account of speckle effects and temporal decorrelation from the water surface motions. Therefore, the presence of a ship is signalled by high values for the cross-correlation of sub-apertures. Good results are shown by Ouchi *et al.* (2004), although it is not always an improvement over the conventional approach (Greidanus 2006).

None of the above methods are used in an "operational" environment yet, as opposed to the more conventional CFAR approach.

Wake detection in SAR images

Line segments are an important part of ship wake SAR signatures, and are relatively easy to automatically detect. Therefore, most existing ship wake detectors look for line segments. This has been implemented in various manners, which are in one way or another variations or subsets of the Hough or Radon transform. Pixel values are summed along possible lines in the image, and a high (low) value of this sum indicates the actual presence of a bright (dark) line in the image. This testing for the presence of line segments is done in local areas, either around previously detected ships, or covering the entire image.

Apart from giving a confirmation on the presence of the ship, the benefit of analysing a ship's wake comes from the information it can give on ship heading and speed. In order to derive ship heading from the wake, it must be determined what type the wake is; this is not always easy. By combining heading with radial speed derived from the SAR-induced azimuth displacement, the ship's vector velocity can be estimated. In case the transverse Kelvin wake is visible, the ship speed can be deduced from its wavelength. However, surface waves are distorted by SAR imaging effects so that this measurement may be unreliable.

Although the (simple) automatic wake detection algorithms based on line detection have been extensively explored and tested, they are often not used in operational circumstances nowadays. One reason is that wakes are not prominent in the (shallow incidence angle, HH) images that are optimised for ship detection. Furthermore, wake signatures in SAR images are in reality more complex than thin straight lines, as they may be broadened, fuzzy, tapered or curved, so that the automatic detection algorithms do not always give satisfactory results.

Detection in optical images

Automatic detection algorithms for ships in optical images are much less developed and widespread than for radar images, a consequence of the advantages of radar over optical as outlined earlier. In addition, ship analysis in optical images may be more complicated. In the first place, a simple thresholding approach quickly runs into trouble as a result of other bright features in optical imagery: small clouds, whitecaps, sun glint and the ship's wake. Secondly, the higher resolution results in much more structure and texture over the target, so that it cannot be simply treated as a 'blob' like is done for radar. A few sophisticated object-oriented and feature-based approaches have been developed to deal with this, but they are not yet mature nor widely used.

2.5   SAR capabilities and performance

Detection

Success of ship detection depends on the contrast between ship and sea clutter. It is, therefore, a function of many variables, related to target, environment and sensor. First, ship RCS increases with size, freeboard and the amount of metal structures fit on the ship. It can be quite low for non-metal (*e.g.* wooden or plexiglas) boats. In addition, a given target has a large intrinsic variability due to interference and SAR motion effects as outlined earlier. Second, concerning environmental effects, sea clutter increases with wind speed, swell and air-sea temperature instability. The presence of surface films reduces radar backscatter, but natural films are often fine-structured, effectively increasing clutter. Local features such as caused by sea ice, intense rain cells, breaking waves, wind or current fronts, etc., can cause false alarms. Third, concerning sensor parameters, ship - sea contrast is increased by higher resolution; shallow incidence; cross- rather than co-polarisation; and HH rather than VV. Finally, contrast is influenced by imaging geometry: aspect angle of target, and angles between radar look direction, wind direction and wave direction.

With all these influences, it is difficult to make quantitative statements about detection performance. To begin with, such statements have to be of complex conditionality: a detection percentage cannot be given as a simple number but only as a function of the above parameters. Secondly, such a large parameter space needs a lot of data to explore; but in practice, the parameter space is only sparsely filled with measurements. Nevertheless, quantitative estimates of detection performance have been published. They give indications on what size ships can be detected per beam mode and wind condition, and generally point to 90–97% detection rates when image modes are matched to targets and adverse weather conditions are avoided (Vachon *et al.* 2002; Wackerman *et al.* 2001; Kourti *et al.* 2005).

The DECLIMS project has compared the performance of a number of contemporary (semi-) operational automatic ship detection systems (Greidanus *et al.* 2004; Greidanus and Kourti 2006a). While on average all give similar results, there can be marked differences on individual scenes. Nevertheless, a number of issues can be recognised that are problematic to all or most of the detectors. These are: accurate land masking; side lobes and azimuth ambiguities from strong targets; and particular types of sea clutter, including sea ice. Furthermore, in general the results of the automatic detectors can still be improved by visual inspection. There are detailed differences in the appearance of ship and clutter signatures that an experienced image interpreter can recognise, but that are not yet captured in automatic detection software. Considering the above points, there is still clearly room for improvements in the automatic detection systems. At the same time, automatic detection algorithms for operational use represent a compromise between accuracy and speed, so one should never expect an extreme level of sophistication.

Classification

With the resolutions presently obtainable from space, classification of ships with SAR is difficult, and identification is essentially impossible. Under favourable circumstances, SAR can show the outline of the ship and the distribution of scattering centres over the ship. Together with the total RCS these are the three available indicators on which classification can be based. The backscatter distribution can be quite variable due to illumination and interference effects. Primarily the size can be used to distinguish *e.g.* fishing vessels (smaller) from tankers and container vessels (larger), especially in particular locations and scenarios where the size ranges of these classes of vessels are known a priori.

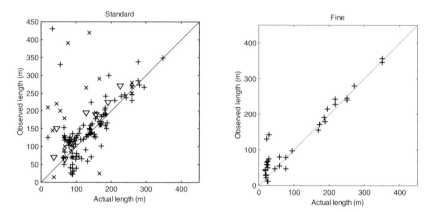

Fig. 4. Comparison of vessel length estimated visually from SAR images (vertical) to actual vessel length (horizontal). The left figure collects 120 ships from 10 RADARSAT Standard and ENVISAT-ASAR images (25 m resolution); the right figure collects 32 ships from 6 RADARSAT Fine images (8 m resolution). Plus is HH, cross is HV, triangle is VV.

Unfortunately, in many cases the outline of the ship is not well defined in a SAR image, due to side lobes, attached wake, strong sea clutter and, most importantly, blurring in azimuth direction caused by ship motions. Often, SAR signatures are point-like in range direction, but elongated in azimuth direction; in such cases, the length estimate cannot be trusted. These problems are confirmed by a benchmark test on classification performed under the DECLIMS project (Greidanus and Kourti 2006a). The comparison between the known ship sizes and results of visual analysis showed that in many cases the size of the SAR signature as seen in the image does not correspond to the actual ship size. The only ships present in the test set that gave reliable size estimates were ships at low sea state imaged by RADARSAT Fine (Figure 4; Greidanus and Kourti 2006b).

3.   Applications

3.1 Concept of use

Four key characteristics of satellite SAR imagery dictate the way they can be used for maritime surveillance: availability under most meteo conditions; snapshot nature; capability of non-cooperative detection but not identification; finite detection and false alarm rates. As a consequence, ship detection results from satellite are in nearly all applications only useful in combination with ship traffic data from other sources. These can come from sensors such as radar on the coast or on patrol boats or aircraft, or from ship reporting systems such as Vessel Monitoring System (VMS), in use on fishing vessels, Automatic Identification System (AIS), for merchant vessels, or Long Range Identification and Tracking (LRIT), also for merchant vessels, from 2008. Satellite ship detections can be used as an independent check on the compliance of vessels with reporting regulations, and to guide the deployment of costly patrol assets, preventing them from having to search large (empty) areas.

Vessel positions obtained from satellite images must be correlated with vessel positions from these other systems, which are widely used in the European waters. Difficulties arise because some systems give ship positions only with large intervals, in which case the most recent position needs to be extrapolated, with consequent position errors. Ship speed and heading, if available, can be an aid to this extrapolation. The need to fuse data from different maritime surveillance systems implies needs for data standardisation and data communication. These pose serious challenges, of practical rather than fundamental nature, considering that the various surveillance systems are operationally run at many different physical locations and by various government branches with different specialised remits and legal constraints including *e.g.* data classification. European level harmonisation and coordination can play a facilitating role.

On the open ocean, survey areas are wide but the ships (mostly) not too small; the ScanSAR Narrow Far mode of RADARSAT presents a good compromise between resolution (50 m) and swath (300 km). Closer to the coast, the smaller ships necessitate the use of higher resolution; RADARSAT Standard beams 4–7 are suitable, or ENVISAT beams IS4-7 in HH and the steeper beams in HV. All these have 25 m resolution and up to 100 km swath. RADARSAT Fine (8 m resolution) is needed for smaller ships (order 10 m) but its 50 km swath is rather restricted. In coastal areas, the denser ship traffic can make it difficult to pinpoint potential threats or infringements; there, AIS data from coastal receivers can be helpful.

Many applications have a real time requirement. The processing chain consists of: data acquisition by the satellite; downlinking to a ground station; image formation ("SAR processing", in the case of SAR); ship detection on the image; fusion between detected targets and ship positions from other sources; and sending the results to the end user. If the acquisition is taken within range of a ground station, then the slowest step of these is the SAR processing. Nevertheless, for a single image this can now be as fast as 5 minutes using a cluster of parallel PCs. The automatic ship detection can be as fast as 1 minute. Acquisition, downlinking and sending the results (*e.g.* by e-mail) are near-instantaneous. Indeed, the fastest time achieved between image acquisition and having the list of detected ships at the end user has been 6 minutes in trials by JRC and the KSAT ground station in 2006 (but in that case without data fusion).

The above applies when all processing steps can be done at the ground station. This may not be the case when *e.g.* the ship detection algorithm is run by a specialist service provider in another location, or when the ship traffic data to be fused is confidential and only available at the end user. In such cases image data may need to be sent around, with consequent additional delays. FTP can be used for this; recent trials have shown that in such set-ups, the final results (fused maritime traffic picture) can be at the end user within 20–30 minutes after acquisition (Lemoine *et al.* 2006).

Automatic detection systems do not perform as well as human operators regarding false alarm rejection. Therefore, several service providers will visually inspect the outputs of the automatic detection before forwarding them to the end user. This practice leads to more reliable results, but seriously increases processing time and operating cost. Improvements in automatic detection algorithms are therefore still much needed.

Not all applications have a real time requirement. Distributions of ship traffic density collected off-line can serve applications in fisheries, traffic safety, risk analysis and environmental safety.

3.2   Application areas

Fisheries control

The main instruments for fisheries monitoring and control at sea are patrol boats and aircraft and the VMS reporting system. VMS, based on on-board GPS receivers and satellite communications transponders, provides the authorities with regular position updates of all the participating fishing ships. In the EU, VMS is mandatory for fishing ships over 15 m in length, bihourly reporting identity, location, speed and heading. While this is obviously a very powerful control system, it will not guard against fishing ships that are active without carrying a transponder, with their transponder switched off or while sending fake position reports. An occasional independent check comparing VMS positions with actual fishing ship positions will reveal these anomalies. This is exactly what satellite imaging can do, in a way that is cost effective in comparison with airborne surveillance or shipborne inspection. The latter are still needed to follow up alarms given by satellite imaging, but need to spend less time searching.

Concerning optical sensors, SPOT-5 has been used together with VMS to monitor small fishing vessels in tropical coastal waters. Very high resolution optical images are being used for monitoring aquaculture sites.

Satellite monitoring is probably most cost effective for outlying areas, where the costs to maintain sufficient patrol assets become too high. A good example are the waters around the Kerguelen Islands, where illegal fishing has been expunged following the set up of a satellite surveillance program with a local receiving station (Losekoot and Schwab 2005).

Pollution control

Oil spill detection is an application of spaceborne SAR in its own right. However, if an oil spill is detected in a satellite image, it can be of value to look for ships in the vicinity: a ship's position and heading may link it to the spill. Although, as mentioned earlier, dual polarisation could cater to oil spill and vessel surveillance at the same time, and the use of the same acquisition for several applications is economically attractive, the preferred mode for oil spill detection is wide swath/low resolution which will miss the smaller ships. Otherwise, it is generally the larger ones (cargo, tanker) that are the main polluters, and they should remain detectable in spill-optimised image modes – except under high wind speeds but then also oil spills are not seen anymore. The use of AIS data can be very helpful to identify the ships in the satellite image.

Maritime security

Under this header falls a wide range of scenarios including anti-terrorism, anti-piracy, maritime border surveillance related to sovereignty, smuggling and illegal immigration, etc. This field is younger than the others and still much in development; the role of satellite surveillance is not yet well defined. One possible role could be open ocean surveillance in combination with LRIT reports, quite similar to how satellite imaging is used to check VMS reports. Satellite observation can capitalise on its wide swath imaging capability of outlying areas, while the limited resolution and low revisit rate are no great obstacles because the targets are relatively large and the role of the imaging is incidental cross-checking, not monitoring.

For maritime border surveillance, forward scenarios have patrol aircraft surveying risk areas of open sea away from the coast. Satellite surveillance can play a supporting role, covering parts of the sea that are out of the footprint of the aircraft's sensors. Scenarios for maritime border monitoring of Europe's *coastal* waters may be less suitable for satellites, since continuous monitoring can be provided there by shore-based sensors. Still, also here satellite surveillance could be used, to probe areas not yet covered by coastal surveillance systems or as a check on their performance.

Maritime safety

Maritime safety is a mature field for which many systems have been developed. Port and coastal authorities use sophisticated Vessel Traffic Services (VTS) systems for maritime traffic control, ships operate on-board radars, AIS has been recently introduced and is in widespread use, and systems are set up to exchange ship traffic data between ports, ships, ship owners and other stakeholders. This infrastructure provides a continuous picture of the maritime traffic. This makes it difficult to find a niche where satellite imaging can still give added value. One possible contribution could be risk assessment by mapping of ship traffic density and patterns (Hajduch *et al.* 2006). Especially new transport routes where no other monitoring is yet in place could benefit from satellite surveillance.

References

Arnesen TN, Olsen RB (2004) Literature review on vessel detection. FFI/Rapport-2004/02619, ISBN-82-464-0859-3

DECLIMS project (2007) Detection and Classification of Marine Traffic from Space, FP5 contr nr EVG2-CT-2002-20002, 2003–2006, http:/declims.jrc.it

Delignon Y, Garello R, Hillion A (1997) Statistical modelling of ocean SAR images. IEE Proc Radar Sonar Navig 144 (6): 348–354

Greidanus H, Clayton P, Indregard M, Staples G, Suzuki N, Vachon P, Wackerman B, Tennvassas T, Mallorquí J, Kourti N, Ringrose R, Melief H (2004) Benchmarking operational SAR ship detection. Proc IEEE Int Geosc Rem Sens Symp IGARSS'04, Anchorage, Alaska

Greidanus H (2006) Sub-aperture behaviour of SAR signatures of ships. In: Proc IEEE Int Geosc Rem Sens Symp IGARSS'06, Denver, Colorado

Greidanus H, Kourti N (2006a) Findings of the DECLIMS project – Detection and classification of marine traffic from space. In: SEASAR 2006: Advances in SAR oceanography from ENVISAT and ERS, ESA-ESRIN, Frascati

Greidanus H, Kourti N (2006b) A detailed comparison between radar and optical vessel signatures. In: Proc IEEE Int Geosc Rem Sens Symp IGARSS'06, Denver, Colorado

Hajduch G, Leilde P, Kerbaol V (2006) Ship detection on ENVISAT ASAR data: results, limitations and perspectives. In: SEASAR 2006: Advances in SAR oceanography from ENVISAT and ERS, ESA-ESRIN, Frascati, Italy

Howell C, Youden J, Lane K, Flett D (2004) Iceberg and ship discrimination with ENVISAT multi-polarization ASAR. In: Proc IEEE Int Geosc Rem Sens Symp IGARSS'04, Anchorage, Alaska

Kourti N, Shepherd I, Schwartz G, Pavlakis P (2001), Integrating spaceborne SAR imagery into operational systems for fisheries monitoring. Can J Rem Sens 27 (4): 291–305

Kourti N, Shepherd I, Greidanus H, Alvarez M, Aresu E, Bauna T, Chesworth J, Lemoine G, Schwartz G (2005) Integrating remote sensing in fisheries control. Fisheries Management and Ecology 12: 295–307

Lemoine G, Indregard M, Cesena C, Thoorens F, Greidanus H, Dörner H (2006) Evaluation of Vessel Detection System use for monitoring of fisheries activities. In: ICES Ann Sci Conf, Maastricht

Lombardo P, Sciotti M (2001) Segmentation-based technique for ship detection in SAR images. IEE Proc Radar Sonar Navig 148 (3): 147–159

Losekoot M, Schwab P (2005) Operational use of ship detection to combat illegal fishing in the southern Indian Ocean. In: 8th Int'l Conf Rem Sens Marine and Coastal Envir, Halifax, Canada

Lyden JD, Hammond RR, Lyzenga DR, Schuman RA (1988) Synthetic Aperture Radar imaging of surface ship wakes. J Geoph Res 93: 12,293–12,303

Margarit G, Fabregas X, Mallorqui JJ, Broquetas A (2004) Analysis of the limitations of coherent polarimetric decompositions on vessel classification using simulated images. Proc IEEE Int Geosc Rem Sens Symp IGARSS'04, Anchorage, Alaska

Ouchi K, Tamaki S, Yaguchi H, Iehara M (2004) Ship detection based on coherence images derived from cross correlation of multilook SAR images. IEEE Geosc Rem Sens Lett 1 (3): 184–187

Tello M, López-Martínez C, Mallorqui JJ (2005) A novel algorithm for ship detection in SAR imagery based on the wavelet transform. IEEE Geosc Rem Sens Lett 2 (2): 201–205

Touzi R, Charbonneau FJ, Hawkins RK, Vachon PW (2004) Ship detection and characterization using polarimetric SAR. Can J Rem Sen 30 (3): 552–559.

Vachon PW, Thomas SJ, Cranton J, Edel HR, Henschel MD (2002) Validation of ship detection by the RADARSAT Synthetic Aperture Radar and the Ocean Monitoring Workstation. Can J Rem Sens 26: 200–212

Wackerman CC, Friedman KS, Pichel WG, Clemente-Colón P, Li X (2001) Automatic detection of ships in RADARSAT-1 SAR imagery, Can J Rem Sens 27 (4): 371–378

Yeremy M, Campbell JWM, Mattar K, Potter T (2001) Ocean surveillance with polarimetric SAR. Can J Rem Sens 27 (4): 328–34

Oil Spill Detection in Northern European Waters: Approaches and Algorithms

Anne H.S. Solberg[1,2] and Camilla Brekke[1,3]

[1] Department of Informatics, University of Oslo, Oslo, Norway
[2] also at Norwegian Computing Center, Oslo, Norway
[3] also at Norwegian Defence Research Establishment, Kjeller, Norway

Abstract. The combined use of satellite-based Synthetic Aperture Radar (SAR) images and aircraft surveillance flights is a cost-effective way to monitor deliberate oil spills in large ocean areas and catch the polluters. SAR images enable covering large areas, but aircraft observations are needed to prosecute the polluter, and in certain cases to verify the oil spill. We discuss the limitations of satellite imaging of oil spills compared to aircraft monitoring. Automatic detection of oil spills has proven to be an interesting complement to manual detection. We present an overview of algorithms for automatic detection, and discuss their potential compared to manual inspection as part of an operational oil spill detection framework. Experimental results show that automatic algorithms can perform comparable to manual detection, both in terms of accuracy in detecting verified oil spills, false alarm ratio, and they can also speed up the image analysis process compared to fully manual services.

1. Introduction

Marine pollution arising from illegal oily discharges from ships represents a serious threat to the marine environment. Oil pollution caused by large accidents like the Prestige event in 2002 capture many headlines, but the majority of the oil pollution cases are caused by operational discharges from tankers. Observed oil spills commonly appear in connection with offshore installations and correlate well with major shipping routes. A combination of aircraft and satellite sensors are currently used to monitor large ocean areas to detect oil spills and catch the polluter. The inclusion of satellite surveillance allows the user to better target the aircrafts used for oil spill surveillance and to cover larger areas.

V. Barale, M. Gade (eds.), *Remote Sensing of the European Seas.*

2.   Remote sensing sensors for oil spill detection

For routine monitoring of illegal oil discharges from ships and offshore installations both aircraft sensors and satellite sensors can be used. Satellite-based Synthetic Aperture Radar (SAR) images can be used to screen large ocean areas, while aircrafts are more suitable to be brought into action to identify the polluter, the extent, and the type of spill.

2.1   Aircraft sensors

Most surveillance aircrafts used for oil pollution monitoring in Europe are equipped with a combination of sensors: Side-Looking Airborne Radar (SLAR), infrared/ultraviolet (IR/UV), Laser Fluoro-Sensor (LFS), Micro-wave Radiometer (MWR). For an overview of aircraft sensors for oil spill detection, see (Goodman 1994; Trieschmann *et al.* 2003). SLAR is the main sensor for long-range detection of oil pollution on the sea surface. The SLAR is used to locate possible spill locations. Then the spill is ins-pected more closely using additional sensors and/or visual inspection. The sensor configuration used on board surveillance aircrafts varies from coun-try to country. An example is the German aerial surveillance, which lo-cates the oil spills by SLAR, IR/UV scanning is used to quantify the extent of the film, a MWR is used to quantify the thickness, and a LFS is used for oil type classification. The SLAR, IR, and LFS can operate at night. A number of different aircraft types are used for oil spill aerial surveillance. They differ in terms of endurance, cruising speed and SLAR sensor equip-ment resulting in different SLAR area coverage during one flight hour. One hour of airborne remote sensing over the sea at a speed of 335 km/h covers an area of 13400 km2 (Tufte *et al.* 2005).

2.2   Satellite sensors

SAR is the main spaceborne remote sensing instrument for oil spill imag-ing, with all-weather and all-day operation capabilities, although it is not capable of oil spill thickness estimation and oil type recognition. The main limitation for spaceborne optical sensors is the need for daylight and cloud-free scenes, but they have a potential to discriminate between oil and algal blooms. A more detailed discussion of other satellite sensors for oil spill detection is given in (Brekke and Solberg 2005).

    SAR is particularly useful for searching large areas. Usually even small volumes of oil cover large areas and thus the need for very high spatial resolution in SAR images is not crucial. SAR has however some limitations, as a number of natural phenomena can produce similar dark objects in the SAR images (see Section 3).

Currently, RADARSAT-1 and ENVISAT ASAR are the two main SAR sensors used for oil spill detection. The best trade-off between spatial coverage and spatial resolution is achieved using RADARSAT-1 ScanSAR and ENVISAT ASAR Wide Swath image modes. Table 1 describes the coverage and resolution of these sensors. The costs of satellite images are much lower than the costs of covering the same area by aircraft. The actual time that the satellite passes over a given location will vary with latitude, but the overflight will be fixed in time.

Table 1. Coverage and resolution for selected RADARSAT-1 and ENVISAT ASAR products.

	RADARSAT-1 ScanSAR Narrow	ENVISAT ASAR Wide Swath
Spatial coverage per scene	300 km × 300 km	400 km × 400 km
Spatial resolution per scene	50 m × 50 m	150 m × 150 m

The number of available RADARSAT-1 or ENVISAT ASAR images of a given area on a given date depends on the geographic latitude and the observation period. Daily coverage is possible in Northern Europe, and a coverage a couple of times a week is possible for all European waters.

2.3   Satellite vs. aircraft – advantages/limitations

The advantages and limitations of satellite-based vs. aircraft monitoring are summarized in Table 2. In order to cover the same area as a RADARSAT-1 ScanSAR Narrow scene or an ENVISAT ASAR Wide Swath scene with an aircraft, 6 or 12 flight hours, respectively, are needed. A limitation with the satellite monitoring is that the images are taken at fixed times of the day. The fate and persistence of oil in seawater are controlled by processes that vary considerably in space and time. The amount of oil spilled, its initial physical and chemical characteristics, and the prevailing climatic and sea conditions have great impact on the lifetime of an oil spill. A reasonable assumption might be that most illegal oil discharges are bilge oil, *i.e.* a mixture of several kinds of oils (fuel oil, hydraulic oil, *etc.*). For instance, in the Finnish surveillance area it is estimated that 1–5 percent of the detected slicks are thicker slicks that persist several days, and 95–99% are bilge oil that persists only some hours (Tufte *et al.* 2005). To cover oil spills occurring at all times of the day, aerial surveillance can be used to supplement the fixed coverage times of the SAR satellites.

Table 2. Advantages and drawbacks with SAR satellite and aerial surveillance. (Adapted from Tufte *et al.* 2005).

Satellite SAR	Aerial surveillance
Advantages	Advantages
Large and well-defined spatial coverage. Less expensive than airborne surveillance. Can be used to cue aircraft to improve aircraft operational efficiency.	Flexible monitoring. High accuracy of oil spill detection. Can be deployed at short notice. Can identify polluter. Can identify additional oil parameters.
Limitations	Limitations
False targets can occur in analysis. Fixed monitoring schedule. Limited to certain wind conditions.	High cost. Smaller spatial coverage. Limited to certain weather conditions.

3. SAR imaging of oil spills

Oil spills dampen the Bragg waves (wavelength of a few cm) on the ocean surface and reduce the radar backscatter coefficient. This results in dark regions or dark spots in a satellite SAR image. A part of the oil spill detection problem is to distinguish oil spills from other natural phenomena that dampen the short waves and create dark patches on the surface. Natural dark patches are termed oil spills look-alikes. Oil spills include all oil related surface films caused by oil spills from oilrigs, leaking pipelines, passing vessels as well as bottom seepages, while look-alikes include natural films/slicks, grease ice, threshold wind speed areas, wind sheltering by land, rain cells, shear zones, internal waves *etc.*

A service for oil spill detection based on SAR images must contain an oil spill detection step where the SAR images are analyzed, and dark regions that might be oil spills are identified. In this process, the factors that can be used to discriminate between an oil spill and a look-alike are important. Due to higher viscosity, oil spills tend to remain more concentrated and provide larger damping to the surrounding sea than natural films (Hovland *et al.* 1994). A newly released oil spill will have reasonably sharp borders to the surrounding sea. As the weathering effects the spill, the borders can become more fuzzy. The shape will be altered by wind and current. Depending on the source of the outlet, certain shapes of the oil spills can be expected. Oil spills from moving ships are thin, linear or piecewise slicks, while oil spills from stationary sources can be wide and regular if a significant amount of oil is released in short time.

Other types of pollution can also cause slicks that are visible in the SAR image Algae can also create dark patches in the SAR image very similar to oil spills. This is in particular a problem in the Baltic Sea, where a certain algae type that dampens the Bragg waves is common during the summer season. If additional information from *e.g.* ocean color sensors is available, it can be used to identify algae.

The wind speed is also important for imaging of oil spills. With very low wind, no backscatter from the sea surface will be seen. Look-alikes are very frequently observed in low to moderate wind conditions (approximately 3 to 7 m/s). As the wind speed increases, the expected number of look-alikes will be lower. For oil spills, the contrast between the spill and the surrounding sea will decrease with higher wind speeds. At high wind speeds (>10 m/s) only larger slicks with thicker oil will be visible. The upper limit for observing oil in the SAR image is not known exactly. In an operational oil spill detection service at Kongsberg Satellite Services (KSAT) in Tromsø, Norway, an upper limit of 15 m/s is used.

4. SAR oil spill detection: manual vs. automatic

Oil spills in a SAR image can be identified by manual inspection, or the image can first be screened by an automatic algorithm for oil spill detection, followed by manual inspection of the suspect alarms only. Manual interpretation of *e.g.* a 400 km × 400 km ENVISAT ASAR image can be a complex and time consuming task because the image is so large that the operator can only view a small part of the scene at a time to be able to detect thin oil spills. Recent benchmarks (Indregard *et al.* 2004; Solberg *et al.* 2006) comparing automatic algorithms to manual detection shows that an automatic algorithm can be a valuable tool when a large number of images are to be inspected.

4.1 Manual oil spill detection

A well-established operational service for oil spill detection is run at KSAT. Trained operators detect oil spills by inspecting the SAR images. In addition to the image, they can use external information about wind speed and direction, oil rig/pipeline location, national territory borders and coast lines. After a possible oil spill has been detected, it is assigned confidence level low, medium or high based on a certain set of rules (Indregard *et al.* 2004).

The location of detected oil spills and their confidence level is then immediately sent to the surveillance aircrafts.

4.2 Automatic approaches

A literature review of automatic techniques for oil spill detection in SAR images can be found in Brekke and Solberg (2005). In this section we give a short update on the state-of-the-art in this field.

Several of the published papers on oil spill algorithms for SAR images (*e.g.* Fiscella *et al.* 2000; Del Frate *et al.* 2000; Solberg *et al.* 1999, 2006) describe a methodology consisting of dark spot detection followed by feature extraction and a classification step (see Figure 1).

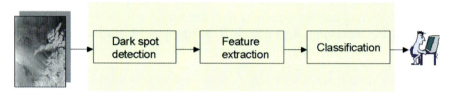

Fig. 1. A framework for oil spill detection algorithms.

Segmentation techniques

As oil spills are characterized by low backscattering levels, the use of thresholding for dark spot segmentation is commonly applied (see *e.g.* Keramitsoglou *et al.* 2006; Solberg *et al.* 1999; Brekke and Solberg 2005). As SAR images tend to become darker with increasing range and as local variations in the wind level and other meteorological and oceanic conditions occur, thresholding algorithms where the threshold is set adaptively based on local statistical estimates should be preferred.

Dark spot feature extraction

Discrimination between oil spills and look-alikes are often based on a number of features computed for each suspicious dark spot on the sea surface (see *e.g.* Topouzelis *et al.* 2002; Solberg *et al.* 1999; Brekke and Solberg 2006; Del Frate *et al.* 2000; Fiscella *et al.* 2000).

Good features are very important for the success of the following classification step. Most of the features applied in the literature are covered by the following types:
- the geometry and shape of the dark spot;

- the physical characteristics of the backscatter level of the dark spot and its surroundings;
- the dark spot contextual features;
- the texture features of both the dark spot and the surroundings.

Slick classification

The purpose of the classification step is to distinguish oil spills from look-alikes. How difficult the classification task is depends on the variability in the feature values for objects in the oil spill class relative to the difference between feature values for objects in the look-alike class. Effective methods for developing classifiers involve learning from example patterns (training). In statistical approaches, the classification decision is based on the probability and the cost of a certain decision (*e.g.* Solberg *et al.* 1999; Brekke and Solberg 2007; Brekke *et al.* 2007).

Various classifiers have been applied to the oil spill detection problem (Fiscella *et al.* 2000; Nirchio *et al.* 2005; Del Frate *et al.* 2000; Keramitsoglou *et al.* 2006). All detection algorithms suffer from false alarms, and dark spots classified as oil spills may be confused with look-alikes (*e.g.* natural film and low wind areas). Applying external data to improve classification and assess the slick nature has been suggested. Girard-Ardhuin *et al.* (2005) combines characteristics of the detected dark spots from the SAR images and meteorological and oceanic data through a multi-sensor approach (including information about surface wind measurements, sea-surface temperature, atmospheric fronts and clouds and chlorophyll).

5.   A benchmark study of oil spill detection approaches

As part of the EC project Oceanides, a benchmark study comparing oil spill recognition approaches was performed. Manual oil spill detection based on SAR images was compared to semi-automatic and automatic approaches. A joint satellite-airborne campaign was performed during 2003. The campaign covered the Finnish and German sectors of the Baltic Sea, in addition to the German sector of the North Sea. The campaign was organized in such a way that a trained operator at KSAT (KSAT1) analyzed the SAR images, and reported possible oil spills to the Finnish and German pollution control authorities. They would check the positions and verify the slicks, and report additional slicks found by the aircraft. This was done for both ENVISAT and RADARSAT-1 images.

For benchmark comparisons, KSAT let another operator (KSAT2) inspect the same SAR images without knowing the aircraft detections or the result of the previous inspection to study the inter-operator variance. The automatic oil spill detection approach developed at Norwegian Computing Center (NR) (Solberg *et al.* 2006) was used to analyze all images without knowing aircraft detections or KSAT results. Figure 2 shows examples of correctly classified oil spills, false alarms, and slicks detected only by aircraft.

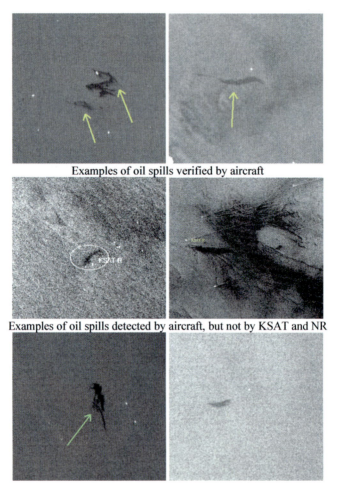

Examples of oil spills verified by aircraft

Examples of oil spills detected by aircraft, but not by KSAT and NR

Examples of false alarms. The slick in the image on the left was verified as algae, while the clearly suspect dark spot in the right image could not be found when the aircraft inspected the position (possibly resolved).

Fig. 2. ENVISAT images with examples of oil spills and false alarms. © ESA/ KSAT/NR.

The benchmark data set contained 27 ENVISAT images and 32 RADARSAT-1 images. The real-time inspection of the ENVISAT images at KSAT (KSAT1) detected 11 oil spills that were verified as oil spills by the aircraft. The repeated inspection by another operator, KSAT2, detected 8 of these verified slicks, the automatic algorithm (NR) also detected 8 of the verified slicks. For RADARSAT-1 data, there was 18 verified oil spills, KSAT2 found 15 of these, while the NR algorithm found 14.

This demonstrates that the inter-operator variance was significant. KSAT has later taken measures to reduce this variability by increasing operator training and harmonising the interpretation process. The performance of the NR algorithm is also almost comparable to KSAT2, so it can be a valuable alternative or supplement to manual inspection.

It is also of interest to study additional slicks reported by the aircraft, but not detected from satellite image analysis. In general, these slicks involved a small amount of oil. For some of them, the time of aircraft pass was several hours after the satellite image acquisition, so the release could be new. For other cases, the satellite image was taken several hours after the aircraft pass, and a small amount of oil could very well be resolved during this time period.

The number of false alarms, the number of slicks reported by satellite, but verified as not oil by the aircraft, was also studied. All false alarms are discussed in detail in (Indregard et al. 2004). Some linear slicks with good contrast were detected in the satellite image but verified as algae.

The confidence levels that are associated with reported possible oil spills from KSAT can be used by the surveillance aircraft crew to prioritize which oil spill positions they will inspect first. Slicks that were assigned confidence level High had very low false alarm ratio, slicks that were assigned confidence Medium had reasonably low false alarm ratio, while slicks that were assigned confidence Low had relatively high false alarm ratio (again the details can be found in (Indregard et al. 2004). However, the confidence assigned by two different operators at KSAT varied. The automatic algorithm can also be used to compute confidence levels using a set of rules that simulate the rules used by the operators at KSAT (see Solberg 2005 for how this is done). By comparing the confidence levels assigned by two KSAT operators and NR's automatic algorithm on a set of 22 selected oil spills, it was found that KSAT1 and KSAT2 assigned the same confidence for 7 out of 22 slicks, while the algorithm and KSAT1 agreed for 13 of 22 slicks, and had a confidence level difference of one (indicating e.g. High vs. Medium) for six additional slicks. What this indicates is that there is still some subjectivity involved in confidence assignment, and using an algorithm to get a "second opinion" to use in confidence assignment might be valuable.

The processing time for manual inspection at KSAT varied between 3 and 25 minutes for RADARSAT-1 images, with an average of 9 minutes. The automatic algorithm had an average processing time of 3 minutes. For ENVISAT images, the average time for manual inspection at KSAT was 10 minutes, while the automatic algorithm had an average processing time of 1.45 minutes.

6. Discussion and conclusions

The combined use of satellite-based SAR images and aircraft surveillance flights is a cost-effective way to monitor large areas and catch the polluters. The coverage in terms of the number of weekly satellite passes of European waters is very good in Northern Europe and decent in the Mediterranean.

Oil spill detections from aircrafts and satellite images were compared in a benchmark study. In general, there were good agreement between aircraft detections and satellite-based detections when the time offset between the image acquisitions was low.

Some information about oil spill statistics and hot spots exist, see *e.g.* (Bauna and Clayton 2004; Tufte *et al.* 2005). Hot spots coincide well with major shipping routes, pipelines, oil rigs *etc.* Tufte *et al.* (2005) discuss sampling requirements as guidelines for operational oil spill monitoring in European waters. They also summarize current monitoring efforts for many countries in Northern Europe, and their experience in using a combination of aerial- and satellite monitoring. The best approach for a specific national authority depends on the size and shape of the area to be monitored, and other resources available. International cooperation with neighboring countries on planning satellite acquisitions and sharing costs is important.

Many different techniques are proposed for automatic detection of oil spills. They often consist of three main parts: segmentation, feature extraction and classification. However, there is still a challenge in reducing the number of false alarms, and automatically assigned confidence levels can be helpful in prioritizing the alarms. Interoperator variance in manual detection should be reduced by better training, or introducing the algorithm as a second information source. This applies to both detection and confidence assignment.

Presently, one of the biggest challenges for operational oil spill detection services is obtaining sustainability in terms of data availability. SENTINEL-1 is a planned two-satellite system (C-band) to be operated as a constellation for maximized coverage/repeat cycle. The first satellite will be launched in 2010 and the second some 12–15 months later (Attema 2005). RADARSAT-2 is a Canadian SAR satellite planned to be launched in the summer of 2007. The SAR instrument will be C-band like RADARSAT-1, but there will be more flexibility in the selection of polarizations.

Acknowledgements

The authors would like to thank the Oceanides project, in particular Marte Indregard, Peter Clayton and Lars Tufte for contributions to the benchmark study on oil spill approaches.

References

Attema E (2005) Mission Requirements Document for the European Radar Observatory Sentinel-1, Requirement Specification. Technical Report ES-RSESA-SY-0007, issue 1, revision 4, European Space Agency

Bauna T, Clayton P (2004) D3c-state of knowledge of potential oil spill hotspots in european waters. Technical report, Oceanides project, European Commission, Archive No. 04-10225-A-Doc, Contract No: EVK2-CT-2003-00177

Brekke C, Solberg A (2005) Oil spill detection by satellite remote sensing. Rem Sens Environ, 95 (1): 1–13

Brekke C, Solberg A, Storvik G (2007) Classifying Oil Spills and Look-alikes in ENVISAT ASAR Images. In Proc. ENVISAT Symposium, Montreux, Switzerland, 23–27 April 2007

Brekke C, Solberg AHS (2006) Segmentation and feature extraction for oil spill detection in ENVISAT ASAR images. Int J Rem Sens (submitted)

Brekke C, Solberg AHS (2007) Classifiers and Confidence Estimation for Oil Spill Detection in ENVISAT ASAR Images. IEEE Trans Geosci Rem Sens (submitted)

Del Frate F, Petrocchi A, Lichtenegger J, Calabresi G (2000) Neural networks for oil spill detection using ERS-SAR data. IEEE Trans Geosci Rem Sens 38 (5): 2282–2287

Fiscella B, Giancaspro A, Nirchio F, Pavese P, Trivero P (2000) Oil spill detection using marine SAR images. Int J Rem Sensi 21 (18): 3561–3566

Girard-Ardhuin F, Mercier G, Collard F, Garello R (2005) Operational Oil-Slick Characterization by SAR Imagery and Synergistic Data. IEEE J Oceanic Engineering, 30 (3): 69–74

Goodman R (1994) Overview and future trends in oil spill remote sensing. Spill Science & Technology Bulletin 1 (1): 11–21

Hovland HA, Johannessen JA, Digranes G (1994) Slick detection in SAR images. Proc. IGARSS'94, 4: 2038–2040

Indregard M, Solberg A, Clayton P (2004) D2-report on benchmarking oil spill recognition approaches and best practice. Technical report, Oceanides project, European Commission, Archive No. 04-10225-A-Doc, Contract No: EVK2-CT-2003-00177

Keramitsoglou I, Cartalis C, Kiranoudis C (2006) Automatic identification of oil spills on satellite images. Environmental Modelling & Software 21 (5): 640–652

Nirchio F, Sorgente M, Giancaspro A, Biamino W, Parisato E, Ravera R, Trivero P (2005) Automatic detection of oil spills from SAR images. Int J Rem Sens 26 (6): 1157–1174

Solberg A, Brekke C, Husøy P (2006) Oil spill detection in Radarsat and Envisat SAR images. IEEE Trans Geosci Rem Sens 45 (2): 746–755

Solberg AHS (2005) Automatic oil spill detection and confidence estimation. Proc Int Symp Rem Sens Environ, St. Petersburg, Russia, pp 943–945

Solberg AHS, Storvik G, Solberg R, Volden E (1999) Automatic detection of oil spills in ERS SAR images. IEEE Trans Geosci Rem SeDns 37 (4): 1916–1924

Topouzelis K, Karathanassi V, Pavlakis P, Rokos D (2002) Oil spill detection: SAR multi-scale segmentation & object features evaluation. Proc SPIE, Remote Sensing of the Ocean and Sea Ice, series 4880, pp 77–87

Trieschmann O, Hunsänger T, Tufte L, Barjenbruch U (2003) Data assimilation of an airborne multiple remote sensor system and of satellite images for the North- and Baltic sea. Proc SPIE 10th Int Symp Rem Sens, pp 51–60

Tufte L, Indregard M, Solberg A, Huseby RB (2005) D11-oil spill monitoring guidelines and methodology to combine satellite oil spill recognition with routine airborne surveillance. Technical report, Oceanides project, EC Archive No. 04-10225-A-Doc

The Use of Satellite Imagery from Archives to Monitor Oil Spills in the Mediterranean Sea

Guido Ferraro, Barbara Bulgarelli, Serge Meyer-Roux,
Oliver Muellenhoff, Dario Tarchi, and Kostas Topouzelis

Institute for the Protection and Security of the Citizen, Joint Research Centre,
European Commission, Ispra, Italy

Abstract. Accidental pollution at sea can be reduced but never completely eliminated, on the other side, deliberate illegal discharges from ships can indeed be reduced by the strict enforcement of existing regulations and the control, monitoring and surveillance of maritime traffic. Notwithstanding, operational oil discharges is a common practice and represents the main source of marine pollution from ships. Consequently, JRC has focused its attention on the need to monitor in the long term the problem of sea-based oil pollution in all European seas. For this reason, JRC has collected aerial surveillance data and satellite imagery from different actors and archives, even if the data for the seas around Europe are not homogeneous. In this paper, JRC intends to present the first results of the analysis of the created database on oil spills for the Mediterranean Sea. In conclusion, the long term oil spill monitoring is the key instrument to assess the implementation of maritime transport regulation such as the Directives on port reception facilities and on ship-source pollution and can be seen as an environmental indicator of the European Seas.

1. Setting the scene

Among the different types of marine pollution, oil is a major threat to the European seas ecosystems. The source of the oil pollution can be on the mainland or directly at sea. Sea-based sources are discharges coming from ships or offshore platforms.

Oil pollution from sea-based sources can be accidental or deliberate. Fortunately, the number of marine accidents and the volume of oil released accidentally are on the decline. On the other side, routine tanker operations can lead still to the release of oily ballast water and tank washing residues. Furthermore, fuel oil sludge, engine room wastes and foul bilge water,

V. Barale, M. Gade (eds.), *Remote Sensing of the European Seas.*

produced by all type of ships, also end up in the sea. In the last decade maritime transportation has been growing steadily, reflecting the intensified co-operation and trade in the European region and a prospering economy. More ships also increase the potential number of illegal oil discharges. Both oil tankers and other kinds of ships are among the suspected offenders of illegal discharges.

In the North Sea regular aerial surveillance to detect oil spills and to catch the polluters started in the eighties. The eight Countries bordering the North Sea work together within the Bonn Agreement and undertake aerial surveillance using aircraft equipped with Remote Sensors (RS). Data of observed oil spills are available from 1986 (Bonn Agreement website, Carpenter 2007). It should be stressed that in the North Sea, there are many off-shore installations which are sources of sea-based pollution. Deliberate illegal oil discharges from ships are regularly observed also within the Baltic Sea since 1988. A complex set of measures known as a Baltic Strategy has been implemented by the nine Contracting Parties to the Helsinki Convention (HELCOM website). These measures include surveillance flights and improved usage of remote sensing equipment.

As a possible result of the flight surveillance in the North and Baltic Sea, a decrease in the number of observed illegal discharges has been identified over the last years despite the rapidly growing density of shipping. Although the number of observations of illegal oil discharges has been decreasing it should be kept in mind that for some areas aerial surveillance is not evenly and regularly carried out and therefore there are no entirely reliable figures for these areas.

For the North-East Atlantic, there are no data available on a regular basis concerning the deliberate oil spills problem. As additional problem, this area is not defined "Special Area" according to Annex I of the MARPOL Convention. Outside "Special Areas", it is difficult to assess if visible oil discharges from ships are illegal. On the other side, in this area there was, for the first time, the operational use of satellite imagery during the Prestige accident in 2002 (Fortuny et al. 2004).

This paper will focus only on the Mediterranean Sea, where there are no data derived from regular aerial surveillance, so the only possible way to analyse trends for this sea is the use of satellite images from archives. The reliability of the satellite image analysis is not yet fully satisfactory and further investigations and validation activities are necessary. However, the use of archive satellite imagery is the only way to extract information for this sea. JRC is carrying out a systematic mapping of the oil spills using satellite imagery in the Mediterranean Sea. This action helps to reveal what is the dimension of the oil pollution problem, thus stressing the need for more concerted international actions.

1.1 Pollution from offshore platforms

Offshore platforms can legally discharge oil at sea, according to detailed parameters. The main discharge associated with an offshore installation is produced water. However oil could come into sea also from oil on cuttings and from produced sand contaminated with oil, well clean-up fluids, releases during well abandonment and pipeline decommissioning. In addition to permitted operational discharges, spillages may occur where systems fail.

Specific rules are set to prevent pollution from offshore activities in different regional agreements, such as the Baltic Convention on the Protection of the Baltic Sea Area of 1992, the Protocol, of the Barcelona Convention, for the Protection of the Mediterranean Sea against Pollution Resulting from Exploration and Exploitation of the Continental Shelf and the Seabed and its Subsoil (adopted in 1994 but not yet in force), and the Convention for the Protection of the Marine Environment of the North-East Atlantic (known as OSPAR Convention) of 1992.

As example, we can make reference to the OSPAR Recommendation 2001/1 for the Management of Produced Water from Offshore Installations. This Recommendation establishes that no individual offshore installation should exceed a performance standard of 40 mg of dispersed oil per litre (*i.e.* 40 parts per million – 40ppm) for produced water discharged into the sea. An improved performance standard of 30 mg/L (30ppm) is to apply by the end of 2006. These discharge limits are based on the total weight of oil discharged per month divided by the total volume of water discharged during the same period. A maximum oil concentration of 100 mg/L (100ppm) is generally applied (OSPAR website).

1.2 The use of satellite imagery to detect oil spills

The different tools to detect and monitor oil spills are vessels, airplanes, and satellites. The vessels, especially if equipped with specialised radars, can detect oil at sea but they can cover a very limited area. The vessel, however, remains necessary in case it is necessary to take oil sampling. As example, in some EU Member States the sampling is necessary to prosecute the polluter.

Aircraft is the most used tool to detect and monitor oil pollution at sea. Observations by experienced aircrew are fully reliable tools for detections, classifications and quantification of observed pollution. Aerial surveillance can be based on the simple visual analysis of the air-crew, using as example the Bonn Agreement Oil Appearance Code, or can be executed with auxiliary RS tools. Among them, the Side-Looking Airborne Radar (SLAR) is the most used.

The main systems to monitor sea-based oil pollution are the use of airplanes and of images from satellites equipped with Synthetic Aperture Radar (SAR). The possibility of detecting an oil spill in a SAR image relies on the fact that the oil film decreases the backscattering of the sea surface resulting in a dark feature that contrasts the brightness of the surrounding spill-free sea. Several studies aiming at oil spill detection have been conducted (Solberg *et al.* 1997; Solberg *et al.* 1999; Del Frate *et al.* 2000; Espedal *et al.* 1999; Espedal *et al.* 2000; Fiscella *et al.* 2000; Pavlakis *et al.* 2001; Topouzelis *et al.* 2002; Karathanassi *et al.* 2007). Most of these studies examined the ability of several algorithms and techniques to detect oil spills using SAR data.

The analysis of this basic fact needs to start from a description of the different mechanisms responsible for the sea surface radar backscattering, which strongly depends on the incidence angle of the radar sensor. In a quite large range of angles, approximately from 20 to 50 deg (the angular span of particular interest for space-borne observations), the main agent of radar backscattering are the wind-generated short gravity-capillary waves. The oil film has a dampening effect on these waves locally decreasing the backscattering. It is implicitly assumed that a light wind field exists in order to activate short gravity-capillary waves. The minimum wind speed is in fact depending on the frequency of observation and the incidence angle. The radar sensors on board of operational satellites used in this study are working in the C band. In this frequency range, a minimum wind field of 2–3 m/s creates sufficient brightness in the image and makes the oil film visible. On the other end, when the wind speed is too high, it causes the spill to disappear. First, because the short waves receive enough energy to counterbalance the dumping effect of the oil film. Then, when the sea-state is fully developed, the turbulence of the upper sea layer may break and/or sink the spill or a part of it.

As a consequence of the above brief discussion the identification of an oil spill in a SAR image includes always as first and basic step the detection of dark features. Typically, a SAR image may show some dark features that are not oil spills (*i.e.*, in most cases due to both meteorological and/or oceanographic effects). These look-alike features pose a fundamental problem to the identification of oil spills and the analysis procedure must include a discrimination phase. The basic functions of a procedure for identification of oil spills can be described as follows:

1. isolation and contouring of all dark signatures, through appropriate threshold and segmentation processing of the image;

2. extraction of key parameters for each candidate signature, which usually are related to its shape, internal structure and radar backscattering contrast;

3. test of the extracted parameters against predefined values, which characterize man-made oil spills, usually determined through phenomenological considerations and statistical assessments;

4. computation of probabilities for each candidate signature. Features falling above a probability threshold are considered to be oil spill with associated a confidence level which is increasing with the corresponding probability. Alternatively, the confidence level can be defined in terms of peculiar characteristics of the identified feature. In this case high confidence spill are the feature having all the characteristics that a real spill usually exhibits in a SAR image. The general approach can be also more sophisticated taking into account relevant environmental parameters having an impact on the spill shape, such as the time history of wind fields and currents.

Finally, it must be recalled that two main limits exist in the use of SAR images: the assessment of the quantity of oil and the identification of the polluter. Of great interest for the competent authorities, is the amount of oil represented by the spillage. However, an accurate estimation cannot be achieved, since it requires accurate knowledge of the spill thickness, which can not be measured by SAR sensors. Moreover, satellites SAR images are unable to identify the pollution culprit (*i.e.* the name of the ship that polluted); satellite can at best detect the position of the probable pollution culprit.

The data used for the Mediterranean Sea are derived from the analysis of satellite imagery. The data type used in these studies was mainly uncalibrated low-resolution image since this is the most targeted and cheap product for the application. A spatial resolution (pixel) of about 200m appeared to be sufficient for statistical investigations of marine oil pollution (Gade and Redondo 1999; Gade *et al.* 2000).

For securing to the maximum possible degree that the detected spills were due to man-made activities and not to look-alike manifestations of natural phenomena, all the images were carefully analyzed using a dedicated semi-automatic detection scheme, which includes, as final step, the decision by a skilled operator. Each identified spill was then registered in a database, together with information concerning its geographic position, the date and time of detection, the spilled area, its average contrast strength, and a vector describing its shape. It is important to underline that oil spills, in the period 1999–2004, have been identified in archive images: the presence of oil at sea has not been confirmed by aerial or vessel surveillance.

For this reason, even though only high confidence features have been taken into consideration, we prefer to term them as 'possible oil spills'.

2. The Mediterranean Sea

Operational pollution from ships is a major problem within the Mediterranean region. While accidental pollution rarely occurs within the Mediterranean waters, operational pollution is a common practice in this basin, representing the main source of marine pollution from ships. Furthermore, the increase of maritime traffic crossing the basin contributes to render the situation even more worrying (REMPEC 2002).

The lack of a regular surveillance service has determined the absence of data on verified spills due to illegal discharges from ships in the Mediterranean. Moreover, this lack of surveillance can encourage the discharge of dirty ballast waters or oily sludge. Due to the large extension of the basin a surveillance system based only on aerial patrolling can be difficult to be implemented for different reasons. On this aspect the satellite surveillance may represent a valid complement because of its ability of providing a global coverage, including remote areas. Satellite surveillance has still a number of limitations, such as an uneven spatial coverage, a quite sparse number of acquisitions in time and a residual number of false alarm cases. Nevertheless comprehensive studies, based on the systematic analysis of space imagery, turned out to be the unique source of information for an overall assessment of the problem. This kind of information, once available, helps to identify the areas at major risk of operational pollution to which particular attention should be given. Considering in fact the small number of accidental pollution events compared to the operational ones within the region, it becomes evident that the use of satellites would be more related to the field of monitoring illicit discharges from ships. Studies carried out by JRC (Bernardini *et al.* 2005, Ferraro *et al.* 2006a, Ferraro *et al.* 2006b, Ferraro 2007, Pavlakis *et al.* 2001, Tarchi *et al.* 2006, Topouzelis *et al.* 2006), based on the analysis of a large number of SAR images, detected a significant number of possible spills within the Mediterranean Sea.

To assess on the distribution of sea-based oil pollution, the number of possible detected spills has to be always compared to the total number of satellite images analyzed. It is also necessary to precisely account for the images which are only partially covering a sea area. To this aim, the coverage is expressed in terms of square degrees observed per year, where only area corresponding to sea is taken into consideration.

Table 1. Yearly coverage and possible oil spills detected over the whole Mediterranean basin in the period 1999–2004.

Year	Coverage (square degrees)	Possible Oil Spills
1999	1382	1638
2000	3642	2297
2001	2495	1641
2002	1840	1401
2003	2289	897
2004	3885	1425
TOTAL	15533	9299

A square degree is a square having approximately 60 nautical miles per side, *i.e.* about 110 kilometres. Due to the fact that all meridians join to the poles, the square degrees are not all equal: their average size tends to decrease towards the north. Table 1 summarizes the results obtained for the whole Mediterranean for the period 1999–2004. For 2003 the results are preliminary and will be further verified. In total 18947 SAR images were analyzed and 9299 possible oil spills were detected. The cumulative result as a point-like map is shown in Figure 1.

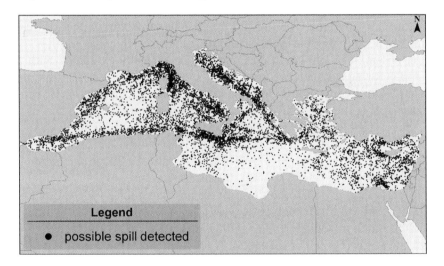

Fig. 1. Possible oil spills detected in the Mediterranean during 1999–2004.

In this map, each possible spill is represented by a dot at the location of the spill centroid (average position). The corresponding coverage of analyzed images for the same period is displayed in Figure 2. The majority of spills are located beyond the 12 nautical miles limit of territorial waters, probably indicating deliberate intention to avoid risks of legitimate actions within the area of jurisdiction of the coastal states. With reference to Figure 2, it should be noted that a low number of images were available for Libyan coastal waters while, by contrast, many images were available for the seas surrounding the Italian peninsular. As a consequence the results are skewed somewhat towards these areas. However, the variations of coverage may be considered quite smooth and the problem almost disappears taking into consideration sub-areas of limited size.

Figure 3 finally displays the oil spill density as obtained merging the information from the two previous maps, *i.e.* by normalizing the number of observed possible oil spills in a given area with the total number of observations available for that area. Such a procedure basically removes any bias effect and accounts for uneven coverage of the area. In summary, the map can be employed to understand the spatial distribution of possible oil spills and to identify hot-spot areas. As could have been easily supposed, the spills distribution appears to be highly correlated with the major shipping routes. Concentrations appear in the Ionian Sea, the Adriatic Sea, the Messina Straits, the Sicily Channel, the Ligurian Sea, the Gulf of Lion and east of Corsica. All over the region, however, the spills show considerable spatial dispersion.

The whole set of detected possible oil spills were then analyzed in terms of seasonal variations. The general trend shows that the number of detections has a maximum during the summer months. Such behaviour has been often observed in similar studies and can be basically explained with the fact that during summer, the mean wind speed is lower, thus determining a higher visibility of oil pollution (Gade *et al.* 2000). However, in particular areas, such as the sea between Corsica, Sardinia and the Italian peninsula, the increased number of summer detections shows a systematic distribution well correlated with local main maritime routes. This fact suggests that the increase in number of detections may also be due to the increment of maritime traffic (ferries) during the tourist season.

It is usually very important to provide an estimation of the oil quantity which is spilled to the sea on a yearly basis. Such a precise estimation would require, in addition to the spilled area, an accurate knowledge of the spill thickness (Brown *et al.* 1995). No information concerning the thickness can be retrieved from radar observations and no direct calculation of the corresponding total volume of oil can be made starting from these data. However they may be used, along with other observational data and other indicators, for an overall estimation. In a recent study (REMPEC 2002) it has been estimated that up to some 100.000 tons of oil and oily waters enter the Mediterranean Sea every year due to operational pollution.

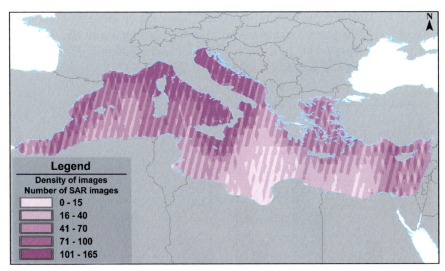

Fig. 2. Coverage of SAR images analyzed in the present study for the Mediterranean Sea during the period 1999–2004.

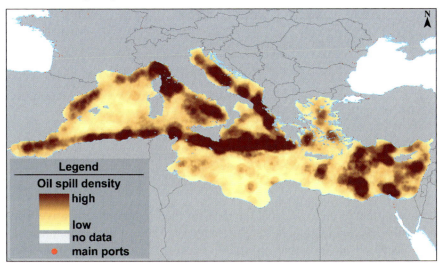

Fig. 3. Oil spill density for the Mediterranean Sea for the period 1999–2004.

Considering these figures, and taking into consideration the distinct hydrological and ecological characteristics of the basin, as well as its extensive coastline (45.000 km) and high concentration of specially protected areas, the situation in the Mediterranean Sea is raising a big concern. In the effort to define a possible indicator of the trend of the detected oil spills in the different years in relation to the area analyzed, a table has been produced

(Figure 4). The "oil spill density" has been calculated dividing the number of detected oil spills per the area coverage. For the Mediterranean Sea, it seems that for the period 2000–2002 the density remains constant but in the years 2003–2004 there is a significant reduction in density.

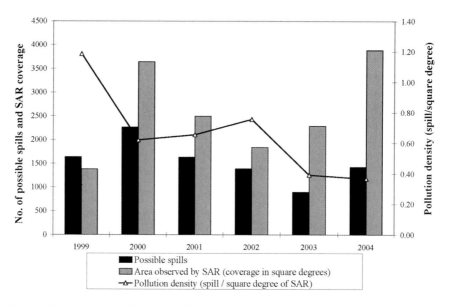

Fig. 4. Total numbers for the Mediterranean Sea: area covered by SAR imagery, possible oil spills (1999–2004) and their ratio.

3. Conclusions

It is important to underline that the data for the Mediterranean derive from oil spills detected in archive satellite imagery. These spills are therefore to be considered as "possible oil spills" because they have not been confirmed by an aircraft and or a vessel. However, the analysis using satellite data allowed the possibility to create density maps of oil spills comparing the area coverage with the number of spills. These maps allow identifying hot spots.

The key positive conclusion of this study is that the operational pollution in the Mediterranean seems decreasing. For the Mediterranean Sea; the ratio spill/area coverage has dropped of approximately 50% comparing 2000–2002 with 2003–2004.

We think that the positive trend in the decrease of oil pollution could be closely related, at least for part of the Mediterranean Sea, to the implementation of the Directive 2000/59 on port reception facilities which entered into force in December 2002. In conclusion, it seems necessary to continue to monitor the trend in the seas around Europe and in particular to reassess the situation after the analysis of the satellite imagery of the year 2005 for the Mediterranean.

References

Bernardini A, Ferraro G, Meyer-Roux S, Sieber A, Tarchi D (2005) Atlante dell'inquinamento da idrocarburi nel Mare Adriatico. European Commission, EUR 21767 IT

Bonn Agreement website: http://www.bonnagreement.org - data concerning aerial surveillance are reported on: http://www.bonnagreement.org/eng/html/aerial-surveillance/welcome.html

Brown C, Fruhwirth M, Fingus M, Vaudreuil G, Monchalin J, Choquet M, Heon R, Padioleau C, Goodman R, Mullin J (1995) Oil Slick Thickness Measurement A Possible Solution to a Long-Standing Problem. Proc Eighteenth Arctic Marine Oil Spill Program Technical Seminar, Environment Canada, Ottawa, Ontario, pp 427–440

Carpenter A (2007) The Bonn Agreement Aerial Surveillance programme: Trends in North Sea oil pollution 1986–2004. Marine Pollution Bulletin 54: 149–163

Del Frate F, Petrocchi A, Lichtenegger J, Calabresi G (2000) Neural networks for oil spill detection using ERS-SAR data. IEEE Trans Geosci Rem Sens 38: 2282–2287

Espedal HA, Wahl T (1999) Satellite SAR oil spill detection using wind history information. Int J Rem Sens 20: 49–65

Espedal HA, Johannessen JA (2000) Detection of oil spills near offshore installations using synthetic aperture radar (SAR). Int J Rem Sens 11: 2141–2144

Ferraro G, Tarchi D, Fortuny J, Sieber A (2006a) Satellite Monitoring of Accidental and Deliberate Marine Pollution. In: Gade M, Hühnerfuss H, Korenowski GM (eds) Marine Surface Films: Chemical Characteristics, Influence on Air-Sea Interactions and Remote Sensing. Springer, Heidelberg, pp 273–288

Ferraro G, Bernardini A, Meyer-Roux S, Tarchi D (2006b) Satellite Monitoring of Illicit Discharges from Vessels in the French Environmental Protection Zone (ZPE) 1999–2004. European Commission, EUR 22158 EN

Ferraro G, Bernardini A, David M, Meyer-Roux, Muellenhoff O, Perkovic M, Tarchi D, Topouzelis K (2007) Towards an Operational Use of Space Imagery for Oil Pollution Monitoring in the Mediterranean Basin: a Demonstration in the Adriatic Sea. Mar Pollution Bull 54: 149–163

Fiscella B, Giancaspro A, Nirchio F, Trivero P (2000) Oil spill detection using marine SAR images. Int J Rem Sens 21: 3561–3566

Fortuny J, Tarchi D, Ferraro G, Sieber A (2004) The Use of Satellite Radar Imagery in the Prestige Accident. Proc International Conference Interspill 2004

Gade M, Redondo JM (1999) Marine pollution in European coastal waters monitored by the ERS-2 SAR: a comprehensive statistical analysis. Proceedings of the International Geoscience and Remote Sensing Symposium, IGARSS 1999, vol 2: 1375–1377

Gade M, Scholz J, von Viebahn C (2000) On the detectability of marine oil pollution in European marginal waters by means of ERS SAR imagery. Proc International Geoscience and Remote Sensing Symposium, IGARSS 2000, vol 6: 2510–2512

Helsinki Convention website: http://www.helcom.fi - Data on aerial surveillance available at: http://www.helcom.fi/shipping/waste/en_GB/surveilance

Karathanassi V, Topouzelis K, Pavlakis P, Rokos D (2007) An object-oriented methodology to detect oil spills. Int J Rem Sens doi:10.1080/0143116060069 3575

North Sea Directorate (1992) The Netherlands, Visibility Limits of Oil Discharges, Rijswijk, 1992. IMO document: MEPC 33/INF.28

OSPAR website: http://www.ospar.org

Pavlakis P, Tarchi D, Sieber A, Ferraro G, Vincent G (2001) On the Monitoring of Illicit Discharges – A Reconnaissance Study in the Mediterranean Sea. European Commission, EUR 19906 EN.
See: http://serac.jrc.it/midiv/pub/jrc_illicit_study.pdf

REMPEC (2002) Protecting the Mediterranean against Maritime Accidents and Illegal Discharges from Ships, Malta

Solberg R, Theophilopoulos NA (1997) Envisys - A solution for Automatic oil spill detection in the Mediterranean. Proc 4th International Conference on Remote Sensing for Marine and Coastal Environments: 3–12

Solberg, Storvik G, Solberg R, Volden E (1999) Automatic Detection of Oil Spills in ERS SAR Images. IEEE Trans Geoscim Rem Sens 37: 1916–1924

Tarchi D, Bernardini A, Ferraro G, Meyer-Roux S, Muellenhoff O, Topouzelis K (2006) Satellite Monitoring of Illicit Discharges from Vessels in the Seas around Italy, 1999–2004. European Commission, EUR 22190 EN

Topouzelis K, Karathanassi V, Pavlakis P, Rokos D (2002) Oil Spill Detection: SAR Multi-scale Segmentation & Object Features Evaluation. Proc 9th International Symposium on Remote Sensing (SPIE), pp 77–87

Topouzelis K, Bernardini A, Ferraro G, Meyer-Roux S, Tarchi D (2006) Satellite mapping of oil spills in the Mediterranean Sea. Fresenius Environmental Bulletin 15: 1009-1–1009-14

Sea Ice Monitoring in the Arctic and Baltic Sea Using SAR

Jan Askne[1] and Wolfgang Dierking[2]

[1] Department of Radio and Space Science, Chalmers University of Technology, Gothenburg, Sweden
[2] Earth Observing Systems, Alfred Wegener Institute for Polar and Marine Research, Bremerhaven, Germany

Abstract. Large regions of the Polar Oceans are covered by sea ice. The ice has a profound impact on the exchange of heat, momentum, and matter between the ocean and the atmosphere, on the solar albedo of the ocean, and on deep ocean circulation. Information about sea ice conditions are needed for ship navigation, fisheries, or oil and gazxs exploration, in geo- and biophysical studies, and in climate research. In this chapter, methods of sea ice monitoring using synthetic aperture radar are addressed. The influence of sea ice properties such as surface roughness or volume structure on the observed radar signatures is explained, also considering environmental effects. We discuss advantages and limitations of different SAR configurations for sea ice observations. The use of SAR imagery for ice type discrimination is examined. The determination of other parameters such as ice drift, timing and length of the melt season, melt pond coverage, sea ice concentration, or extension of polynyas are briefly summarized. The usefulness of SAR data for validation and improvements of numerical models for simulating the dynamics of the sea ice cover is emphasized.

1. Introduction

Satellite remote sensing is regarded important for increasing our knowledge about the Earth's oceans, their biological and physical properties, and variations of these properties, which may be indications of global climate change. Data acquisitions from space are essential in particular for the vast sea ice areas of the Arctic and Antarctic since their accessibility is limited. For seasonally ice covered regions close to more densely populated coasts, such as *e.g.* the Baltic Sea, regular information from space-borne sensors about physical and biological conditions is useful for supporting marine

V. Barale, M. Gade (eds.), *Remote Sensing of the European Seas.*
© Springer Science+Business Media B.V. 2008

operations and fisheries. Because the visibility of sea ice regions from space is limited by short time intervals of daylight during winter, and because these regions are often shrouded in clouds and fog in particular during summer, severe gaps exist in the information that can be obtained from optical satellite sensors. Because of their independence from cloud and light conditions, passive microwave radiometers and radar systems are intensively used for sea ice monitoring. Passive microwave sensors have provided regular coverage of sea ice in the Polar Regions for almost 30 years. These are characterized by a relative coarse spatial resolution on the order of kilometers but a wide swath which makes it possible to cover large region in short time intervals. For many applications, however, sea ice variations on spatial scales of meters are of interest. In such cases, Synthetic Aperture Radar (SAR) with its much better spatial resolution is utilized. The drawback is that the swath widths are smaller. For sea ice monitoring, the launch of the European Earth Remote Sensing Satellite, ERS-1, in July 1991 was an important milestone, and similarly the launches of RADARSAT in 1995 and of ENVISAT in 2002.

In this chapter, methods and applications of sea ice monitoring using SAR are addressed, focusing on the Arctic and the Baltic Sea. Although European countries have also a large interest in the Antarctic, we refrain from a detailed discussion of SAR remote sensing of the Antarctic sea ice regions for the sake of brevity of this chapter.

1.1 The need for sea ice monitoring

Between 11% and 15% of the Earth's surface are covered by sea ice with characteristics changing seasonally and in response to meteorological and oceanographic conditions. The presence of sea ice on the ocean surface changes the solar albedo and the exchange of heat, momentum, and matter between the ocean and the atmosphere. Recent investigations have been focusing on the role of sea ice formation for deep ocean circulation. Increased atmospheric concentrations of greenhouse gases have a larger effect on climate in the Arctic than anywhere else on the globe (Hassol 2004). In order to gain a more detailed understanding of the various interactions and feedback mechanisms that have an influence on Earth's climate, various parameters characterizing the sea ice cover need to be monitored over years and decades. Such parameters are ice extent, concentration, thickness, drift, and duration of the melt period.

Continuous observations of the sea ice cover are also required for marine traffic and operations. If the average extent of Arctic sea ice during summer continues to decrease as observed during the last years, shipping and exploration of oil and gas along the Northern Sea Route will be possible. In regions such as the Baltic Sea, undisturbed ship traffic is fundamental all year round. For operational tasks ice information is required at short notice with a relatively high spatial resolution.

1.2   The potential of SAR to observe sea ice properties

The backscattered intensity which determines the grey tone of a single pixel in a SAR image is composed of direct reflections and diffuse scattering (in variable proportions) from the ice surface and the ice bulk. It depends on the surface characteristics (roughness, snow or melt water coverage) and volume structure (cracks, inclusions, layering), and on the dielectric properties. The pixel-to-pixel changes of the grey tone seen in a SAR image are related to ice type distribution, ice floe dimensions, and the presence of ridges, open water leads, polynyas, melt ponds or frost flowers on the ice surface. The potential of SAR to image different sea ice properties with sufficient contrast and detail depends on SAR parameters such as frequency, polarization, incidence angle, and noise level. The recognition of details characterizing the ice cover is influenced by the spatial resolution of the SAR. A high spatial resolution is coupled with a small swath width and vice versa. Different aspects of sea ice remote sensing are discussed in the literature (*e.g.* Carsey 1992; Tsatsoulis and Kwok 1998).

Sea ice properties affecting scattering

Sea ice properties that have a strong impact on the radar signature are: mm to dm undulations on the ice surface ("small-scale" roughness); number, size, shape and orientation of volume scatterers such as air bubbles and brine inclusions; and density, wetness, grain size, and surface roughness of the snow cover. The dielectric constant depends on salinity and temperature. New and young ice types reveal a larger salinity than older ice types due to drainage of brine from the ice to the upper ocean layer.

The typical sea ice roughness spectrum decays with spatial wavelength (Carlström and Ulander 1995). A surface appears rougher at shorter radar wavelengths (and therefore brighter in the radar image), until a limit is reached for which the scattering mechanisms get more complex. Surface scattering intensity is more sensitive to the radar incidence angle than volume scattering. Radar signature variations occur over ridges, hummocks, and rubble fields ("large-scale" roughness) because of incidence angle variations, but also because of an increased number of ice volume inhomogeneities (air voids, cracks).

The intensity of volume scattering depends on the penetration depth into the ice, which decreases with increasing radar frequency. At typical dimensions of most sea ice inclusions (mm-scale), volume scattering increases as the radar wavelength gets smaller relative to the size of the inclusions (Rayleigh scattering). In effect, the scattering contribution from the ice volume increases with increasing radar frequency (considering radar bands from P to K_u)[1].

Snow accumulation on sea ice is highly variable in time and space. Snow insulates the ice surface from the lower atmosphere and influences the temperature profile in the ice bulk. It also depresses the ice surface, occasionally causing the ice-snow interface to be flooded if the ice is thinner. A dry snow layer causes refraction of the radar waves, hence the incidence angle on the ice surface is changed. Considerable scattering from the snow can be observed only at higher frequencies (K_u-band), in particular in case of old, coarse-grained snow layer. A wet ice or snow surface layer changes the backscatter signature radically, since it acts like a shield that is not penetrated by the radar waves. Thin ice (thickness less than 30 cm) is often covered by a very saline slush (snow completely saturated with water) layer or with frost flowers (small, distinct ice crystals or bumps of ice crystals). The wet, saline and rough surface layer of the ice or snow can produce strong and highly variable backscattering intensities in a SAR image. Frost flowers and wet saline snow on thin Arctic ice may affect surface and backscatter even more drastically. Changes by 5 to 15 dB over a period of 1 to 14 days, caused by temperature drops and snow fall, were observed (Ulander *et al.* 1995). Besides temporal varying effects (melting, refreezing), radar signature analysis has to take into account also regional variations. Arctic and Baltic Sea ice differ considerably, and in the Arctic a number of ice regimes can be distinguished, *e.g.* relatively smooth level ice at the Russian coast or rugged and ridged ice north of Greenland.

Many parts of the Arctic Ocean and adjacent regions are dominated by thick Multi-Year Ice (MYI, thickness > 2 m) that has survived at least one summer season. The surface layer of MYI is typically rough and almost completely desalinated. During fall and winter, the radar backscatter originates from the rough surface and the air bubbles in the upper ice layer. During summer, the surface is wet, and melt ponds change the surface roughness. The backscatter is dominated by the ice surface and is typically very low. Sea ice formed during the recent winter season is called

---

[1] The radar bands are associated with frequencies: P-band (420–450 MHz), L-band (1215–1400 MHz), S-band (2.2–2.5 GHz, 2.7–3.7 GHz). C-band (5.25–5.925 GHz), X-band (8.5–10.68 GHz), Ku-band (13.4–14 GHz 15.7–17.7 GHz).

First-Year Ice (FYI, thickness 0.3–2 m). In the Arctic it is characterized by a salinity of 5 to 10 psu. The radar waves penetrate only into the upper ice layer, and the backscattered signal is dominated by surface contributions. Ice volume contributions are low but increases at larger incidence angles and may also affect scattering from thin ice with a very smooth surface (nilas). Examples of typical Arctic sea ice SAR signatures, under various environmental conditions at different frequencies, polarizations and incidence angles, are given in Onstott 1992 or Onstott and Shuchman 2005.

Conditions are different in the Baltic Sea. The ice cover melts completely in summer. The ice salinity varies from about 2 psu in the south to about 0.2 psu or less in the north because of the low salinity of the brackish water. The thickness ranges from 0.2–1 m. Field measurements and results of theoretical scatter modeling show that the backscattering response at shorter radar waves (X-band) is mainly caused by volume inhomogeneities in the uppermost part of the ice. At moderate wavelengths (C- and S-band), the scattering is dominated by the roughness of the snow-ice interface. At L-band, strong radar returns can originate also from scattering close to or at the ice-water interface (Dierking *et al.* 1999).

In order to improve and test our understanding of the scattering phenomena, theoretical backscatter modeling has been combined with *in situ* measurements of ice properties. Most models developed hitherto assume a statistically distributed small-scale surface roughness and small sizes of the volume inclusions (Winebrenner *et al.* 1992).

SAR configurations

In order to find the optimum radar configurations for sea ice mapping, a number of measurements using ground based, airborne, and space borne radar have been carried out (see *e.g.* Onstott 1992). The major conclusion from these studies with regard to the separation of sea ice types over all seasons is that the selection of C-band radar is a reasonable compromise between the use of shorter wavelengths (X- and K_u-band) during winter and longer wavelengths (L-band) during summer. This assessment is based on observations and theoretical modeling of sea ice signatures and addresses in particular the discrimination of FYI and MYI in the Arctic. For specific task of monitoring, other frequency bands than C-band may be more useful. Shorter wavelengths (X- and K_u-band) are better suited to distinguish different thin ice classes and FYI, longer radar waves are more helpful in discriminating deformed and level ice. Also the polarization influences the discrimination of ice classes and the detection of certain sea ice features such as ridges, although to a minor degree compared to the frequency. For example, the optimum polarization for discrimination of

FYI and MYI at C-band is HH- or HV. Compared to VV-polarization, open water has a lower backscattering coefficient at HH-polarization. The difference increases at larger incidence angles. For thick or rough FYI and MYI ice, the polarization difference is smaller. Dual-polarization radar measuring HH and VV simultaneously is therefore useful for recognizing open water areas in the ice cover. The choice of the incidence angle also influences the detection of deformation features. Ridges are more easily recognized at larger incidence angles than at angles close to nadir. Comparing C- and X-band radar data at different polarizations, the best ice type discrimination for Baltic Sea ice was obtained at C-band VH-polarization at an incidence angle of 45° (Mäkynen and Hallikainen 2004). The utilization of fully polarimetric SAR improves sea ice type discrimination further (polarimetric SAR modes are available on RADARSAT-2 and on ALOS). It was noted, however, that the accuracy of sea ice classification is significantly enhanced by using different frequencies (Dierking *et al.* 2004).

Other radar parameters of importance for sea ice observations are the radar swath width, spatial resolution, temporal coverage, and noise level. For certain cases (*e.g.* shipping routes close to harbors, highly dynamic key regions such as polynyas) imaging of limited areas with high resolution (30 m or better) and hence narrow swath (100 km and less) is useful. In general, a wider coverage (400 to 500 km with spatial resolutions of 100 to 150 m, available from RADARSAT and ENVISAT) is preferred. For a complete coverage of the Arctic, the satellite needs to follow a polar orbit. At large incidence angles, latitudes up to 88°N can be covered. With the 400 km wide swath of ENVISAT ASAR (repeat cycle 35 days), areas above latitude 60°N are covered at least every third day. The SAR noise level determines how well thin ice types that are often characterized by a low backscatter can be discriminated in the imagery.

2.   Extraction of large-scale sea ice properties

Large scale properties of the sea ice cover are of major relevance for climate studies as well as for operational monitoring. These properties include, *e.g.* ice type distribution, ice concentration and thickness, frequency, size, and distribution of leads and polynyas, ice deformation, and ice motion. Examples of the retrieval of such large scale properties follow.

2.1 Ice type discrimination

The discrimination of ice types and ice vs. open water is important not only for marine operations. Ice classification permits indirect conclusions on the ice thickness distribution and on the degree of deformation in a given area. It is needed for the determination of ice concentration and (in a sequence of consecutive images) changes of ice type composition. Information on areal coverage and evolution of thin ice and pancake ice (small ice floes with a rim that develop on the wind-roughened sea) is useful for the estimation of vertical heat exchange between ocean and atmosphere and salt fluxes into the upper ocean layer.

Since the backscattering intensity of a given ice type varies dependent on physical and dielectric ice properties and on environmental conditions, ice classification in SAR imagery can be quite complicated, in particular for single-frequency, single-polarization SAR. Also the discrimination of open water leads and ice floes is difficult, since a water surface shows varying backscattering intensities dependent on wind speed and direction. The RADARSAT Geophysical Processor System, RGPS (Kwok 1998) includes an algorithm based on ice backscatter characteristics in winter combined with ice motion information and records of near-surface air temperature. It works with five classes: MYI, deformed FYI, undeformed FYI, smooth younger ice types and calm open water (Fetterer *et al.* 1994). Potential and limitations for ice type classification are illustrated by

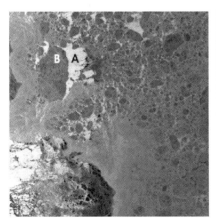

Fig. 1. *Quasi*-simultaneous images, centered at 82°N 12°E (image length and width ~80 km; pixel spacing ~25 m, left; and ~100 m, right), collected by ERS (C-band VV), left panel (© European Space Agency), and RADARSAT (C-band HH), right panel (© Canadian Space Agency), 19/09/1996 (Dokken *et al.* 2000).

Figure 1 demonstrates the sensitivities of radar signatures to polarization, and environmental effects. An ERS SAR image from the Arctic Ocean (left panel) is compared to a RADARSAT image acquired 97 minutes later (right panel). The wind speeds measured on an ice breaker in a distance of approximately 50 km from the centre pixel were 8 and 14 m/s at the times when the images were acquired, respectively. The temperature was around −1°C, and snow was wet. The images demonstrate the influence of polarization and incidence angle on the sea ice signature. In the ERS-1 image (VV-polarization, 23° incidence), the wind-roughened open water appears bright (A), thin ice almost black, smooth ice floes dark, and rougher ice (ice floe size smaller than SAR resolution) grey. Floe B is MY ice with ridge clusters on it. Because of the wet snow, the volume scattering may be reduced, and therefore the floe appears darker as it is typical for MY ice. At HH-polarization and ≈30° incidence angle, the backscattering from the water surface is much lower (it has also to be taken into account that the look-direction of images *a* and *b* are different because of ascending and descending orbits). The ice floes stand out less clearly.

Figure 2 illustrates the sensitivity of the sea ice signature to radar frequency. It shows (left panel) an ERS-1 SAR image from the coast of East Greenland and (right panel) a JERS-1 SAR image acquired about 25 minutes earlier.

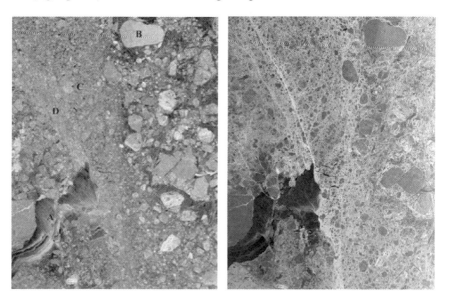

Fig. 2. *Quasi*-simultaneous images of the coast of East Greenland (image width ~50 km; pixel spacing ~100 m), collected by ERS (C-band VV), left panel (© European Space Agency), and JERS-1 (L-band HH), right panel (© NASDA), in 2004 (Dierking and Busche 2006).

Zones of new ice formation such as the one marked by "A" reveal a much larger signature variation in the ERS data. MYI floes (*e.g.* at "B") stand out more clearly. In the JERS images, the intensity contrast between brash ice and moderately sized floes of smooth level ice is larger (zone "C"). A shear zone, marked by "D," separates belts of ice drifting with different velocities. The shear feature is difficult to identify in the ERS-1 image but obvious in the JERS image.

In Figure 3, differences between backscattered intensities of typical Baltic Sea ice types and intensity variations due to changing environmental conditions are illustrated. The ERS-1 image shown has dimensions of 30×22 km^2.

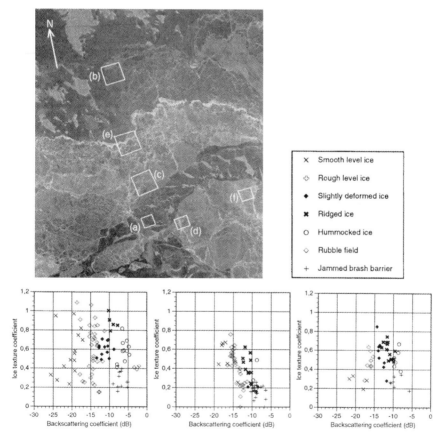

Fig. 3. ERS-1 SAR image (17/03/1992) showing different ice types: (a) smooth level ice, (b) rough level ice, (c) ridged ice, (d) hummocked ice, (e) rubble fields, (f) jammed brash barrier. Graphics: radar signatures of sea ice from the Baltic Sea, from 1992 (top) with cold dry weather, 1993 (center) mild rainy weather, and 1994 (bottom) cold dry weather (Dammert *et al.* 1994).

Ice types were identified on helicopter photographs (Dammert *et al.* 1994). The graphs on the right show partial overlap of the signatures which complicates automatic classification (Mäkynen and Hallikainen 2004). The preceeding Figures demonstrate that a robust classification of sea ice types has to take into account the SAR configuration and the environmental conditions of data acquisitions. They also show that a combination of images acquired at different polarizations and in particular different frequencies enhances the discrimination performance between ice types.

2.2   Ice melt and freeze-up

The length of the melt season (the time between spring melt onset and autumn freeze-up) strongly influences sea ice decay and hence the freshwater fluxes. The sensitivity of SAR signatures to ice surface melt can be used to identify the melt and freeze transitions (Figure 4). Melt onset is marked by a steep decrease in the radar backscattering from multi-year ice, which is related to the appearance of liquid water in the snow cover.

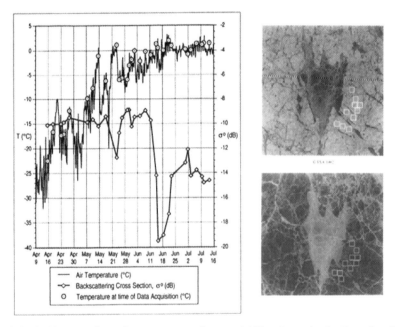

Fig. 4. Left: Near surface air temperature from a drifting buoy in the Beaufort Sea, and the backscattered radar intensity of multi-year ice at C-band, observed during a period from April to July. Right: ERS-1 SAR images from June 2 (above) and from June 15 (below). Equal grey levels in separate images connote equal backscattered intensities (Winebrenner *et al.* 1998).

At freeze-up, backscatter increases significantly, related to increased volume scattering from air bubbles in the ice (Winebrenner *et al.* 1998). The onset of melt influences also first year ice radar backscattering, but the effect is less visible at C-band. Algorithms based on histogram variations of the backscattered radar intensity have been developed using RADARSAT imagery. It was shown that the melt signal is detectable over multiyear ice, first-year ice, and mixtures of these ice types (Kwok *et al.* 2003).

Meltponds are the result of summer melt processes on sea ice, and they play a significant role in the heat exchange between atmosphere, ice, and ocean, and in the disintegration of the ice. Since the extent of melt ponds is often smaller than the SAR resolution (Perovich *et al.* 2002), and because of the wind effect on the radar backscatter of open water, no operational algorithms are available for the retrieval of meltpond concentration. During an Arctic expedition a 7 km^2 area at latitude 84°N was covered by a video mosaic simultaneously with ERS-1 SAR imaging. The melt pond concentration varied between 12 and 31%, and the backscatter was a mix between open water (-7.0 dB) and old ice (-13.5 dB) signatures. It was found that the backscattered intensity increases with melt pond concentration, as in Figure 5 (Askne *et al.* 1994; also Yackel and Barber 2000).

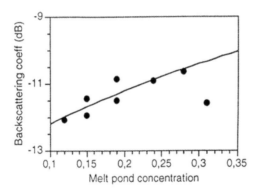

Fig. 5. Backscatter as function of meltpond concentration (500×500 m^2 averages).

2.3 Ice motion

Information about ice motion is needed, *e.g.* for calculations of ice volume or mass transport, for analyses of ice deformation patterns and for studies of the effect of large scale atmospheric wind fields on sea ice. For the observations of ice drift patterns, a number of satellite sensors have been used including SAR. The estimated error of the drift velocity derived from consecutive SAR images is at least smaller than half of the error achieved with passive microwave sensors, as validated by buoys (Stern and Moritz 2002). SAR is in particular useful when detailed information is needed over smaller regions such as the Baltic Sea, or over highly dynamic areas such as Fram Strait. The temporal sampling has to be of the order of a few days (depending on the magnitude of the ice drift velocity in a particular area). When ice concentration is low (as in the marginal ice zone) and wind speed high, the ice motion can be rather complex. In such cases, the results of unsupervised drift retrieval algorithms are highly erroneous.

There is a number of approaches to determine the ice drift from successive radar images that differ in their capability to deal with rotational motion and variations of image features: correlation of image segments (Fily and Rothrock 1987); evaluation of the similarity between features such as ice floes (McConnell *et al.* 1991); derivation of motion vectors using changes of the local brightness both in the spatial and the time domain, i.e. the optical flow method (Sun 1996); and application of invariance transformation (Sun 1994). The RGPS includes products for high-resolution ice motion (Kwok 1998). The ice motion routine is based on a Lagrangian approach *and* uses area matching techniques to follow an array of points from a source image to a set of target images. This product has been validated by comparison with buoys showing displacement errors (RMS) of the order of 0.8 km for separations up to 20 km.

2.4   Other applications of SAR imagery

Sea ice concentration (the fraction of a unit area on the ocean surface covered by sea ice) has been continuously monitored since 1978 using passive microwave sensors. The accuracies of ice concentration retrievals from passive microwave data have been evaluated by comparisons to SAR imagery, aerial photograpy and ship observations (Dokken *et al.* 2000; Pettersson *et al.* 1996). However, since the ice-water discrimination in one-channel SAR imagery is difficult at certain environmental conditions, SAR is hitherto not used for operational observations of ice concentration.

Leads and polynyas are areas of considerable heat flow from the ocean to the atmosphere during winter and of increased solar radiation uptake during spring and summer. During ice formation on the leads and in the polynyas, salt is released to the upper ocean layer. This process contributes to the maintenance of the Arctic Ocean halocline and possibly to deep water formation. For typical sizes of leads and polynyas, SAR images with their comparatively high spatial resolution are used to delimit the polynya area, to distinguish open water and new ice types, and to estimate opening and closing velocities. With one-channel SAR data, the identification of the polynya border may be difficult (see *e.g.* Dokken *et al.* 2002).

Sea ice deformation is a result of divergent and convergent sea ice motion, by which leads, pressure ridges, and near-shore shear zones are formed. Consecutive SAR images show the ice deformation which in further analysis can be related to wind and ocean drag. An automatic algorithm for computation of statistical parameters characterizing the deformation state is described in (Li *et al.* 1998). The forces acting on the ice due to wind depend on the number and height of ridges (besides further parameters such as ice concentration) and need to be known in models simulating atmosphere-ice interactions. Ridges and rubble fields are severe hazards to marine operations and offshore structures, an item that is important for ship traffic in the Baltic Sea. Scattering from ice ridges has been studied by (Carlström 1997; Manninen 1996). Typical widths of the ice ridges are smaller than typical spatial SAR resolution of 30–150 m. In such cases, the backscatter value of an image pixel is a mixture of contributions from the ridge and the adjacent level ice. Ridges or clusters of ridges are easier to recognize at lower frequencies and larger incidence angles.

Ice thickness is one of the most important ice properties for climate research and for operational mapping. As part of the RGPS areas of new ice formation are found by searching for areas of divergent ice motion (Kwok 1998). The evolution of these areas is followed in subsequent SAR images. The thickness of the newly developed ice is estimated from the accumulated freezing days. The thickness of older ice can only be roughly determined by separating FYI and MYI. The thickness of thin ice (up to 0.3–0.5 m) is highly correlated with radar signatures such as intensity and co-polarization ratio as was found in a number of laboratory studies and analyses of SAR imagery. Only under specific conditions, a sufficiently robust relationship between radar backscatter and thickness may exist also for thicker ice. In general, however, the radar backscatter variations are only slightly correlated with thickness for FYI and MYI.

3. Operational SAR mapping and model validations

SAR images are nowadays routinely used by operational ice services. The objective of ongoing projects is to implement a coherent European operational marine system for the high latitudes covering the Baltic Sea, European Arctic and Canada, and Antarctica[2]. The Canadian Ice Service[3]

---

[2] http://www.polarview.org/
[3] http://ice-glaces.ec.gc.ca/

has since long been using SAR data together with other sources of information Usually, different satellite products are employed depending on availability (optical for cloud-free conditions, wide swath RADARSAT or ENVISAT, or ENVISAT Global Monitoring mode). Numerical models simulating sea ice dynamics (ice growth, decay, motion, and deformation) are used in climate research and for ice forecasting. In comparison to data from satellites and from ground based sensors, which provide information about the actual ice conditions, different model assumptions can be initialized, tested and verified (Baltic Sea examples in Leppäranta *et al.* 1998).

4. Conclusions

The launch of ERS-1 was a breakthrough for sea ice monitoring using SAR. Today we have a wider range of data products available, considering the various missions in space (C- L-, and X-band, and imaging modes that differ in polarization, incidence angle, swath width and spatial resolution). For the operational sea ice services, a high rate of data acquisitions is important for most ocean regions, making wide swath modes their first choice. The continuation of C-band SAR missions is highly prioritized, considering the experience gained and the algorithms developed for the ERS-1, ERS-2, RADARSAT and ENVISAT SAR systems. Although the spatial resolution is of great help, single-frequency, single-polarization SAR is not sufficient for accurate classification due to various environmental effects on the radar signatures of the sea ice as we showed in this chapter. In the coming years, sea ice specialists have enhanced possibilities to work with different frequencies and with polarimetric data products. The development of robust algorithms for different applications is often time consuming, considering the logistic difficulties for validation campaigns. But we expect that this problem will be alleviated because of new in-situ measurement technologies and the increased awareness and interest in sea ice observations using SAR. The latter is mainly driven by the pressing need to cover the sea ice regions over the globe for understanding climate developments and for extended operational use.

References

Askne J, Ulander LMH, Birkeland D (1994) Accuracy of ice concentration derived from ERS-1 SAR images during the late melt period in the Arctic. EARSeL Advances Rem Sens 3: 44–49

Carlström A (1997) A microwave backscattering model for deformed first-year sea ice and comparisons with SAR data. IEEE Trans. Geoscience and Remote Sensing GRS-35: 378–391

Carlström A, Ulander LMH (1995) Validation of backscatter models for level and deformed sea ice in ERS-1 SAR images. Int J Rem Sens 16: 3245–3266

Carsey FD (ed) (1992) Microwave remote sensing of sea ice. AGU Geophysical Monograph Series. American Geophysical Union, Washington

Dammert PBG, Ulander LMH, Larsson B (1994) Radar signatures of sea ice and leads. In: Ulander LMH (ed) Baltic Experiment for ERS-1 (BEERS). National Maritime Administration, Norrköping, Sweden, pp 71–98

Dierking W, Busche T (2006) Sea Ice Monitoring by L-Band SAR: An assessment based on literature and comparisons of JERS-1 and ERS-1 imagery. IEEE Trans Geosci Rem Sens 44: 957–970

Dierking W, Pettersson MI, Askne J (1999) Multifrequency of scatterometer measurements of Baltic Sea Ice during EMAC-95. Int J Rem Sens 20: 349–372

Dierking W, Skriver H. , Gudmandsen P. (2004) On the improvement of sea ice classification by means of radar polarimetry, *Proceedings of the 23rd EARSeL Symposium, Remote Sensing in Transition*, Ghent, Belgium, ed. By R. Goossens; Rotterdam: Millpress, pp 203–209

Dokken ST, Håkansson B, Askne J (2000) Inter comparison of Arctic Sea ice concentration using Radarsat, ERS, SSM/I and in-situ data. Can J Rem Sens 26: 521–536

Dokken ST, Winsor P, Markus T, Askne J, Björk G (2002) ERS SAR characterization of coastal polynyas in the Arctic with SSM/I and numerical model investigations. Rem Sens Environ 80: 321–335

Fetterer F, Gineris D, Kwok R (1994) Sea ice type maps from Alaska Synthetic Aperture Radar Facility imagery: An assessment. J Geophys Res 99: 22443–22458

Fily M, Rothrock DA (1987) Sea ice tracking by nested correlations. IEEE Trans Geosci Rem Sens GE-25: 570–580

Hassol SJ (ed) (2004) Impacts of a warming Arctic: Arctic Climate Impact Assessment. Cambridge University Press, Cambridge

Kwok R (1998) The RADARSAT Geophysical Processor System. In: Tsatsoulis C and Kwok R (ed) Analysis of SAR data of the Polar Oceans. Springer-Verlag, Berlin, pp 235–258

Kwok R, Cunningham G, Nghiem S (2003) A study of the onset of melt over the Arctic Ocean in RADARSAT synthetic aperture radar. J Geophys Res 108: 3363–3376

Leppäranta M, Sun Y, Haapala J (1998) Comparisons of sea-ice velocity fields from ERS-1 SAR and a dynamic model. J Glaciology 44: 248–262

Li S, Cheng Z, Weeks WF (1998) Extraction of intermediate scale sea ice deformation parameters from SAR ice motion products. In: Tsatsoulis C and Kwok R (ed) Analysis of SAR data of the Polar Oceans. Springer-Verlag, Berlin, pp 69–90

Manninen AT (1996) Surface morphology and backscattering of ice-ridge sails in the Baltic Sea. J Glaciology 42: 141–156

McConnell R, Kwok R, Curlander JC, Kober W, S. PS (1991) Ψ-S correlation and dynamic time warping: Two methods for tracking ice floes in SAR images. IEEE Trans Geosci Rem Sens 29: 1004–1012

Mäkynen M, Hallikainen M (2004) Investigation of C- and X-band backscatter signatures of Baltic Sea ice. Int J Rem Sens 25: 2061–2086

Onstott RG (1992) SAR and scatterometer signatures of sea ice. Chapter 5. In: Carsey FD (ed) Microwave remote sensing of sea ice. American Geophysical Union, Washington D.C., pp 73–104

Onstott RG, Shuchman RA (2005) SAR measurements of sea ice. In: Jackson CR and Apel JR (ed) Synthetic aperture radar marine user's manual. US Department of Commerce, 81–115

Pettersson MI, Cavalieri DJ, Askne J (1996) SAR Observations of Arctic freeze-up as compared to SSM/I during ARCTIC-91. Int. J. of Remote Sensing 17: 2603–2624

Perovich DK, Tucker WB, Ligett KA (2002) Aerial observation of the evolution of ice surface conditions during summer. J. Geophy. Res. 107: 8048–8064

Stern HL, Moritz RE (2002) Sea ice kinematics and surface properties from RADARSAT SAR during the SHEBA drift. Journal of Geophysical Research (for the SHEBA Special Section) 107: 1–10

Sun Y (1994) A new correlation technique for the ice motion analysis. EARSeL Advances of Remote Sensing 3: 57–63

Sun Y (1996) Automatic ice motion retrieval from ERS-1 SAR images using the optical flow method. Int. J. Remote Sensing 17: 2059–2087

Tsatsoulis C, Kwok R (ed) (1998) Analysis of SAR data of the polar oceans: recent advances. Springer, Berlin

Ulander LMH, Carlström A, Askne J (1995) Effect of frost flowers, rough saline snow and slush on the ERS-1 SAR backscatter of thin Arctic Sea ice. Int. J. of Remote Sensing 16: 3287–3306

Winebrenner DP, Bredow J, Fung AK, Drinkwater MD, Nghiem S, Gow AJ, Perovich DK, Grenfell TC, Han HC, Kong JA, Lee JK, Mudaliar S, Onstott RG, Tsang L, West RD (1992) Microwave sea ice signature modeling. Chapter 8. In: Carsey FD (ed) Microwave remote sensing of sea ice. American Geophysical Union, Washington D.C., pp 137–176

Winebrenner DP, Long DG, Holt B (1998) Mapping the progression of melt onset and freeze-up on Arctic sea ice using SAR and scatterometry. In: Tsatsoulis C and Kwok R (ed) Analysis of SAR data of the Polar Oceans. Springer-Verlag, Berlin, Germany, pp 129–144

Yackel JJ, Barber DG (2000) Melt ponds on sea ice in the Canadian archipelago 2. On the use of Radarsat-1 synthetic aperture radar for geophysical inversion. J. Geophys. Res. 105: 22061–22070

SAR Observation of Rip Currents off the Portuguese Coast

José C.B. da Silva

Institute of Oceanography & Department of Physics, University of Lisbon, Lisbon, Portugal

Abstract. Signatures of rip currents can be identified in SAR images and aerial photographs of the near-shore. An area on the Portuguese West coast (bathymetry and wave climate of which are briefly described) was studied. An image corresponding to low wind speed conditions is shown, where a series of rip-like cell features are consistent with rip current morphology. A simple model is presented to explain the signatures, based on wind contrast due to the relative water motion within rip currents. The model is discussed for near-threshold wind conditions of Bragg wave excitation.

1. Introduction

Rip currents are near-shore cell circulations usually consisting of two converging longshore feeder currents which meet and turn seawards into a narrow, jet-like and fast-flowing rip-neck that extends through the surf zone, decelerating and expanding into a rip-head past the line of breakers (Figure 1). Rips are relatively easy to observe, due to increased turbulence generated by wave-current interaction, sediment plumes and/or foam or bubble patches seaward of the breakers (Aagaard and Masselink 1999).

Rip currents exist as a response to an excess of water built up onshore by breaking waves, so called set-up, and often display a periodic longshore spacing. They increase in intensity and decrease in number as wave height increases, and can vary in location and intensity over time. Rips have been observed to persist for periods of two days to several months (MacMahan *et al.* 2006). Typical rip velocities are between 0.3–1 m/s and can easily exceed 1 m/s during storm conditions and short periods of time (of the order of 10 min), due to their non-stationary character (Short 1985; MacMahan *et al.* 2006). The flow associated to rip currents is maximum in the rip neck, and flows faster at low tide, having also a considerable vertical shear near the surface. A review of rip current formation mechanisms,

V. Barale, M. Gade (eds.), *Remote Sensing of the European Seas.*

rip current hydrodynamics and associated morphology, as well as on observations and measurements is given in MacMahan *et al.* 2006.

Fig. 1. Rip current feature observed in aerial photograph (*Vila do Bispo*, Portugal). Rip morphology can be identified as an elongated rip-neck and a mushroom type rip-head. Photo credits: *Esquadra 401*, Portuguese Air Force.

Observation of rip currents is important in coastal engineering studies because they can cause a seaward transport of beach sand and thus change beach morphology. As rip currents are an efficient mechanism for exchange for nearshore and offshore water, they are important for across-shore mixing of heat, nutrients, pollutants and biological species. Rip currents are particularly dangerous for beachgoers, as rips pose a potential threat to swimmers (Short and Hogan 1994). This is because swimmers can be caught in the strong offshoredirected currents and flushed out to sea, since the current speeds usually exceed the swimmers ability to swim back to shore (National Weather Service, Rip Current Service). According to the United States Lifesaving Association (USLA), over 100 drownings due to rip currents occur every year in the United States alone, and statistics show that most of water rescues on surf beaches are due to rip currents. It is therefore important to know where the most threatening rips can be found so that life-guards may be informed.

Studies of rip currents have relied mainly on numerical modelling and video camera observations (Yu and Slinn 2003). Although satellite remote sensing, in particular by Synthetic Aperture Radar (SAR), lacks good temporal resolution, it provides single "snap-shots" that in principle are optimal for rip current observations. Because *in situ* measure-ments of sur-face waves and currents in the near-shore regime remains a significant challenge, remote sensing techniques can complement such measurements in an efficient and cost-effective way. High resolution SAR observations can improve significantly our capabilities to study rip currents, providing per-iodic observations of rip current morphology and distribution, monitoring the occurrence/migration of the most threatening rip current occurrences and under which circumstances rip currents are most prominent.

2.   Study region: bathymetry and wave climate

Aerial photography and SAR data (ERS and ENVISAT) reveal that rip currents can be found almost everywhere in the near-shore Portuguese beach environment. This is not surprising since the Portuguese coast is an exposed Atlantic coast line, with most of the beaches facing the eastern boundary of the North Atlantic Ocean. Here we select an area in the West coast of Portugal, South of Figueira da Foz and to the North of Lisbon (Figure 2), to illustrate an example of rip current observations by SAR.

The wave field is characterized by wave directions mainly contained in the NW quadrant, with angles of incidence varying between 270° and 340°, the most frequently observed being those between 280° and 290° (Gomes *et al.* 2005; Silva 2006). The significant wave height is largely dominated by the class between 1 and 2 meters, but the class between 2 and 3 meters is also observed for a significant fraction of time (Silva 2006).

The near-shore bathymetry South of Figueira da Foz down to Nazaré (extending for some 60 km alongshore) is characterized by a quasi-continuous under-water ondulatory bar some 700 m offshore, as detected in SAR images The bar is continuous and quasi-linear, with a variable height of 2–6 meters (Gomes *et al.* 2005). The bottom is mostly sandy and according to Oliveira *et al.* (2004) the mean sediment size is 0,71 mm, and the area presents a mean tide range of approximately 2,2 meters.

3.   SAR observations

Rip cell features were identified in one radar image of the West coast of Portugal (near Figueira da Foz). Figure 3 shows an ENVISAT ASAR image in IMP VV polarization mode dated 30 of August 2004 acquired at

1054 UTC. The image corresponds to very low wind conditions, as can be seen from the low backscatter background (dark clutter). The satellite acquisition time corresponds to near low tide conditions, which occurred at 0955 UTC. This favours rip observations because rip current flow fastest at low tide, suggesting that the radar observability mechanism is linked to some sort of interaction involving currents, wind and waves.

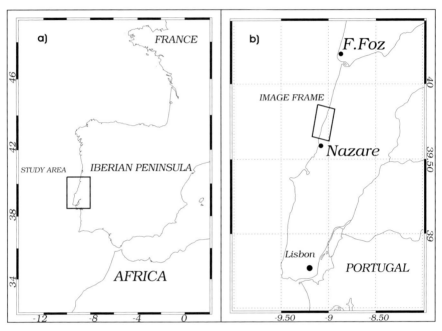

Fig. 2. (a) Study region in the West coast of Portugal. (b) Rectangle indicates location of SAR image frame shown in Figure 3, acquired on 30/08/2004.

In Figure 3 a "jig-saw tooth"-like structure is observed close to (and along) the coast line, and rip currents are observed as bright features in the form of rip-cells with enhanced backscatter on a dark background. The features interpreted here as rip-cells are periodically observed for some 60 km alongshore, and 34 of these features have been identified along the coast in one single SAR overpass. The characteristic lengths of these rip cells are of some 650 m (perpendicular to the coast) and the average widths of the rip-necks are approximately 200 m. The challenge now is to explain why rips appear as enhanced backscatter features on a dark background.

Da Silva *et al.* (2006) described the wave conditions close to the time of image acquisition. The significant wave height was $H_s = 1.13$ m and the peak wave period was $T_p = 6.2$ s. The wave direction was 321° (from NW) and the breaking wave angle estimate was 12° (angle of wave propagation relative to the normal to the coast when waves start to break). Wind measurements were available some 40 km to the North of the SAR image extract presented in Figure 3. The wind speed was $V_w = 2.7$ m/s and direction was 270° (from W) at Figueira da Foz (Figure 2).

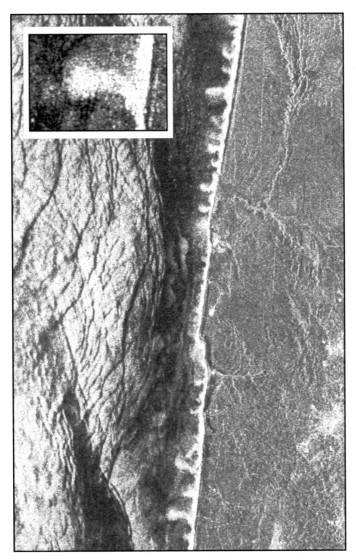

Fig. 3. ENVISAT ASAR VV polarization image (16 km × 27 km), 30/08/2004 (1054 UTC), corresponding to the area in Figure 2. The bright elongated cells extending offshore from the coast are thought to be rip current signatures. One of these cells is shown in full resolution (top left corner), revealing the typical rip current morphology (length of the rip cell ~1.4 km).

4.   Rip SAR signature interpretation: a simple model

The background backscatter level in the SAR image shown in Figure 3 is near the noise floor level. This usually occurs when the wind speed is below some threshold of wind wave excitation (which for C-band Bragg waves is approximately $V \approx 2.0$ m/s). In the region where rip features are observed, we thus assume that wind speed is below this threshold ($V < 2$ m/s), but in the average direction registered at the Figueira da Foz meteorological station.

A wind contrast model based on the growth of waves at low wind speeds can be used to explain the enhanced backscatter signatures of the near-shore rip cells in Figure 3. It takes into account the modulation of short-scale surface (Bragg) waves by the effect of relative variations of wind velocity and the surface rip currents. In this simple model, the wind wave spectrum variations are described by the balance equation for the spectral density of wave action $N(k,x,t)$ (Hasselmann, 1968; Keller & Wright, 1975; Willebrand, 1975)

$$\frac{dN}{dt} = (\beta - \gamma) - \delta N^2 \qquad (1)$$

with the right-hand side in the form suggested by Hughes (1978). Here $N(k,x,t) = \rho \omega(k) S / k$, where S denotes the surface wave height spectrum, $\omega(k)$ the intrinsic frequency of surface waves, k is the wavenumber of surface waves, β and γ the wind wave growth rate and the wave damping coefficient, respectively, and δ is a phenomenological coefficient in the last term of Eq. 1, describing nonlinear limitation of the wind wave spectrum. The right hand side of Eq. 1 neglects non-linear wave-wave interaction and wave breaking terms (see e.g. Hasselmann 1968). Although we assume that these terms should be small compared to wind wave input and film damping, the assumption should be tested for particular cases.

Considering the relaxation times of cm and dm-scale waves, which are of the order of $(\beta - \gamma)^{-1}$, to be small compared to rip current events, we can neglect the straining effect due to rip currents and consider the wave spectrum to be close to the local equilibrium spectrum. The equilibrium spectrum is found from Eq. 1 by setting $(dN/dt) = 0$. Then the relative spectrum variation can be written as the wind contrast K_w, given by (see e.g. Ermakov et al. 1998)

$$K_w = \left(\frac{N(k,x,t)}{N_0} - 1 \right) = \frac{\beta - \beta_0}{\beta_0} \qquad (2)$$

where β and β_0 are the wind wave growth rates affected and unaffected by the rip currents, respectively. The growth rate used here is in the form presented by Hughes (1978), and is based on several sets of data of different authors, including cm-scale wave growth data. It is given by

$$\beta = \omega\left(U_* \cos\theta/c_p\right)\left(0.01 + 0.016\left|\cos\theta\right|U_* /c_p\right)$$
$$\left\{1 - \exp\left[-8.9\left(U_* /c_p - 0.03\right)^{1/2}\right]\right\} \tag{3}$$

where c_p is the wave phase speed and ω the wave frequency, given by $\omega = (gk + k^3\sigma/\rho)^{1/2}$, g is the acceleration of gravity, ρ the water density and σ the surface tension. In Eq. 3, U_* is the wind friction velocity and θ is the angle between the surface wave propagation and the wind directions. The value of $U_* \cos\theta$ is the component of U_* in the propagation direction of the surface waves. The cosine term in Eq. 3 assumes therefore anisotropy in the growth rate of waves, extending the growth rate equation to any relative angle of waves and wind. The relation between wind speed V_w at the standard height of 10 m and U_* can be obtained by the empirical formula $U_* = 0.034V_w$ for $V_w < 7$ m/s (see *e.g.* Amorocho and De Vries, 1980).

The effective wind velocity relative to the water surface, V_e, moving with characteristic offshore current velocity U_c, is given by (in vector form) $V_e = V_w - U_c$, where

$$|V_e| = \sqrt{V_w^2 - 2V_w U_c \cos\theta_v + U_c^2} \tag{4}$$

and where θ_v is the angle between V_w and U_c.

Figure 4 presents results of the wind contrast model at low wind conditions. The simulations are for rip currents with speeds of $U_c = 0.75$ m/s. Different wind directions relative to the current (θ_v) were considered. Essentially, two types of contrasts are possible: positive contrasts, when there is a component of the wind velocity opposing the rip current; and negative contrasts, when there is a component of the wind velocity in the rip current direction. Note that, for different wind-current configurations (different θ_v), the wind contrast curves in Figure 4 have different lower limits for the wind speed. This is, because, in the model, we only compute contrasts when the effective wind speed is above the threshold within the rip current area.

For the case corresponding to the image in Figure 3, taking $V_w=1.5$ m/s (direction 270°) as the background wind velocity and $U_c = 0.75$ m/s, this gives an effective wind speed of $V_e = 2.5$ m/s over the rip surface currents. This is above the threshold for Bragg wave generation ($V \approx 2.0$ m/s). It thus explains the high backscatter contrast of the rip features, and why they appear as enhanced backscatter signatures on a dark image background. The wind contrast K_w near threshold wind conditions is large, producing strong positive contrast signatures (of the order of $\beta/\beta_0 \approx 1.7$, for the present case).

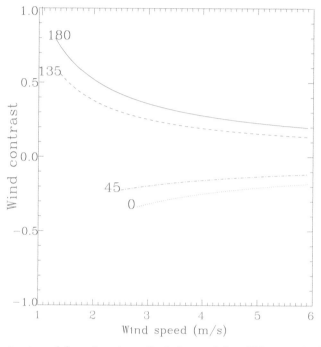

Fig. 4. Contrast model as function of wind speed for different wind directions relative to rip current. The curves show different angles of wind and current.

5. Discussion

At very low wind speeds, the wind is below the threshold for producing any measurable radar backscatter at all. But as soon as the "effective" wind speed increases, due to the presence of strong rip currents, it may rise above the wind threshold for producing measurable SAR backscatter levels (above the radar noise floor). A similar wind threshold mechanism has been used to explain strong contrast internal wave signatures at very low wind speeds. When the background backscatter is near the noise floor, internal wave signatures are characterized by bright bands on a dark radar background (see da Silva *et al.* 2000). In the above discussed model, when the rip current surface velocity opposes some component of the wind velocity ($90° < \theta_v < 270°$), the effective wind is enhanced over the rip current feature area. Generation of Bragg waves occurs inside the rip current feature area, whereas outside the rip current area the wind may still be below the wind threshold for Bragg wave generation. This would explain the high contrast at such near-threshold wind speeds.

The bright elongated signatures extending from the shore line (see inset in Figure 3 for detail of one of the rip cells) is typical of rip current morphology. The SAR bright signature exhibits a rip-neck that extends offshore for some 100 s of meters (through the surf zone), and encloses a wider rip-head feature also typical of rip currents (Figure 1). We thus based our image interpretation of rip currents not only on the wind contrast model presented above, but also on the morphology of the features.

Studies of wave breaking on an opposing current, such as rip currents, are based on one-dimensional conservation of wave action. Smith (1999) has established a criterion for wave breaking based on laboratory data. The wave parameters corresponding to the SAR observation of 30/08/2004 are within the scaled conditions that were used in Smith (1999). We have used Smith (1999) criterion to estimate the depth of breaking in the presence of a rip current. Figure 5 shows a simulation for a moderately-strong rip current ($U_c \approx 1$ m/s), revealing that wave breaking occurs for water depths lower than 3 m. The bright continuous band along the beach visible in the SAR image is probably due to wave breaking (see *e.g.* Wackerman and Clemente-Colón 2004), because waves are expected to break within 300 m of the shore (Figure 5b). Note however that the sand-bar depth is generally greater than 3.5 m, so that in the frame of this model, it is unlikely that wave breaking occurs over (and near) the bar. The sand-bar is located some 600–900 m offshore (Silva 2006), and some of the observed rip current features extend more than 1 km offshore. Although waves are not expected to break before reaching some 300 m off the coast, the wave steepness is significantly increased, due to wave-current interaction (Figure 5a).

Wave-slopes of shoaling waves within rip currents will probably produce significant tilt modulation. Tilt modulation is a mechanism that explains SAR imaging of long waves interacting with a short (Bragg) wave field (Holt 2004). It is effective for range-traveling waves (wave crests moving toward or away from the antenna) as is the case for the SAR image in Figure 3. When the varying slope of the long wave changes the local tilt (or orientation) of the short waves, these tilting waves act as reflecting mirrors or facets to the incoming radar waves. As the facets change tilt along different phases of the longer propagating wave, the radar return will vary and a characteristic SAR signature is produced on the fine resolution image. This signature is usually in the form of bright and dark lines, but in the case of Figure 3 these lines are almost imperceptible due to the reduced wavelength of the long wave field (about 50 m at 10 m depth, considering monochromatic waves with a period of 6.2 s).

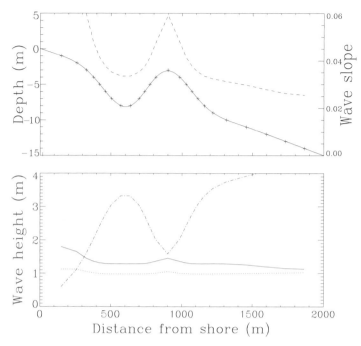

Fig. 5. (a) Idealized bathymetric cross section of the study region (solid line with crosses). Sand bar is located some 800–1000 m offshore and is 3 m deep at its shallowest depth. Wave slopes (dashed line) were computed taking into account wave interaction with opposing rip current. (b) Wave height estimated with (solid line) and without (dotted line) the opposing current. The dash-dot line represents the upper limit of wave heights for breaking waves, according to Smith (1999). Waves do not break until they reach some 300 m from the coast.

In addition, the wind contrasts should be strongest at the crest of the long wave as compared to the trough. This is a direct consequence of the airflow over the long waves, which is distorted and preferentially forms short (Bragg waves) at the crest of the long wave (Holt 2004). In fact, the air streamlines are displaced closer to the water surface on the upwind side than on the downwind side of the crest of the long surface wave (see Belcher and Hunt 1998). This leads to an asymmetric drag force on waves relative to the crest. In turn, this should affect the Bragg cm-scale roughness, that combined with the tilt modulation can enhance radar backscatter. The local incident angle varies significantly when waves encounter an opposing current and the wave slopes increase. For strong currents in the near shore, such as rips, the radar backscatter is enhanced due to tilt. We thus suggest that the rip current signatures observed in the SAR image of Figure 3, may be the result of wind modulation combined with tilt modulation due to wave-current interaction.

6. Conclusions

The SAR rip signatures presented here are consistent with rip current morphology, and we have been able to explain them with a simple wind contrast model based on the wind speed threshold of Bragg wave generation. However, other mechanisms should be investigated to explain rip current observation by SAR, particularly at higher wind speed conditions. Backscatter from wave breaking due to wave-current interactions is a possible candidate to explain the observability of rip currents in SAR images. One major difficulty that has hampered rip current observations has been lack of adequate SAR spatial resolution (for spaceborne sensors). The TERRASAR-X mission will offer much smaller pixel sizes than ENVISAT and RADARSAT modes, which correspond to larger numbers of independent looks and thus less speckle noise at a given spatial resolution. We believe that TERRASAR-X is thus specifically suited for observation of near-shore processes such as rip currents.

Image data was provided by ESA, projects AOPT-2423 & AOE-563. This is a contribution to projects AMAZING (PDCTE/CTA/49953/2003) and SPOTIWAVE-II (POCI/MAR/57836) funded by FCT, & EU SIMP. Wind data has been provided by the Institute of Meteorology, Portugal. The Author gratefully acknowledges the Woods Hole Oceanographic Institution where this paper was prepared, during a Sabbatical Leave and FCT.

References

Aagaard T, Masselink G (1999) The surf zone, In: Short AD (ed), Handbook of Beach and Shoreface Morphodynamics, John Wiley & Sons, Chichester, Chapter 4, pp 72–118

Amorocho J, De Vries JJ (1980) A new evaluation of the wind stress coefficient over water surface. Journal of Geophysical Research 85: 433–442

Belcher SE, Hunt JCR (1998) Turbulent flow over hills and waves. Annu. Rev. Fluid Mech. 30: 507–538

Da Silva JCB, Ermakov SA, Robinson IS (2000) Role of surface films in ERS SAR signatures of internal waves on the shelf, 3. Mode Transitions. Journal of Geophysical Research 105: 24089–24104

Da Silva JCB, Sancho F, Quaresma L (2006) Observation of Rip Currents by Synthetic Aperture Radar. Proceedings of SEASAR 2006, Advances in SAR Oceanography from ENVISAT and ERS missions, ESA publication SP-613, Frascati, Italy

Ermakov SA, da Silva JCB, Robinson IS (1998) Role of surface films in ERS SAR signatures of internal waves on the shelf. 2. Internal (tidal) waves. Journal of Geophysical Research 103: 8032–8043

Gomes F, Bessa Pacheco M, Jorge da Silva A, Silva R, Rusu E (2005) A Utilização dos SIG na estimativa da corrente de deriva litoral (Aplicação à costa oeste de Portugal continental entre a Figueira da Foz e a Nazaré), Instituto Hidrográfico, internal manuscript with pp 9 in Portuguese

Hasselmann K (1968) Weak-interaction theory of ocean waves. In: Holt M (ed) Basic Developments in Fluid Dynamics, vol 2. Academic Press, San Diego, pp 117–182

Holt B (2004) SAR imaging of the ocean surface, Chapter 2. In: Jackson C, Apel J (ed) Synthetic Aperture Radar Marine User's Manual. U.S. Department of Commerce and NOAA, Washington DC, pp 25–79

Hughes BA (1978) The effect of internal waves on surface wind waves, 2. Theoretical Analysis. Journal of Geophysical Research 83: 455–465

Keller WC, Wright JW (1975) Microwave scattering and the straining of wind-generated waves. Radio Science 10: 139–147

MacMahan JH, Thornton EB, Reniers AJHM (2006) Rip current review. Coastal Engineering 53: 191–208

Oliveira F, Oliveira T, Silva R, Larangeiro S (2004) Dinâmica sedimentar do trecho litoral Praia da Vieira - Praia Velha. VII Congresso da Água, Lisboa, Portugal, cd-rom. pp 15–30, in Portuguese

Short AD (1985) Rip current type, spacing and persistence, Narrabeen Beach, Australia. Marine Geology 65: 47–71

Short AD, Hogan CL (1994) Rip currents and beach hazards, their impact on public safety and implications for coastal management. In: Finkl CW (ed) Coastal Hazards, Journal of Coastal Research, Special Issue 12. pp 197–209

Silva FS (2006) Rip Currents Identification with Synthetic Aperture Radar. Lic. Dissertation Thesis, Faculdade de Ciências do Mar e do Ambiente, University of Algarve, Faro. pp 73, in Portuguese

Yu J, Slinn DN (2003) Effects of wave-current interaction on rip currents. Journal of Geophysical Research 108: 3088 doi:10.1029/2001JC001105

Smith JM (1999) Wave breaking on an opposing current. Coastal Engineering Technical Note CETN IV-17. U.S. Army Engineer Research and Development Center, Vicksburg, MS, pp 9

Wackerman CC, Clemente-Colón P (2004) Wave Refraction, Breaking and Other Near-Shore Processes. In: Jackson C, Apel J (ed) Synthetic Aperture Radar Marine User's Manual. U.S. Department of Commerce and NOAA, Washington DC, pp 171–187

Willebrand J (1975) Energy transport in a non-linear and inhomogeneous random gravity wave field. Journal of Fluid Mechanics 70: 113–126

Current Measurements in European Coastal Waters and Rivers by Along-Track InSAR

Roland Romeiser[1] and Hartmut Runge[2]

[1] Institute of Oceanography, University of Hamburg, Germany
[2] Remote Sensing Technology Institute, German Aerospace Centre (DLR), Oberpfaffenhofen, Germany

Abstract. The Along-Track Interferometric Synthetic Aperture Radar (Along-Track InSAR, ATI) technique permits a direct high-resolution imaging of ocean surface current fields from aircraft and satellites. With several airborne ATI experiments and a first demonstration of current measurements from space with data from the Shuttle Radar Topography Mission (SRTM) in February 2000, European scientists have built up a leading expertise in this field. The German satellite TerraSAR-X, which was launched on June 15, 2007, will be the first to offer ATI capabilities during a longer period. We give an overview of ATI fundamentals, SRTM results, predicted ATI capabilities of TerraSAR-X, the potential for further improvements, and promising applications in coastal waters and rivers.

1. Introduction

The technique of current measurements by Along-Track Interferometric Synthetic Aperture Radar (Along-Track InSAR, ATI) was first proposed by Goldstein and Zebker (1987) and demonstrated with an airborne ATI system of NASA-JPL (Thompson and Jensen 1993). Later, European scientists too became involved and developed a numerical imaging model (Romeiser and Thompson 2000), carried out several experiments of their own (*e.g.* Romeiser 2005) and demonstrated current measurements from space with data from the Shuttle Radar Topography Mission (SRTM) (Romeiser *et al.* 2005b). In the near future, similar activities will be possible with the German satellite TerraSAR-X (Romeiser and Runge 2007).

ATI is the only existing technology for a direct imaging of current fields from satellites at spatial resolutions on the order of 1 km and higher. Demand is expected mainly for applications in coastal areas. European scientists have made significant contributions to recent developments in these fields, and the use of ATI data will further enhance their capabilities.

V. Barale, M. Gade (eds.), *Remote Sensing of the European Seas.*

2.   ATI fundamentals

The ATI technique exploits the fact that two complex SAR images of the same scene, acquired with a short time lag on the order of milliseconds, exhibit phase differences proportional to Doppler shifts of the back-scattered signal and, thus, to line-of-sight target velocities. This permits a direct imaging of line-of-sight velocity fields. To obtain two SAR images with a short time lag from a moving platform, one needs two antennas separated by a corresponding distance in flight direction. Accordingly, the technique is called *along-track* interferometry; not to be confused with (single-pass) *cross-track* interferometry or *repeat-pass* interferometry, which are topographic mapping techniques.

2.1   Imaging mechanism

The high resolution of SAR images in azimuth (flight) direction is obtained by tracking the Doppler history of received signals and mapping targets to azimuthal positions where the Doppler shift of their contribution to the total received signal becomes 0. This method of synthesising a long antenna with high azimuthal resolution works well for stationary, non-moving targets, but it causes artefacts in the positioning of moving targets, and it does not preserve information on velocities at full spatial resolution. Only the phase differences of ATI images permit velocity estimates on a pixel-by-pixel basis, since they are basically proportional to mean Doppler shifts of all signal contributions mapped onto an individual pixel.

However, ATI images of the ocean are affected by the same mapping artefacts as conventional SAR images, and detected "Doppler velocities" include contributions of sub-resolution-scale wave motions, which may vary within an image. The resulting nonlinearities of the imaging mechanism must be taken into account in the data interpretation. Furthermore, temporal decorrelation of the backscattered signal and instrument noise cause uncertainties of the retrieved velocities.

2.2   Current retrieval technique

Despite the imaging artefacts and phase noise, one can usually apply a simple phase-to-velocity conversion to ATI data to get a first solution for the current field, which can then be refined by corrections based on model computations (Romeiser 2005). Usually, the retrieval of current fields from ATI data is much easier and less ambiguous than the interpretation of SAR intensity signatures of current features, which requires a priori assumptions on basic properties of the observed features and quite accurate information on wind speed and direction.

Note that the standard ATI concept only permits measurements of a single current component. With an airborne system, one can perform perpendicular overflights to obtain fully 2-D current fields. This is more difficult with satellite data from ascending and descending overpasses, which are usually separated by hours to days. In principle, vector current measurements can be obtained during a single overpass if two pairs of ATI antennas with different look directions are used, as demonstrated with an airborne system by Toporkov *et al.* (2005). However, for the time being, we have to construct fully 2-D current fields from spaceborne ATI data on the basis of model simulations or a statistical analysis of data from a number of ascending and descending satellite overpasses. Fortunately, some applications, such as the bathymetric monitoring (Sect. 4.2), do not depend on fully 2-D current measurements.

2.3 Instrument parameters and data quality

A crucial ATI parameter is the time lag τ between the two images, which is determined by the along-track antenna separation L and the platform velocity V. Depending on transmit/receive antenna settings, the time lag is $\tau = L/V$ or $\tau = L/2V$. For current measurements, τ needs to be sufficiently long to obtain significant phase signatures of the current variations of interest and sufficiently short to avoid phase ambiguities or a decorrelation of the backscattered signal. The decorrelation time depends on the radar frequency and on wind and wave conditions. According to Romeiser and Thompson (2000), decorrelation times at X-band (10 GHz) and L-band (1 GHz) are on the order of 5 to 15 ms and 50 to 150 ms, respectively, where the lower (higher) values are for high (low) wind speeds.

While the temporal decorrelation leads to increasing absolute phase noise and, for very long time lags, to a quasi-uniform phase distribution that does not contain recoverable velocity information anymore, the problem at short time lags is that phase variations associated with current variations become small compared to the instrument noise. However, the phase noise is a zero-mean contribution which can be reduced by (complex) averaging of a sufficient number of independent full-resolution pixel values. The number of phase samples that need to be averaged to obtain velocity estimates with a given accuracy is a good measure of the data quality of an InSAR system, since it describes the relation between measuring accuracy and effective spatial resolution.

Two diagrams showing the theoretical behaviour of this parameter as function of effective ATI baseline and instrument noise level, for a spaceborne ATI (V = 7000 m/s) at X-band, VV polarisation, an incidence angle of 30°, wind speeds of 5 and 15 m/s are presented in Figure 1. Ideal baselines for the given parameters are in the range of about 20 to 40 m. Black solid circles indicate that the parameters of SRTM and TerraSAR-X, whose current measuring capabilities are examined below, are clearly suboptimal. Improvement could be obtained by increasing baselines or reducing noise levels. For TerraSAR-X, a baseline increase by a factor of 2 would have a similar effect as a noise reduction by about 4 dB.

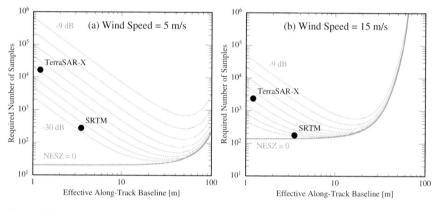

Fig. 1. Theoretical number of ATI phase samples to be averaged to obtain current estimates with an RMS error of 0.1 m/s vs. effective along-track baseline, for instrument noise levels (NESZ) of 0 and –30 to –9 dB in steps of 3 dB and for wind speeds of 5 and 15 m/s. Radar frequency = 9.65 GHz, polarisation = VV, incidence angle = 30°. Solid circles: SRTM and TerraSAR-X.

As shown by Romeiser and Thompson (2000), the dependence of the ATI imaging mechanism on radar frequency is mainly characterised by a proportionality of the ideal baseline range to the wavelength and by slightly varying nonlinear contributions to the imaging mechanism, but there is no fundamental preference for high or low radar frequencies. Since VV polarisation yields the largest backscattered power from the ocean surface and thus the highest signal-to-noise ratio (SNR), VV polarisation is more desirable than HH or cross polarisation. Regarding the incidence angle, a tradeoff between good SNR values at steep incidence angles and a more linear imaging mechanism of horizontal current variations at larger incidence angles must be found. Depending on instrument parameters, wind speed, and the current field to be imaged, ideal incidence angles are usually in the range of 30° to 60°.

3.   Spaceborne ATI missions

Until now, there has been no civilian satellite with routine ATI capabilities
However, a successful demonstration of the technique has been possible
with data from SRTM, TerraSAR-X is expected to permit similar measurements in an experimental mode of operation, and there are plans and
and concepts for further improved future instruments.

3.1   SRTM

The main objective of this Space Shuttle mission in February 2000 was a
high-resolution topographic mapping of the earth's land surfaces. A set of
C- and X-band antennas was installed in the cargo bay of the shuttle and at
the end of an expandable mast with a resulting cross-track separation of
60 m (Rabus *et al.* 2003). The long cross-track baseline was essential for
the topographic measurements. Just by chance there was an additional effective along-track baseline of 3.5 m, resulting in a time lag of 0.5 ms. Although this time lag is very short and there were no dedicated data takes
for ocean applications, it has been possible to exploit a few X-band images
of coastal areas for current retrievals. Figure 2 shows a result for the Dutch
Wadden Sea from Romeiser *et al.* (2005b). The effective spatial resolution
of the SRTM-derived line-of-sight currents is on the order of 1 km, and the
RMS difference between SRTM-derived currents and reference currents
from a numerical circulation model is better than 0.1 m/s.

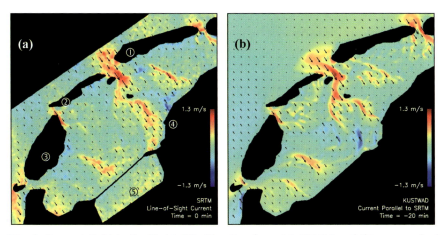

Fig. 2. Current field in the Dutch Wadden Sea from (a) SRTM and (b) a numerical
circulation model. Area size = 70 × 70 km², grid resolution = 100 × 100 m². Map:
① Terschelling, ② Vlieland, ③ Texel, ④ Harlingen, ⑤ Lake IJssel.

3.2   TerraSAR-X

TerraSAR-X is a new German satellite mission with a highly programmable, high-resolution X-band SAR. First routine data acquisitions are expected in late 2007. Similar to SRTM, TerraSAR-X is dedicated mainly to land applications, but the instrument design offers experimental modes of operation in which separate sections of the phased array antenna panel with a total length of 4.8 m can be used as individual receive antennas for polarimetry and ATI. The effective along-track baseline between two antenna halves is 1.2 m, which is even less favourable than the 3.5 m of SRTM. Furthermore, the instrument noise level of TerraSAR-X is more than 8 dB higher than the –29 dB of SRTM. However, advantages of TerraSAR-X lie in its high single-look resolution of about 3.0 m for stripmap data, which permits an averaging of many individual phase samples within reasonable effective resolution cells, and in the possibility to improve the SNR by selecting steeper incidence angles than the 55° of SRTM.

Romeiser and Runge (2007) have shown by numerical simulations that the current measuring capabilities of TerraSAR-X will be clearly better than those of SRTM in the Wadden Sea case (see Section 3.1) if stripmap data with a nominal swath width of 30 km are used. The effective spatial resolution can be 400 m under favourable conditions (moderate wind speed, incidence angle around 30°). The performance in ScanSAR mode with a nominal swath width of 100 km will be slightly better or worse than in the SRTM/Wadden Sea case; predicted effective resolutions at an incidence angle of 30° range from about 600 to 1500 m.

3.3   Future missions

Although the predicted current measuring capabilities of TerraSAR-X are promising, one can see from Figure 1 that the instrument parameters are far from the theoretical optimum. Accordingly, a major performance improvement would be obtained with an increased baseline or a reduced instrument noise level. A concept for increasing the along-track baseline of a follow-on system by a factor of 3 with two passive antenna extensions is discussed in the paper by Romeiser and Runge (2007).

A more complex yet more mature project, which has been approved for full implementation and a planned launch in 2009, is TanDEM-X (TerraSAR-X add-on for digital elevation measurement). TanDEM-X uses a second TerraSAR-X type satellite in close formation flight with the first one. Main mission objective is an advanced topographic mapping capability with inter-satellite cross-track baselines of about 300 to

500 m (Moreira *et al.* 2004). The two satellites will be in slightly different orbits in such a way that they move in a helical pattern with alternating large vertical and horizontal cross-track baselines. The along-track baseline does not vary very much and can be selected individually. We hope to be able to use TanDEM-X for current measurements with an along-track baseline close to the theoretical optimum of 20 to 40 m.

While TanDEM-X, the proposed TerraSAR-X follow-on upgrade, and some other developments will mainly lead to an improved data quality and, hopefully, an improved availability of ATI data, current measuring capabilities will still be limited to a single component. As already mentioned, it is technically feasible to obtain fully 2-D vector current measurements during a single overpass with a dual-beam InSAR (Toporkov *et al.* 2005). This may be a logical further step.

4. Spaceborne ATI applications

Open-ocean circulation studies consider 3-D flow patterns and do not require very high spatial resolutions. The use of altimeter data in this field will not be challenged very much by ATI systems. In coastal regions, where a high spatial resolution is appreciated and surface current measurements are often sufficient, the use of spaceborne radars is limited mainly by coarse temporal sampling. While a single satellite permits repeated measurements every few days only, ground-based radars can provide data with comparable spatial coverage and resolution every few minutes. In our opinion, the most promising spaceborne ATI applications are applications in coastal areas that require measurements within areas of some tens to thousands of square kilometres at high spatial, but not temporal resolution – applications for which in-situ measurements are not sufficient, altimetry will not work, and HF radar or airborne InSAR measurements are not practical for logistical, technical, or economic reasons.

4.1 Coastal oceanography

The general circulation patterns in European coastal seas are usually well known and reproducible by numerical circulation models. However, processes on short spatial and temporal scales, such as the formation of mesoscale eddies, the response of upper layer dynamics to rapid changes in the wind field, or effects of changes in bottom topography, river discharges, or other boundary conditions are usually not understood in full detail and, difficult to predict. Despite the coarse temporal sampling, repeated high-resolution current measurements by ATI can be very valuable for basic research and operational monitoring applications and for the development of further improved theoretical models. In combination with ocean colour data, ATI-derived currents can also complement studies on ecosystem dynamics or sediment dynamics, which until now have had to rely on point measurements and numerical modelling. Near real time delivery of ATI data and ATI-derived current fields is feasible.

4.2   Bathymetric mapping and monitoring

The mapping and monitoring of underwater bathymetry in coastal waters with strong tidal currents (for example, the Wadden Sea off the coast of the Netherlands, Germany, and Denmark) on the basis of conventional SAR intensity imagery was demonstrated by Calkoen et al. (2001) and has been available as an operational commercial service for several years. This technique exploits that bathymetric features become visible in SAR images due to a modulation of the tidal flow by the spatially varying water depth and a corresponding surface roughness modulation via wave-current interaction. Since the imaging mechanism is quite indirect and includes several nonlinearities and dependencies on parameters that are not well known, the inversion is done through an iterative optimisation scheme for water depths and model parameters, which are modified until best possible agreement between observed and simulated radar image intensity variations is obtained with correct depths at known calibration points. This method is sufficient for the identification of major bathymetric changes, which can then be examined in more detail by conventional echosoundings. This way, the use of SAR data improves the cost efficiency of ship operations of the responsible monitoring agencies.

Even more improvement can be expected from the use of ATI data instead of conventional SAR images, since the relation between water depths and surface currents is clearly more direct than the one between water depths and SAR image intensities. Romeiser et al. (2002) demonstrated a bathymetric mapping on the basis of airborne ATI data using a very simple approach, exploiting just linearised surface current - water depth relations derived from reference data at a few locations in the test area by a regression analysis. An example result from an experiment at the German island of Sylt is shown in Figure 3. In combination with a full physics-based flow model, such as the one used by Calkoen et al. (2001), bathymetry retrievals from ATI data should be more accurate and reliable than the conventional SAR-based approach, and the method should be applicable to more complex scenarios since the data interpretation is less ambiguous.

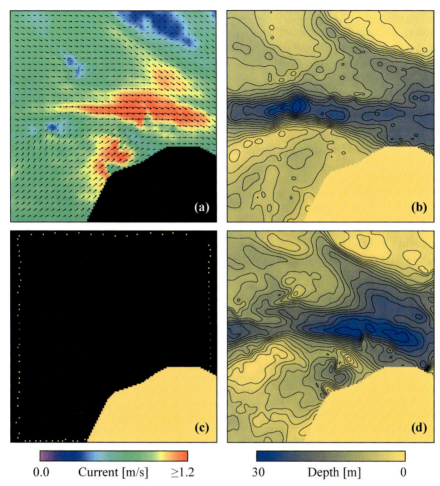

Fig. 3. Bathymetry retrieval: (a) airborne ATI-derived current field north of the island of Sylt; area size = 3.5 × 3.5 km², grid resolution = 25 × 25 m²; (b) depth map from echosoundings with an effective resolution of 200 m; (c) 78 selected reference depth points near the boundaries; (d) depth map derived from the reference depths and the ATI-derived current field.

4.3 Coastal and offshore engineering

Coastal and offshore engineers are interested in effects of currents, waves, and winds on shore and water based structures, as well as on effects of modifications of these structures. For example, British and German Engineers have recently formulated requirements for the site selection of electric power generators in tidal waters and for further investigations on variations in the 3-D flow around such generators (European Commission 2005). While it is obvious that the generators should be placed in areas with long periods of strong and uniform tidal currents, data from a British prototype system indicate that currents acting on the rotor exhibit strong variations on short time scales, which had not been expected, and which affect efficiency and wear of the machine quite strongly. To study and optimise the relevant dynamic processes systematically, high-resolution measurements in combination with theoretical investigations are required. ATI data can make a significant contribution to this specific research as well as to many similar engineering tasks.

4.4   River runoff monitoring

Another promising field of application is the monitoring of river runoff, which is important for coastal oceanography, hydrology, and climate research. Furthermore, the redistribution of water due to climate changes and changes in population, industrialisation, and land use can have major effects on the earth system and on economical and political developments. At present, many rivers are monitored locally, and data from stations throughout the world are collected and archived at the Global Runoff Data Centre (GRDC) in Koblenz, Germany. However, in some regions measurements are practically impossible for various reasons, and many countries do not publish existing runoff data. The development of a satellite-based monitoring system is highly desirable (Alsdorf *et al.* 2003).

British scientists have already demonstrated the monitoring of water levels in rivers on the basis of reprocessed conventional radar altimeter data (Berry 2002). A concept for a more specific high-resolution altimeter mission for river applications has been proposed to ESA and NASA. The use of stationary microwave Doppler scatterometers for current measurements in rivers has been demonstrated by Plant *et al.* (2005).

Romeiser *et al.* (2005a) derived currents in the Elbe river (Germany) from another SRTM image. Results are shown in Figure 4. Since the dominant flow direction can be assumed to be parallel to the river bed, it is possible to construct a fully 2-D surface current field from the ATI-derived component and the shape of the river bed. The SRTM-derived currents in the Elbe river agree quite well with results of a hydrodynamic model. Like in the Wadden Sea case (Sect. 3.1), a similar or even better data quality is expected from TerraSAR-X. The University of Hamburg is currently studying concepts for a data synthesis system for optimal river runoff assessments on the basis of spaceborne ATI and altimeter data, other available data, and numerical model computations.

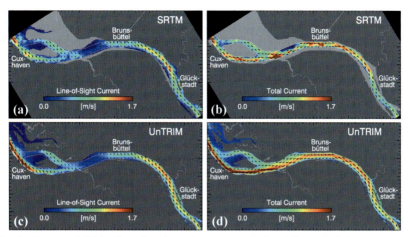

Fig. 4. Current field in the Elbe river: (a) SRTM-derived line-of-sight currents, (b) SRTM-derived quasi-2-D total surface currents, (c) model-derived component parallel to the look direction of SRTM, (d) model-derived total 2-D current field. Area size = 55 × 30 km², grid resolution = 100 × 100 m².

5   Conclusions

Despite some technical limitations, the upcoming spaceborne ATI systems will be attractive for a number of applications in coastal waters and rivers. The European remote sensing community has a leading position in this field. Although TerraSAR-X was not specifically designed for ocean applications, its predicted current measuring performance is good, and its utilisation for this purpose will be limited by the availability of the experimental modes of operation rather than by the achievable data quality. We are optimistic to obtain a sufficient amount of data for a demonstration of repeated high-resolution current measurements in a few test areas. Since a demand for such data is obvious, we hope that positive results will trigger the interest of many potential users and help to pave the way towards an implementation of further improved and fully operationalised spaceborne ATI systems for ocean applications.

Acknowledgements

We thank H. Breit, M. Eineder, and S. Suchandt (DLR), J. Sprenger and D. Stammer (Univ. Hamburg), K. de Jong and J. Vogelzang (RWS), A. Sohrmann, H. Weilbeer, and N. Winkel (BAW), and W. Sent (BSH) for their valuable contributions to the research presented in this work.

References

Alsdorf D, Lettenmaier D, Vörösmarty C, NASA Surface Water Working Group (2003) The need for global, satellite-based observations of terrestrial surface waters. EOS Trans AGU, 84 (269): 275–276

Berry PAM (2002) A new technique for global river and lake height monitoring using satellite altimeter data. Int J Hydropower Dams 9 (6): 52–54

Calkoen CJ, Hesselmans GHFM, Wensink GJ, Vogelzang J (2001) The Bathymetry Assessment System: Efficient depth mapping in shallow seas using radar images. Int J Rem Sens 22: 2973–2998

European Commission (2005) SEAFLOW pilot project for the exploitation of marine currents. Final report JOR3-CT98-0202, European Commission, Brussels

Goldstein RM, Zebker HA (1987) Interferometric radar measurement of ocean surface currents. Nature 328: 707–709

Moreira A, Krieger G, Hajnsek I, Hounam D, Werner M, Riegger S, Settelmeyer E (2004) TanDEM-X: A TerraSAR-X add-on satellite for single-pass SAR interferometry. In: Proc IGARSS 2004, IEEE, Piscataway, NJ, 4 pp

Plant WJ, Keller WC, Hayes K, Spicer K (2005) Streamflow properties from time series of surface velocity and stage. J Hydr Eng 131: 657–664

Rabus B, Eineder M, Roth A, Bamler R (2003) The shuttle radar topography mission – a new class of digital elevation models acquired by spaceborne radar. ISPRS J Photogrammetry Rem Sens 57: 241–262

Romeiser R (2005) Current measurements by airborne along-track InSAR: Measuring technique and experimental results. IEEE J Ocean Eng 30: 552–569

Romeiser R, Runge H (2007) Detailed analysis of ocean current measuring capabilities of TerraSAR-X in several possible along-track InSAR modes on the basis of numerical simulations. IEEE Trans Geosci Remote Sens 45: 21–35

Romeiser R, Thompson DR (2000) Numerical study on the along-track interferometric radar imaging mechanism of oceanic surface currents. IEEE Trans Geosci Rem Sens 38-II: 446–458

Romeiser R, Seibt-Winckler A, Heineke M, Eppel D (2002) Validation of current and bathymetry measurements in the German Bight by airborne along-track interferometric SAR. In Proc IGARSS 2002, IEEE, Piscataway, NJ, 3 pp

Romeiser R, Sprenger J, Stammer D, Runge H, Suchandt S (2005a) Global current measurements in rivers by spaceborne along-track InSAR. In Proc IGARSS 2005, IEEE, Piscataway, NJ, 4 pp

Romeiser R, Breit H, Eineder M, Runge H, Flament P, de Jong K, Vogelzang J (2005b) Current measurements by SAR along-track interferometry from a Space Shuttle. IEEE Trans Geosci Rem Sens 43: 2315–2324

Thompson DR, Jensen JR (1993) Synthetic aperture radar interferometry applied to ship-generated waves in the 1989 Loch Linnhe experiment. J Geophys Res 98: 10,259–10,269

Toporkov JV, Perkovic D, Farquharson G, Sletten MA, Frasier SJ (2005) Sea surface velocity vector retrieval using dual-beam interferometry: First demonstration. IEEE Trans Geosci Rem Sens 43: 2494–2502

Wave and Current Observations in European Waters by Ground-Based X- Band Radar

Friedwart Ziemer

Institute for Coastal Research, GKSS, Geesthacht, Germany

Abstract. Radar remote sensing in Earth Observations means not only the global and regional survey of geophysical parameter by satellite or airborne radar, but also the local observation by ground based radar techniques. The information acquired by ground based instruments mounted at fixed coastal or offshore stations provide the potential for repeated or even permanent observation. The restriction of the limited insight of the ground based radar can be overcome by the use of ship borne radar techniques to extend the observation area along the ship's track to a regional scale. This article gives examples using the cross section and the Doppler shift of X-band radar signal to observe the wave and the current field.

1. Introduction

Enclosed by seas the continent of Europe is structured by islands and peninsulas, where typically the distance to the sea is not more than some hundred kilometres. Vice versa, many activities at European seas take place in vicinity of a coastline. In European ecology and economy the sea plays an important role. Ship accidents clustering along the coasts are frequently caused by waves and currents. The risk of accidents close to the coasts is even higher as the local interaction between the propagating ocean waves with the current or the bathymetry or both may steep up single waves. Current shear zones, as well as the bathymetry, may cause wave refraction in a way that the wave energy is focused to certain sites by the interference of waves with nearby periods or wavelengths entering the site from different directions. At focal points this may result in extreme wave height and wave steepness whereas the surrounding region shows a normal situation. Another danger is evoked by the crossing wave situation by forcing the ships hull into a periodic response which can be critical for the ships stability. In the case where the surrounding wave field and the ships response are known, a well controlled speed and direction may mitigate the danger

423

V. Barale, M. Gade (eds.), *Remote Sensing of the European Seas.*

for the ship dramatically. Another source for risks is related to the off shore industry. The production and transport of oil and gas at sea needs a lot of accompanying ship operations such as supply actions and tanker docking. What is needed for a high level of security in all these operations is the knowledge on the directionality of the actual sea state surrounding the site of operation.

For the routine management of *offshore*-work, wave and current monitoring is necessary. A series of activities such as rig supply, docking of tank ships, crane manoeuvre or exploration drilling and in Vessel Traffic Control (VTC) need the warning during critical sea states close to ship traffic ways. In some cases there is a need for warning of extreme high and steep wave crests and even though the absolute height of a wave is not so important for a ship. The local steepness or the mean forcing period can be critical to the ships response. This is different for fixed constructions, which are not lifted by the buoyancy. Here the absolute "crest height" is important, when the lowermost part of the construction is not far enough above the water surface. Actual wave and current information is needed for the cost and time optimisation in ship routing, for security in fishing activity, for the information back flow to weather services for data assimilation to increase the validity of forecast- models, for data networks like EURO-GOOS, for the coastal management to mitigate coastal erosion, for the harbour authorities for the minimisation of wave induced ship movement at the quay and, last but not least, for scientific research in oceanography, meteorology or more general in environmental research.

By gathering actual information on the sea state, radar instruments get an increasing importance as radar has the unique feature of providing observations on global, regional and local scales. For the European seas and their coasts, the monitoring of regional and local scales is essential. Within the chain of mapping instruments the '*ground based*' radar fills the gap providing information on the local scale with high resolutions. The feature ground based means radars are mounted either at a coastal stations, onboard ships or onboard offshore rigs. The observations acquired by a ground based radar cover a scale with a diameter of up to some nautical miles. Ground based radar, even when it is mounted onboard a ship, remains close to the site of observation. This guarantees the correlation in time between succeeding radar observations, an important feature to acquire time series of images scanned from the same section of the ocean surface. The hardware of the radars discussed in this chapter is basing on ordinary commercially available marine X-band radar with additional digital radar signal storage. For the detection of the Doppler frequency shift, discussed in the last chapter of this article, the transmitted signal was stored as well to acquire the phase shift of received signal.

2. Wave observation by marine X-band radar

The best possible description of a propagating wave field is provided by a three dimensional sample that normally is acquired as a matrix of a digitised electronic signal $g^{(3)}(x_i, y_j, t_n)$. This signal is related to the wave field at the ocean surface $\zeta^{(3)}(x, y, t)$ by any transfer function. The index $^{(3)}$ indicates the three dimensions spanning the sample space. For the analysis we consider that within each sample of the surface, ζ is known to be Gaussian distributed in space and time. If the radar backscatter field is Gaussian distributed, we assume the radar imaging to be linearly linked with the wave field. Under this precondition we define the imaging:

$$\zeta^{(3)}(x, y, t) \rightarrow g^{(3)}(x_i, y_j, t_n) \tag{1}$$

to be a linear process for the long gravity waves ($\lambda \geq 1\,m$).The radar backscatter can be described as the convolution of the electromagnetic background noise (speckle) received from each resolution cell with a function describing the local modulation evoked by the long surface gravity waves. The spectral decomposition of a radar image (signal spectrum) equals the product of the spectra resulting from the modulating wave field with the background noise multiplied by the spectral modulation (Alpers and Hasselmann, 1982). Within a spectrum of a single radar image the signal and the noise share the same wave number spectrum. The definition of noise is more clear in the spectrum of a three dimensional sample as the signal is distributed over a series of spectra with different frequencies. Under linear transfer conditions for a bandwidth of wave numbers or frequencies the principal of super positioning of individual components holds for the wave field and for its image. Vice verse, this allows the description of the wave field by its Fourier components.

An operational system and the data acquisition are described elsewhere in the present volume. We discuss an observation matrix covering 32 time steps with 128×128 radar pixels each, which is decomposed into its Fourier components:

$$g(x, y, t) \xrightarrow{FFT} G(k_x, k_y, \omega) \tag{2}$$

The resulting coefficients $G(k_x, k_y, \omega)$ are as well defined over an orthogonal system with the size and resolution defined by the sampling theorem: $\Delta k_x = 2\pi/L_x$, $\Delta k_y = 2\pi/L_y$, $\Delta \omega = 2\pi/T$ and the Nyquist limits: $k_{x, Nyquist} = \pi/\Delta x$, $k_{y, Nyquist} = \pi/\Delta y$, $\omega_{Nyquist} = \pi/\Delta t$. For the coordinates in the Fourier domain we use as well the vector: $\vec{\Omega} = (k_x, k_y, \omega)$.

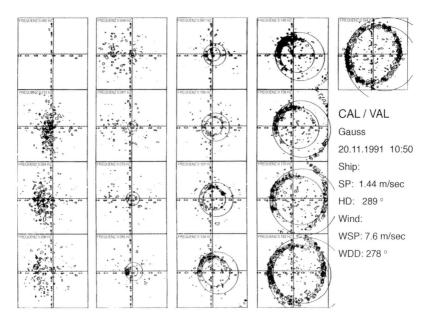

CAL / VAL

Gauss

20.11.1991 10:50

Ship:

SP: 1.44 m/sec

HD: 289 °

Wind:

WSP: 7.6 m/sec

WDD: 278 °

Fig. 1. Three dimensional sample spectrum presented at 16 positive frequencies. Zero frequency in the upper left corner. The frequencies downwards increase in row direction. Isolines give variances of the radar sample. The intersection lines of the dispersion shell with the neighbouring frequencies are plotted. This sample shows aliasing in the uppermost five frequencies.

Figure 1 shows a three dimensional sample spectrum (2) presented as 17 two-dimensional k-spectra $\left| G^{(2)}(k_x, k_y)^2 \right|_{\omega = \omega_i}$ with $i = 0,1,2,3,\ldots,16$ derived from a radar image series with 32 time steps. Isolines give the variances of the radar sample. The frequencies downwards increase by $\Delta\omega = 2\pi/T$ in row direction. For each frequency the intersection lines with the actual wave dispersion for the two neighbouring frequencies are given. These lines define the filter to separate wave induced variance from the background noise. The filter is defined by two models. The first model describes the isotropic dispersion relation for surface gravity waves propagating over varying water depths h:

$$\omega_0^2 = gk \tanh(kh). \tag{3}$$

In most of the cases the conditions are not isotropic, meaning the frequency ω becomes a function of the angle between a relative movement, for example, of a ship $\vec{u}_{ship}$ and the wave travel direction. We define the *current of encounter* $\vec{u}_e$ as the movement between the observing antenna and the wave travel direction, Θ_k as the wave direction, Θ_u as the current direction and Θ_e as the angle of encounter. The Doppler shift of the wave frequencies $\omega_{Doppler} = |\vec{k}||\vec{u}_e|\cos\Theta_e$ may be written as dot product $\omega_{Doppler} = (\vec{k} \bullet \vec{u}_e)$. For the frequency of encounter ω_e we thus attain the sum of the isotropic part (3) corrected by the Doppler impact:

$$\omega_e = \omega_0 + (\vec{k} \bullet \vec{u}_e).$$

(4)

Equation (3) and (4) define a surface within the three dimensional Fourier domain where we consider to find the energies of linear gravity waves even if the propagation is under the influence of shallow water or a current of encounter. In the case of an actual measurement, we localize wave energies in the vicinity of the dispersion shell. Under the use of Dirac's Delta function δ we define a filter function: $\delta_{signal} = \delta\{\omega_0(\vec{k}) - \omega_e(\vec{k})\}$ and the inverse δ_{noise}. Thus we get the $\vec{k}$-spectrum of the imaged wave field by integrating along all frequencies the product of the transfer function with the quotient signal power spectrum divided by the noise spectrum:

$$E(\vec{k}) = \frac{k^\beta}{|M|^2} \frac{\displaystyle\int_0^{\omega_{Nyquist}} \delta_{signal} |G(\vec{\Omega})|^2 d\omega}{\displaystyle\int_0^{\omega_{Nyquist}} \delta_{noise} |G(\vec{\Omega})|^2 d\omega}.$$

(5)

In (5) we see the square of the modulation transfer function: $k^\beta / |M|^2$. If we assume the local surface tilt as the main modulation, the value of β will be -2 to produce a heave spectrum. Ground based radars image the waves under grazing incidence and shadowing becomes important and the value of β differs from -2. From a series of experiments (Ziemer 1995; Nieto-Borges and Soares, 2000) it was concluded that for ground based radars the exponent β approaches -1.2. Following the convention the wave energy in (5) is normalized by: $g\rho_0/2$. The scaling factor $|M|^2 = m_{0\,radar}/m_{0\,buoy}$ can be calibrated by any well proved wave staff. The retrieved values for β and $|M|^2$ must be validated by independent measurements. Parameters such as the significant wave height can be calculated from the spectral moments of (5). Figure 2 shows results attained during the campaigns for calibration (black dots) and an independent validation (open rhombs). The correlation coefficient that was calculated from the validation data set has the value k = 0.978 at the 95% confidence interval.

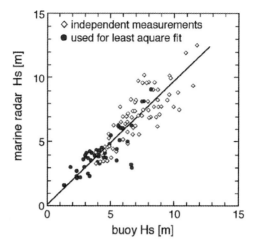

Fig. 2. Scatter diagram of significant wave heights HS comparing radar and buoy measurements. The results (black dots) were acquired during a calibration phase in the fall 1994. Independent measurements (open rhombs) have been acquired during a validation phase in January 1995.

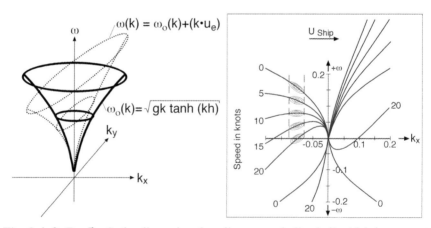

Fig. 3. left: For $\vec{u}=0$ the dispersion describes a parabolic shell which is symmetric to the frequency axis. For $\vec{u}\neq 0$ the frequency of the wave package is lowered or lifted by the Doppler - effect depending on the angle of encounter. Right: Cut through the three dimensional space along the frequency axis and along one of the wave number components. The intersection lines give the placement and displacement of energies of dispersive waves observed under 5 different conditions of encounter.

Three dimensional sampling overcomes the limitations of the 180° directional ambiguity of static radar measurements. As point symmetry is valid, power with negative frequency is redundant. Leaving off the redundant negative part we produce a wave number spectrum that is free of the 180° ambiguity. However before cutting energies with negative frequencies, we have to make sure not to cut those which have been observed to be Doppler shifted to negative frequencies. An example for this is given on the right graph in Figure 3. The graph on the left compares the isotropic dispersion, for which $\vec{u} = 0$ is valid, with a non-isotropic case, where $\vec{u} \neq 0$. The symmetric parabolic dispersion shell describing Equation (3) is distorted by a current of encounter as described by Equation (4). The energies of dispersive waves observed under five different conditions of encounter are given by the five cuts through the three dimensional space on the right of Figure 3. The cut direction is that of the current of encounter. We see already at a relative speed of 15 *kts* virtual negative wave frequencies may be observed.

Each sampling of a periodic process underlies the risk of aliasing. Aliasing occurs when the chosen sampling interval is too long compared to the periods or lengths occurring within the sampled process. To avoid aliasing the Nyquist theorem, stating that the sample interval has been chosen with half of the shortest period or length of the expected oscillation, must be satisfied. A violation of the sampling criteria may shift the frequency or wave number of an observed process to false (aliased) lower values. Aliasing occurs in multidimensional sampling as well, but it is possible to use properly sampled oscillations in the one domain to de-fold aliased oscillations in the other domain. Here an example:

Following the sample theorem the highest properly resolved wave number results from $k_{Nyquist\,(space)} = \pi/radar\ resolution$; and the highest properly resolved frequency from $\omega_{Nyquist\,(time)} = \pi/antenna\ rotation\ time$. Figure 1 shows a case with aliased wave energies that result from a better spatial resolution compared to the time resolution. We consider our sampling intervals in space with $\Delta r = 7.5\,m$ and in time with $\Delta t = 2\,s$. Using the dispersion we can define the highest dissolvable frequency by: $\omega_{Nyquist(space)} = c\pi/\Delta r = c\,k_{Nyquist}$, resulting in a 2.5 times higher frequency than by the time sampling $\omega_{Nyquist} = \pi/\Delta t$. To define a filter covering the full observation spectrum including the aliased parts of the dispersion shell a reordering of the observed dispersion is needed. For this the features of the Fourier domain *point symmetry* and *periodicity* are used to mirror and shift the dispersion branches into those positions where the frequencies are not aliased (Senet *et al.* 2001). The resulting dispersion shell reaches a higher Nyquist frequency limit that corresponds to the use of an antenna with virtually faster rotation.

3.   Spectra of encounter

The two dimensional wave energy spectrum (Equation 5) is without directional ambiguity, free of aliasing and calibrated. As the wave number does not underlie any Doppler effects, the integration along all frequencies result in an image spectrum that is free of any Doppler effect. All these features hold, even if the radar measurements have been acquired from on-board a travelling ship. Vice versa, the resulting spectrum is that of the natural sea and not the one actually acting on the ship. However the ship responds by rolling and pitching to the trespassing of a wave field in the time domain. Thus we have to discuss the spectrum of encounter between the wave field and a ship in the frequency domain.

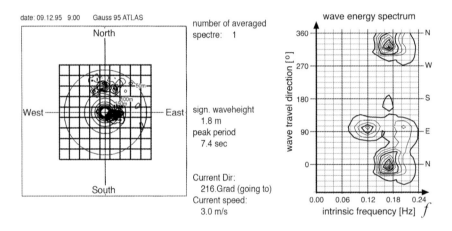

Fig. 4. Transformation of the two dimensional wave energy spectrum $E(k_x,k_y)$ from the k-space to $E'(f_0,\theta)$ in the frequency direction space, an intrinsic spectrum free of encounter effects. For convenience wave energies going to the angular window 300^0 and 360^0 are given to be redundant.

To construct the spectrum of encounter we have to transfer the wave energy from the wave number domain into the frequency direction domain $E(f_0,\theta) = \Im_1 E(k_x,k_y)$ (see Figure 4). Here we substituted $E(f_0) = 1/\pi\, E(\omega_0)$.

The Jakobian $\Im_1 = \left\{ \dfrac{1}{2\pi} |\vec{k}| \dfrac{d|k|}{d\omega} \right\}$ transfers the spectral energies from the

(k_x,k_y)-space into the (f,θ)-space.

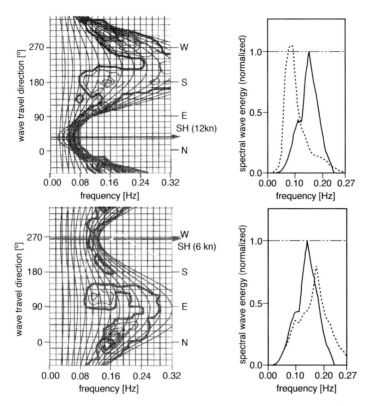

Fig. 5. Two dimensional and one dimensional spectra of encounter under the assumption sailing with different speeds and directions in the wave field of figure 4. The tow "Trials", are tests of other combination of the ships velocities within the same sea state. Examples are given for a speed 12 *kts* heading towards 35° and 6 *kts* and 270°. The one dimensional spectra show the spectra of encounter (dashed) compared to the intrinsic situation continuous line. The energy values have been normalised to the maximum of the intrinsic spectrum.

For a system observing the wave field with the velocity of encounter $\bar{u}_e$ we have to assess the spectrum of encounter. Using Equation (4) we get for the Doppler shift: $f_{Doppler} = (2\pi f_o)^2 |\bar{u}_e| \cos(\Theta_u - \theta)/g$. To transfer the wave energy $E(f_o,\theta)$ to the spectrum of encounter we have to use the Jakobian: $\mathfrak{I}_2 = \left\{ 1/\left(1 + 8\pi^2 f_0 |\bar{u}_e| \cos\Theta_e / g\right) \right\}$. The intrinsic frequency spectrum of the wave field shown in Figure 4: $E(f_0) = \int_{0^0}^{360^0} E(f_0,\theta)d\theta$ and the corresponding spectra of encounter $E(f_e) = \int_{0^0}^{360^0} \mathfrak{I}_2 E(f_e,\theta)d\theta$ are shown at the right of Figure 5. It is evident that the ship's direction in the waves plays the key role. The one dimensional spectra show the spectra of encounter (*dashed*) compared to the intrinsic situation (*continuous*). The total energy for the intrinsic and the encounter spectra is the same. The energy values of both spectra have been normalised to the maximum of the intrinsic spectrum. "Trials" are testing possible combinations of the ships velocity under the same sea state. First the ship heads towards 35° with 12 *kts* speed. The spectrum of encounter is dilated in the swell part and the wind sea part is compressed into a small frequency interval. The integrated frequency spectrum shows the same maximum power but at a lower frequency. The second trial shows the situation under 6 *kts* heading to 265^0. The energy of wind sea and swell is dispersed over a wider bandwidth. The peak power is lowered but the peak is shifted to a higher frequency.

4.   Current mapping by ship based coherent radar

In this paragraph experimental radar will be presented that was developed to map horizontal current profiles. This development is a cooperation between the Electrotechnical University St. Petersburg, Russia, and the Institute for Coastal Research/GKSS Germany on the base of nautical X-band radar (Braun *et al.* 2007). Phase detection and an automatic antenna steering enable the measurement of the Doppler shift for each of 256 radial bins along a radial beam with about 1° beam width. To acquire the full surface current vector each surface element has to be scanned twice under 90° different antenna view directions using two synchronised radars. The range resolution of the radar system is about 7.5 m. Using the Doppler relation we calculate the radial velocities from the backscattered signal for each range bin. Integrating during about 1 *s* over 1000 pulses the radial velocities are detected with an accuracy of 0.03 m/s covering the range of ±7.5 m/s.

To produce geo-coded current maps a navigational system was synchronised with the radar during the scanning to track the antenna position and its momentary view direction (Ziemer and Cysewski 2006). During acquisition, a precise GPS navigation system tracks the ship's position and its North orientation to collocate the two current components in a post processing step. Additionally the navigation data are needed to correct the instantaneous antenna movements due to the pitch and roll movements of the ship. Another correction is necessary to minimise movements, which are directly impacted by wind friction at the sea surface. Each resulting component is written into a geo-coded grid. The last step in the post processing procedure is to compose the full surface current vector by merging the two components into a common grid. Used together with a vertical current scanner as the Acoustic Doppler Current Profiler (ADCP) a three dimensional current observation may be carried out.

Figure 6 shows a surface current map that was composed as described above. The data were acquired with an antenna height of 9 m during a 20 minutes ship's cruise from North to South. The radars scanned towards the west producing a strip of about 500 m width. The current situation is ebbing. Thus the water flows out of the bight west of the track, guided by the bathymetry it meets the northward going flow within the main gully. At the North-East end of the island the gully narrows due to decreasing depth and the tide is accelerated. Further to the north, where the water has passed the island, the cross section is deepening and the speed decreases considerably. At the east end of the islands, the northward current meets an east going return flow and locally strong eddies were observed. The current features and eddies were verified by ADCP observations.

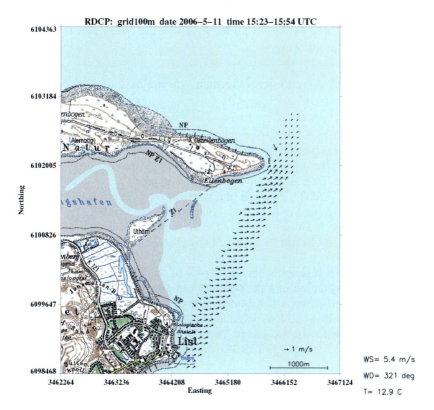

Fig. 6. Surface current map acquired during ebbing by two synchronised Doppler radars. For further details see text.

The coherent radar described in this chapter is under development and was only used in areas sheltered from long waves. To adapt this system for the open ocean, the rolling and pitching of the ship must be tracked by motion sensors by at least 5 Hz to correct the wave induced antenna motions. In addition the signal analysis has to be expanded to interpret the wave impact to the Doppler measurements.

5. Summary and conclusions

During the last decade, a new generation of ground based microwave techniques managed the transfer from research to operational use. Especially the combination of modern hardware with intelligent analysis produced a totally new perspective in monitoring, allowing insight into coastal and offshore processes that could not been observed before in addition the predicted raise of the mean sea level raises the demand for monitoring. Still, new techniques are under development and it is to be foreseen that they will take their places in the monitoring of the European coasts enlarging the efficiency of monitoring.

References

Alpers W, Hasselmann K (1982) Spectral signal to clutter and thermal noise properties of ocean wave imaging synthetic aperture radars. Int J Rem Sens 3: 423–446

Braun N, Ziemer F, Bezuglov A. (2007) Sea-Surface Current Features Observed by Doppler Radar. IEEE Trans Geosci Rem Sens (in press)

Nieto JC, Soares CG (2000) Analysis of directional wave fields using X-band navigation radar. Elsevier Science Coastal Engineering 40: 375–391

Senet CM, Seemann J, Ziemer F (2001) The Near-Surface Current Velocity Determined from Image Sequences of the Sea Surface, IEEE Trans Geosci Rem Sens 39 (3): 492–505

Ziemer F (1995) An Instrument for the Survey of the Directionality of the Ocean Wave Field. Workshop on Operational Ocean Monitoring Using Surface Based Radars, Geneva, WMO/IOC Report No. 32: 81–87

Ziemer F, Cysewski M (2006) High Resolution Sea Surface Maps Produced by Scanning with Ground Based Doppler Radar, eProceed. Intern. Geosci. Remote Sens. Sympos. (IGARSS) 2006, http://ieeexplore.ieee.org/Xplore Denver, Colorado

Nautical Radar Measurements in Europe: Applications of WaMoS II as a Sensor for Sea State, Current and Bathymetry

Katrin G. Hessner[1], José Carlos Nieto-Borge[2], and Paul S. Bell[3]

[1] OceanWaveS GmbH, Lüneburg, Germany
[2] Dpt. Of Signal Theory and Communications, Escuela Politécnica Superior, Madrid, Spain
[3] Proudman Oceanographic Laboratory, Liverpool, Great Britain

Abstract. This paper presents the remote sensing techniques of measuring sea states, currents and bathymetry by using an X-band nautical radar. It briefly describes the fundamental methods to infer sea state information (*e.g.* ocean wave and current parameters) from nautical radar imagery. In addition, this work describes in detail the performance of the Wave Monitoring System WaMoS II (a commercial system for real-time monitoring of wave fields based on nautical radar technology). Two examples of nautical radar applications are presented: the first application is an example of the standard WaMoS II installation for sea state measurements, and the second application shows results of a research project aiming at the determination of shallow water bathymetry by means of nautical radar imagery.

1. Introduction

Wave information is usually derived from time series of the sea surface elevation measured at a certain location in the open sea. These measurements are carried out by in-situ sensors such as buoys, lasers, and pressure sensors with high temporal resolution (*e.g.* sampling frequencies of about 2Hz). However, deployments of such sensors are limited by the local water depth, as well as the mooring facilities. For instance buoys can be easily damaged by ships or during severe meteorological conditions. Furthermore, the use of point measurements assumes that the obtained wave information is representative for a particular area, which is often not the case, particularly in coastal waters, where coastal effects like wave refraction, diffraction, shoaling *etc.* take place. Under these conditions, the sea state can vary significantly in the area of interest.

435

V. Barale, M. Gade (eds.), *Remote Sensing of the European Seas.*
© Springer Science+Business Media B.V. 2008

Complementing point measurements, the imaging of the sea surface based on remote sensing techniques provides information about the spatial variability of the sea state in the area of interest. One of these techniques is based on the use of ordinary nautical X-band radars to analyse ocean wave fields.

Under various conditions, signatures of the sea surface are visible in the near range (less than 3 nautical miles) of nautical X-band radar images. These signatures are known as sea clutter, which is undesirable for navigation purposes. Therefore, the sea clutter is generally suppressed by filtering algorithms. Sea clutter is caused by the backscatter of the transmitted electromagnetic waves from the short sea surface ripples in the range of half the electromagnetic wavelength (*i.e.* ~1.5 cm). The longer waves like swell and wind sea become visible as they modulate the backscatter signal mainly via hydrodynamic modulation of the ripples by the interaction with the longer waves, tilt modulation due to the changes of the effective incidence angle along the long wave slope, and the partial shadowing of the sea surface by higher waves (Keller and Wright 1975; Alpers *et al.* 1981; Plant 1990; Wenzel 1990; Lee *et al.* 1995).

Since standard X-band nautical radar systems allow the sea surface to be scanned with high temporal and spatial resolution, they are able to monitor the sea surface in both time and space. The combination of the temporal and spatial wave information allows the determination of unambiguous directional wave spectra. In addition to point measurements techniques nautical radar imagery also permits the observation of spatial variations in the wave field. Furthermore, the use of nautical radar as a remote sensor enables to measurement of wave field features from moving vessels.

In the past different systems using nautical radars for measuring sea states have been developed. This paper focuses on one of these devices, the German system WaMoS II (Wave Monitoring System), which is described in the following section.

2. The WaMoS II system

WaMoS II is a high-speed video digitising and storage device that can be interfaced to any conventional navigational X-band radar and a software package running on a standard PC. The software controls the radar and data storage. In addition, the WaMoS II software carries out the wave analysis and displays the results (see Figure 1).

The system was developed at the German GKSS Research Centre and the equipment was first tested in 1991. In 1994 the technology was transferred to OceanWaves GmbH in order to market and commercialise the system. Since then WaMoS II has been improved and expanded to cover ship and shallow water applications. Since 2001 the system has been type approved by the Germanischer Lloyd and Det Norske Veritas. During several applications WaMoS II proved to be a powerful tool to monitor ocean waves from fixed platforms as well as from moving vessels, especially under extreme weather conditions (Young *et al.* 1985; Ziemer and Rosenthal 1987; Ziemer and Günther 1994; Nieto-Borge *et al.* 1999; Hessner *et al.* 2001).

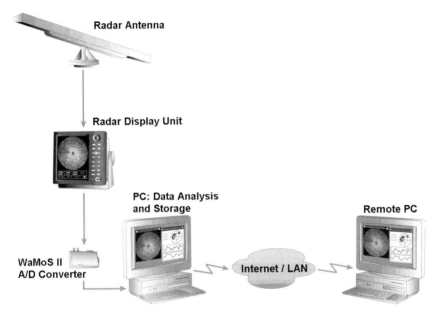

Fig. 1. Schematic of the different parts of WaMoS II and the data flow from the radar antenna to the user display.

A typical WaMoS II wave measurement consists of the acquisition of a radar image sequence and the subsequent wave analysis. The sea clutter image sequence is transformed into the spectral domain by means of a three dimensional Fast Fourier Transform (FFT). First the surface current is estimated from the image spectrum by means of a least-squares-method using the dispersion relation for linear water waves as reference. Then the dispersion relation including the current component (Doppler) is used to separate wave related spectral information from noise. Finally, by applying a modulation transfer function the wave spectrum is determined. From this wave spectrum all kinds of commonly used types of wave spectra may be derived (wave number spectrum, frequency direction spectra, frequency spectra, etc) and various standard spectral wave parameters can be inferred.

The standard WaMoS II software delivers unambiguous directional wave spectra and time series of the integrated standard wave parameters significant wave height (H_s), peak wave period (T_p), peak wave direction (θ_p) and peak wave length (λ_p) in real time. These data can be made available to the user on the WaMoS II PC and can also be transferred to other stations via Internet, LAN, NMEA *etc.*

Recent developments allow WaMoS II to obtain sea surface elevation maps (Nieto-Borge *et al.* 2004), individual wave parameters (Reichert *et al.* 2005), wave groups (Dankert *et al.* 2003), near surface current fields (Dankert *et al.* 2004), bathymetry (Bell 2005; Bell *et al.* 2005; Hessner *et al.* 1999), and high-resolution ocean wind fields (Dankert *et al.* 2005).

3.   Limits, resolution, and accuracies

The standard WaMoS II analysis uses a continuous sequence of 32 radar images. Each radar image represents one antenna revolution. The sampling time depends on the radar antenna rotation rate. Using standard radars with rotation rates ranging from 1.5–3 s, WaMoS II can detect waves in the range of 0.025 Hz to 0.35 Hz. The actual range, resolution and accuracy can vary for each WaMoS II installation, depending on the particular radar and set-up geometry. Table 1 shows typical values of standard output parameters, corresponding resolution, and accuracies.

4.   Applications

This section describes some of the applications of the WaMoS II technology for wave and current monitoring. For that purpose, two different locations are described: The WaMoS II station at the FINO[1] 1 platform in the North Sea and an installation at Teignmouth located on the British coast. These two locations have been selected because of the two different geographic and oceanographic conditions, as well as the specific application (*e.g.* FINO 1 station delivers wave field and meteorological information in real time and Teignmouth is a coastal station, which delivers bathymetry information from the analysis of the wave shoaling imaged by the radar).

---

[1] Forschungsplattformen in Nordsee und Ostsee (Research Platforms in the North Sea and Baltic Sea).

Table 1. WaMoS II parameters.

Wave Spectra	Resolution	Range	
2-d frequency-direction	0.005 Hz	0.02 Hz – 0.35 Hz*)	
spectrum $S(f, \theta)$	4°	0 – 360°	
1-d spectrum $S(f)$	0.005 Hz	0.02 Hz – 0.35Hz*)	
Wave parameters	Accuracy*)**)	Range	Resolution
Significant Wave Height H_s	+/– 10% or +/– 0.5 m	0.5 – 20m***)	0.1 m
Peak direction θ_p	+/– 5°	0 – 360°	1°
Peak period T_p	+/– 0.5 s	3.5 – 40s*)	0.1 s
Peak wave length λ_p	+/– 10%	15 – 600m*)	1 m
Current parameters			
Current speed U	+/– 0.2 m/s	0 – 40 m/s	0.01 m/s
Current direction U_θ	+/– 2°	0 – 360°	1°

*) Typical ranges. The numbers depend on the radar hardware, the total time of measurement and therefore can vary for each individual installation.
**) Based on comparative wave measurements assuming an equally distributed error.
***) There is no limit in estimating the wave heights, but up to now, H_s of 20 m was the highest value measured with WaMoS II.

4.1 Offshore sea-state measurements in the North Sea

Within the framework of the German FINO programme the first platform FINO 1 has been in operation since September 2003 (see Figure 2). The platform is located in the North Sea, approximately 45 km off the island of Borkum. The aim of the project is to record precise measurements of the meteorological conditions in the lower atmospheric boundary layer. A WaMoS II was installed to investigate the load and stability of the structure due to surface waves and currents.

The atmospheric and hydrographic measurements provide important input for the design of offshore wind turbines and for improving atmospheric and oceanographic models thereby forming the basis for the safe and economical operation of wind turbines on the open sea (Herklotz 2007).

Figure 3 and 4 show examples of radar images showing sea clutter and the corresponding wave spectra determined by WaMoS II. The first example shows a bimodal sea state with a dominant wind sea (red) and secondary swell (green), while the second example shows a well developed swell.

Fig. 2. The FINO 1 Platform. The nautical radar antenna used for the WaMoS II measurements is installed below the helicopter deck.

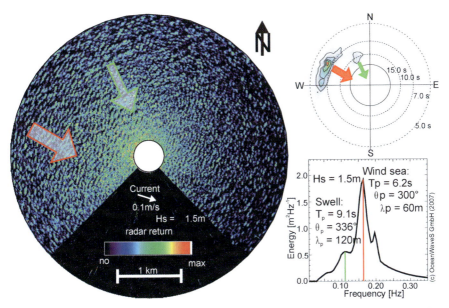

Fig. 3. Radar images measured onboard FINO 1 on 7 Feb, 2006 0009 UTC. On the right the corresponding wave spectra and spectral wave parameter as obtained by WaMoS II are given.

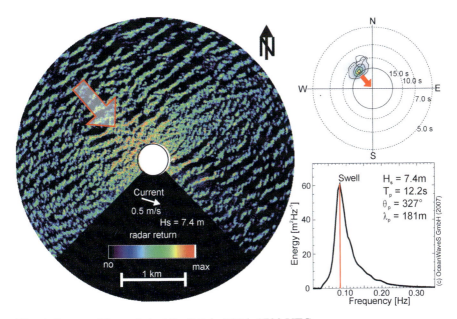

Fig. 4. Same as Figure 3, but for 8 Feb, 2006, 1700 UTC.

The time series of H_s, T_p and θ_p shown in Figure 5 were obtained by WaMoS II (red) and a Waverider buoy (blue) deployed next to the FINO 1 platform. A good agreement between the WaMoS II and the buoy measurements can be seen. Slight deviations can be expected, as WaMoS II delivers spatial mean wave parameters while the buoy delivers wave parameters measured at a point.

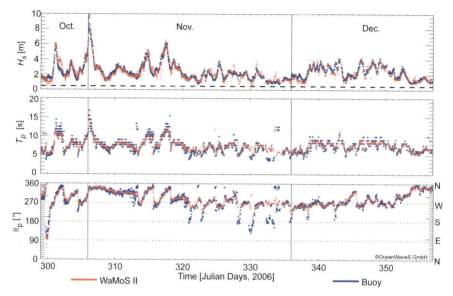

Fig. 5. Comparison of the significant wave height (H_s; top), peak wave period (T_p; middle) and peak wave direction (θ_p; bottom) time series obtained by WaMoS II (red) and a Waverider buoy (blue) at FINO 1.

4.2   High resolution bathymetry application: Teignmouth, UK

Within the EU funded COAST3D project a radar and digital recorder were installed at Teignmouth in order to obtain high resolution bathymetry from the nautical radar imagery. Within the study a validation of radar derived bathymetry by means of echo sounder surveyed bathymetry was performed.

Teignmouth is a small town on a locally south-east facing coastline in the south west of the UK. The study area included a working port and tidal inlet to the Teign Estuary, a rocky headland and a length of straight groined beach. The tidal inlet is surrounded by a complex system of sand banks that evolve on a cyclic basis over the period of a few years.

The nautical radar was deployed on the shoreward end of Teignmouth Pier, approximately in the middle of the study area. A tide gauge was located on the same pier, and echo sounder surveys were carried out at the start and end of the experiment, but only to a distance of 1 km offshore due to limits on the availability of the survey boat.

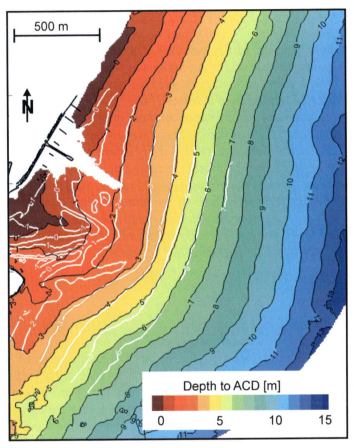

Fig. 6. The amalgamated bathymetry data for Teignmouth using one tidal cycle of hourly records. Radar derived bathymetric contours are shown in black, with the contours from the echo sounder survey shown in white for comparison. Depths are to Admiralty Chart Datum.

Sequences of 64 radar images with an antenna rotation rate of 2.4 seconds were recorded hourly during the deployment. Each image sequence was processed to map the spatial distribution of wavelengths at a range of wave frequencies, and the water depth inferred using the best fit to a wave dispersion equation that includes a correction for the amplitude dispersion of finite amplitude waves in shallow water (Hedges 1976). A set of these water depth maps spanning one tidal cycle were corrected to be relative to chart datum using the tide gauge data and were averaged to give the amalgamated bathymetry shown in Figure 6. The radar derived bathymetry is shown in colour with depth contours marked in black at 1m intervals. The corresponding echo sounder survey contours are shown in white (Bell 2005).

The complex system of sand bars around the mouth of the inlet is clearly visible from the survey contours. The maximum depth at the limit of the echo sounder survey is approximately 8m, although the radar derived water depths extend to almost twice the range of the survey and show maximum depths at this range of almost 14m.

The agreement between the depth contours of the radar derived bathymetry and the survey is excellent, particularly in this northern part of the area, with differences being on the order of centimeters.

It should be noted that to perform the echo sounder survey of the area represented by the white contours in Figure 6. required calm sea state conditions and approximately three days of ship time and man power. In contrast, the radar technology can produce a bathymetric map of almost double that range during wave events using only a few minutes of radar data although with slightly lower spatial resolution and accuracy.

5.   Summary

The principles of sea state measurement using an X-band nautical radar have been described. In addition, the Wave Monitoring System WaMoS II has been also described giving some technical features and accuracies.

Two examples for the use of nautical radar data were presented. The first example shows results of a standard sea state monitoring application at the platform FINO 1 in the German Bight. The second example shows results of a deployment at Teignmouth in the UK with the objective of inferring shallow water bathymetry by means of nautical radar images. These two examples represent just a section of possible WaMoS II applications and show that nautical radar is a reliable tool for monitoring wave fields and related phenomena, such as the wave shoaling due to variable bottom topography.

Nowadays about 40 WaMoS II systems have been installed on different platforms at different locations world-wide. These installations include moving vessels, coastal stations and off-shore platforms and use the capabilities of the WaMoS II for real time monitoring. The data from these stations are mainly used to support safe off-shore and harbour operations and to provide weather services with sea state data.

Acknowledgements

The Teighnmouth experiment was carried out within the COAST3D project which was funded by the Marine Science and Technology (MAST) RTD programme of the European Union under contract number MAS3-CT97-0086, and partly through national funding from the UK Ministry of Agriculture Fisheries and Food research project FD0803, the UK Environment Agency R&D Programme, the UK Environment Research Council and the Netherlands Rijkswaterstaat. The material presented for the FINO 1 Platform was kindly provided by the Federal Maritime and Hydrographic Agency (BSH), Germany.

References

Alpers W, Hasselmann K (1982) Spectral Signal to Clutter and Thermal Noise Properties of Ocean Wave Imaging Synthetic Aperture Radars. Int J Rem Sens 3: 423–446

Alpers W, Ross DB, Rufenach CL (1981) On the Detectability of Ocean Surface Waves by Real and Synthetic Aperture Radar. J Geophys Res 86: 6481–6498

Bell PS (2005). Remote determination of bathymetric changes using ground based radar. University of Wales Bangor, School of Ocean Sciences, PhD Thesis

Bell PS, Williams JJ, Clark S, Morris BD, Vila-Concejo A (2005) Nested radar systems for remote coastal observations. J Coastal Res SI 39 : 483–487

Dankert H (2004) Retrieval of surface-current fields and bathymetries using radar-image sequences, Proc. Int Geosci Rem Sens Symp IGARSS '04

Dankert H, Rosenthal W (2004) Ocean Surface Determination from X-Band Radar-Image Sequences. J Geophys Res 109: doi:10.1029/2003JC002130

Dankert H, Horstmann J, Rosenthal W (2005) Wind and Wave Field Measurements using Marine X-Band Radar-Image Sequences. IEEE J Ocean Engineer 30: doi: 10.1109/JOE.2005.857524

Dankert H, Horstmann J, Lehner S, Rosenthal W (2003) Detection of Waves Groups in SAR Images and Radar-Image Sequences, IEEE Trans Geosci Rem Sens 41(6): 1437–1446

Hedges TS, (1976) An empirical modification to linear wave theory. Proc Inst Civ Eng 61: 575–579

Herklotz K (2007) Oceanographic Results of Two Years Operation of the First Off-shore Wind Research Platform in the German Bight - FINO1 DEWI Magazin Nr. 30, February 2007

Hessner K, Reichert K, Dittmer J, Nieto-Borge JC, Günther, H (2001) Evaluation of WaMoS II Wave data. Proc WAVES 2001, San Francisco, 1–6 Sep 2001.

Keller W.C, Wright J.W. (1975) Microwave scattering and the straining of wind-generated waves. Radio Science 10: 139–147

Lee PHY, Barter JH, Beach KL, Hindman CL, Lake BM, Rungaldier H, Shelton JC, Williams AB, Yee R, Yuen HC (1995) X-Band microwave backscattering from ocean waves. J Geophys Res 100: 2591–2611

Nieto-Borge JC, Rodríguez GR, Hessner K, González PI (2002) Inversion of nautical radar images for surface wave analysis. J Atmos Ocean Tech 21: 1291–1300

Nieto-Borge JC, Hessner K, Reichert K (1999) Estimation of the significant wave height with X-band nautical radars. Proc 18th Int Conf Offshore Mech Arctic Eng (OMAE), St. John's, Newfoundland, Canada, 1999, number OMAE99/OSU –3063.

O'Reilly WC, Herbers THC, Seymour RJ, Guza RT (1996) A Comparison of Directional Buoy and Fixed Platform Measuremtns of Pasific Swell. J Atmos Ocean Techno 13: 231–238

Plant WJ, Keller WC (1990) Evidence of Bragg scattering in microwave Doppler spectra of sea return. J Geophys Res 95: 16,299–16,310

Reichert K, Hessner K, Dannenberg V, Tränkmann I, Lund B (2005) X-Band Radar as a Tool to Determine Spectral and Single Wave Properties. Proc 8th Wave Hindcasting Workshop

Senet CM, Seemann J, Ziemer F (2001) The near-surface current velocity determined from image sequences of the sea surface. IEEE Trans Geosci Rem Sens 39: 492–505

Trizna DB (2001) Errors in bathymetric retrievals using linear dispersion in 3-d fft analysis of marine radar ocean wave imagery. IEEE Trans Geosci Rem Sens 39: 2465–2469

Wetzel LB (1990) Electromagnetic scattering from the sea at low grazing angles. In: Geernaert G, Plant WJ (eds) Surface Waves and Fluxes, vol II, Remote Sensing. Kluwer Academic Publishers, Doordrecht, pp 109–171

Young IR, Rosenthal W, Ziemer F (1985) A Three-dimensional analysis of marine radar images for the determination of ocean wave directionality and surface currents. J Geophys Res 90: 1049–1059

Ziemer F, Günther H. (1994) A system to monitor ocean wave fields. Proc 2nd Int Conf Air-Sea Interaction Meteorol Oceanogr Coastal Zone. Lisboa, 22–27 September 1994

Land-Based Over-the-Horizon Radar Techniques for Monitoring the North-East Atlantic Coastal Zone

Klaus-Werner Gurgel and Thomas Schlick

Universität Hamburg, Institut für Meereskunde, Hamburg, Germany

Abstract. Land-based radar remote sensing techniques for measuring oceanic parameters have been developed for some 30 years. This paper describes the fundamentals and possible applications of shore-based High-Frequency (HF) radars, which are operated in the 3–30 MHz frequency range and use ground-wave propagation mode of the electromagnetic waves. Depending on the operating frequency selected, working ranges up to 200 nautical miles or a spatial resolution down to 300 m can be achieved. The parameters measured include surface current fields, wave directional spectra, and wind direction. The performance of these systems has been evaluated within several experiments, and after the demonstration of its operational capabilities, these systems now start to be integrated within monitoring services. In addition to the oceanographic applications, an HF radar can also be used to track ship locations at far ranges.

1. Introduction

Satellite sensors provide global coverage combined with poor temporal resolution, because it may take days until the same area is observed again. In contrast, land based radars do not provide a global coverage, but offer continuous measurements, *e.g.* every 10 min. Applications like coastal management, pollution monitoring, ship guidance, and rescue operations, require this high temporal resolution to track the tidal and wind driven oceanographic features. All remote sensing radars, including satellite borne systems, have the advantage to avoid instrumentation to be moored in the open sea, where they may be damaged by bad weather conditions or ships passing by.

HF radars are operated in the 3–30 MHz frequency band. Crombie (1955) discovered that electromagnetic waves in the HF band are interacting with the ocean surface due to Bragg scattering. Barrick (1971) further developed the theory and describes a radar for mapping of ocean surface

447

V. Barale, M. Gade (eds.), *Remote Sensing of the European Seas.*
© Springer Science+Business Media B.V. 2008

currents (Barrick *et al.* 1977). Electromagnetic waves in the microwave frequency range propagate by line-of-sight. In the HF frequency range, there are two more propagation modes. Both modes allow to monitor areas behind the horizon. This is why the HF radars are also called Over-The-Horizon Radars (OTHR). First of all, there is ground-wave propagation, where the electromagnetic wave is following a conductive layer. The ocean surface is a good conductor due to the salty water and provides an effective waveguide. Depending on the radar frequency and the transmitted power used, radar working ranges up to 200 nautical miles can be achieved (Ponsford *et al.* 2003). Skywave propagation is the second possibility in the HF frequency range. In this case, the ionosphere acts as a mirror, reflecting the electromagnetic wave back to the earth. As these skywave systems tend to require large areas of land to install all the antennas, up to 0.1 km × 1.5 km, most of these systems are operated by military organisations and they are not only used to measure oceanographic parameters (Georges *et al.* 1998). Skywave radars are normally installed far off the coast, *e.g.* the JINDALEE radar in the centre of Australia, while ground-wave radars require an installation of the antennas close to the water.

The University of Hamburg recently developed a new HF radar for coastal applications, called WEllen RAdar (WERA) (Gurgel *et al.* 1999b). This system is now commercially available and is used by several research institutes as well as by the French Hydrographic Office (SHOM).

2.   Basic physics and performance of HF radars

2.1   Propagation and scattering of electromagnetic waves

HF radar remote sensing is based on the scattering of electromagnetic waves from the rough sea surface. In case of ground-wave propagation and a monostatic set-up, the strongest backscattered signal is due to Bragg scattering at ocean waves of half the electromagnetic wavelength: A radar frequency of 15 MHz (20 m electromagnetic wavelength) couples to 10 m long ocean waves. The phase speed C of these ocean waves is given by the dispersion relationship

$$C = \sqrt{\frac{g\lambda}{2\pi} \tanh \frac{2\pi h}{\lambda}} \qquad (1)$$

where g is the acceleration due to gravity, λ is the ocean wavelength, and h is the water depth. The shallow water term, $\tanh(2\pi h/\lambda)$, approximates to 1 and can be neglected, if the water depth is larger than the ocean wave length.

Low radar frequencies, *e.g.* 5 MHz, require the shallow water term, if the water is less than 30 m deep. The phase speed leads to a Doppler shift of the backscattered signal. As there are always ocean waves travelling towards and away from the radar, two strong first-order Bragg peaks can be identified. Figure 1 shows an example of a backscatter spectrum from a specific patch of the ocean. The deviation between the measured phase speed and the theoretical value due to (1) is attributed to an underlying surface current, which can be estimated by evaluating this difference. Around the two first-order peaks, second-order side bands can be observed, which contain information on the ocean wave directional spectrum.

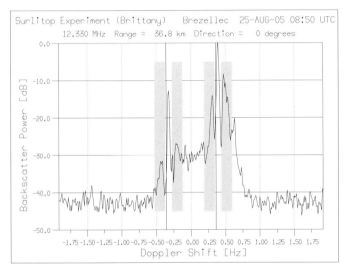

Fig. 1. Power spectrum of sea echoes, normalized to the strongest signal, as measured by WERA. The two 1st-order Bragg peaks and the 2nd-order sidebands (gray areas around the 1st-order peaks) are indicated. The two vertical lines mark the location of the 1st-order Bragg peaks at ±0.358 Hz, when no surface current is present. This spectrum has been measured on a selected sea surface patch, 36.8 km off the shore, perpendicular to the antenna array.

The working range of an HF radar at a given transmit power strongly depends on the conductivity of the ocean surface and the radar frequency (Gurgel *et al.* 1999a). The conductivity is a function of salinity and temperature of the water: High salinity or high water temperature result in low attenuation and high ranges. The salinity of the ocean is about 35 psu[1], which corresponds to a conductivity of 45 mmho/cm. Radar working ranges given in literature normally refer to these conditions. In the Baltic Sea, the salinity

---

[1] practical salinity unit: 1 psu is equivalent to 1 g of salt per liter of water

varies between 14 PSU in the west to 7 PSU around Bornholm and turns to 0 PSU in the eastern part. At 7 PSU, the conductivity is around 7 mmho/cm. In this case, the working range of the HF radar is reduced by about 50% (Gurgel 1995). East of Bornholm, the use of HF radars appears to be questionable due to a dramatic reduction in working range. At the Dead Sea, where the water contains about 300 g of salt per litre, a decreased transmitter power was sufficient to cover the complete area (Essen *et al.* 1995). Sea ice reduces the working range due to the high content of fresh water at the surface and the ice floes damping the Bragg-resonant short ocean waves.

The second major parameter affecting the working range is the radar frequency. High frequencies are stronger attenuated than low ones. The long-range ground-wave radars are operated in the 2.5–5.0 MHz frequency band (Ponsford 2003). However, due to a stronger impact of radio interference by other radio services, the available bandwidth and thus the spatial resolution is lower, compared to frequencies in the upper HF band around 25.0 MHz. Table 1 gives an overview on these characteristics.

2.2   Measurement of the surface current

The surface current is measured by exploiting the Doppler shift of the first-order Bragg peaks: The presence of ocean current moves both peaks by the same amount and to the same direction relative to the value given by the dispersion relationship (Equation 1). This additional Doppler shift can be converted to the radial component of an underlying surface current. Due to the decrease of the orbital motion of the Bragg resonant ocean waves, the current measured represents the average from the surface down to about 16% of the ocean wave length, *e.g.* the top 80 cm of the ocean at 30 MHz radar frequency. Since one HF radar measures the radial component, a second HF radar some 10 km apart is required to calculate the 2-dimensional surface current from the radial components. Figure 2 shows an example of a surface current field measured off Brest, France.

The accuracy of the 2-dimensional surface current depends on the errors of the radial components itself and on the angle between them. The angle leads to a factor called Geometrical Dillution Of Precision (GDOP), which is well known from navigation systems like the Global Positioning System (GPS). Basically, an angle of 90° leads to the lowest errors, while angles around 0° or 180° dramatically increase the error perpendicular to the radial components. If more than two radial components are available, a least-squares algorithm can be applied to further decrease the total error.

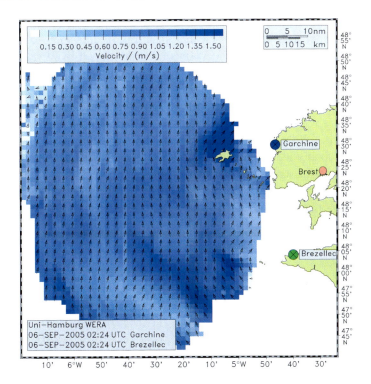

Fig. 2. An example of a surface current field measured west of Brest. Two WERA HF radars are installed at Garchine and Brézellec. Note the strong tidal driven current signatures around Ushant Island, west of Garchine.

Several comparisons between currents measured by HF radar and measurements by means of current meters or ADCPs have been made. The total RMS error observed is in the range 5–15 cm/s (Essen *et al.* 2000, Graber *et al.* 1997). Note, that both systems by principle can not measure the same: The HF radar averages the current over an area of about 1 km^2, while the *in situ* instruments always provide point measurements. Especially in areas with strong current shear due to topography or geostrophic effects, large deviations can be observed. When comparing *in situ* measurements to numerical model results, a similar problem arises. In this aspect, the characteristics of HF radar current measurements and numerical model results are quite similar (Essen *et al.* 2003).

2.3 Measurement of ocean wave spectra

Most of the algorithms described for HF radar wave processing are based on the inversion of the Barrick-Weber equations (Barrick *et al.* 1977).

They describe the interaction of the ocean wave directional spectrum with the second order sidebands in the backscatter spectrum (Figure 1). The most advanced approach based on this theory has been developed by Wyatt (2000), allowing for bi-modal sea states. However, when the length of the Bragg-scattering wave raises to the order of the significant waveheight Hs, this theory reaches its limitations. A solution is to move to lower radar frequencies, which is not always possible. An empirical approach has been developed by Gurgel *et al.* (2006), aimed to overcome this limitation. However, due to problems to reliably identify the first-order Bragg peaks during these high sea state conditions, this empirical solution is currently limited in a similar way. At 27 MHz radar frequency, the wave measurement saturates at $Hs \approx 7.5$ m, at 12.5 MHz, the limit is at $Hs \approx 16$ m. Figure 3 shows an example of a wave field measured off Brest, France.

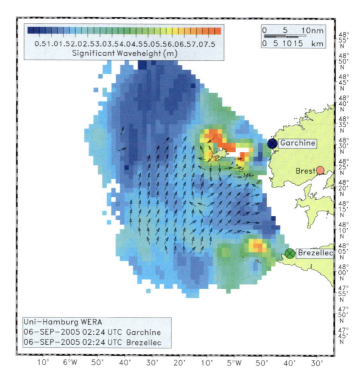

Fig. 3. A wave field measured by HF radar using the empirical algorithm. The colour scale represents the significant waveheight, the arrows shows the mean wave direction. The patches of irregularly high sea state around Ushant Island are due to high local current variability.

Significant progress in the inversion technique has been achieved by Wyatt *et al.* (2003) within the EuroROSE[2] project. Besides further development of the algorithm itself, the accuracy of HF radar wave measurements has been evaluated by comparison to a number of directional wave buoys and examples of wave directional spectra are shown in this paper.

2.4 Measurement of the wind direction

HF radar wind measurements can be done by exploiting the first-order Bragg peaks of the backscatter Doppler spectrum. Depending on the radar frequency, they give information on the wind-induced short ocean waves of 5–10 m length. The height and direction of these ocean waves follow meteorological events with some 10-15 min delay. Figure 4 shows an example of the wind direction measured off Brest, France.

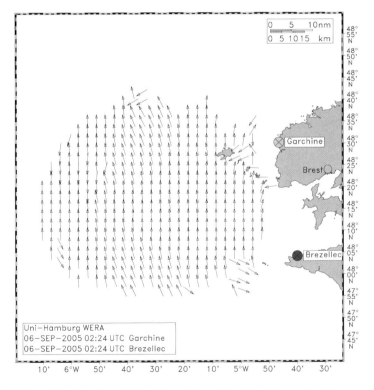

Fig. 4. The wind direction derived from the ratio of the first-order Bragg peaks.

---

[2] European Radar Ocean Sensing, EU Mast-3, CT98-0168

The wind direction can be calculated from the ratio of the two first-order Bragg peaks by fitting their amplitudes to a $\cos^s(0.5\theta)$ or $\text{sech}^s(1.0\theta)$ angular distribution model (Gurgel *et al.* 2006). This solution is ambiguous, giving two possible wind directions, if one Bragg line ratio is available. Data from a second radar site at a different angle to the measurement area resolves this ambiguity. Ocean current measurements require two radar sites, anyway.

In contrast to microwave scatterometers, wind speed measurements in the HF frequency band are much more complicated and still under investigation. At low wind speed, there is an increase in wave energy of the Bragg-scattering waves with increasing winds, which can be used to derive the wind speed. However, at stronger winds, there is saturation in wave energy because of the nonlinear wave-wave interaction and energy transfer to longer waves. This saturation prevents a simple algorithm for wind speed measurements.

2.5   Ship detection and tracking

Besides the echoes from the rough ocean surface, echoes from ships and islands can be observed in the Doppler spectra. As the land is not moving, these echoes appear at zero Doppler shift, but echoes from sailing ships can be mapped into the part of the Doppler spectrum, which is used for oceanographic measurements. Because the wave measurements rely on the weaker second order sidebands, they are mainly affected, and an irregularly high significant waveheight can be observed at these locations. If it is known that this signal is related to a ship, the wave algorithm can handle the situation correctly by interpolation from surrounding areas. Work on this topic has been started, and first results have been published by Gurgel *et al.* (2005).

On the other hand, applications for ship detection and tracking exist, especially because an HF radar has a much higher working range compared to microwave radars. The Canadian Navy is now monitoring the Exclusive Economic Zone (EEZ) up to 200 nautical miles, using radar systems provided by Raytheon Canada (Ponsford *et al.* 2003). First tests with a standard 12.5 MHz WERA system at 30 W transmit power have shown, that larger ships can be detected and tracked up to 120 km range.

Currently, an 8 MHz WERA system is installed near Lisboa, Portugal, to develop and evaluate WERA's ship tracking capabilities. The software consists of two major parts: A detection algorithm based on Constant False Alarm Rate, CFAR (Rohling 1983), and a tracker implemented with a nearest-neighbour search and an alpha-beta filter.

3. HF radar systems

Table 1 summarizes the Bragg resonant ocean wave and the expected working range for different radar frequencies. The ranges are given for surface current measurements at 50W average transmit power. Processing of ocean wave parameters requires an increased signal-to-noise ratio, which reduces the working range by about 30%. The detection range for ship targets depends on the ship size. R_{opt} and R_{min} are given by Shearman (1987), the other values have been calculated using a propagation model (Gurgel et al. 1999a). The maximum possible range resolution depends on the feasible radar bandwidth, which is a function of radio frequency interference and the ionospheric conditions.

Table 1. Bragg scattering ocean waves, working range for current measurements at 50 W average transmit power, and the maximum possible range resolution as a function of radar frequency.

f_0 (MHz)	P_{Bragg} (s)	λ_{Bragg} (m)	R_8 (km)	R_{16} (km)	R_{35} (km)	R_{opt} (km)	R_{min} (km)	R_{resol} (km)
8.00	3.47	18.75	107.0	158.5	218.0	300.0	200.0	2.00
12.00	2.83	12.50	65.0	100.0	143.0	180.0	115.0	1.50
16.00	2.45	9.38	45.5	70.5	102.5	130.0	75.0	1.20
20.00	2.19	7.50	34.0	53.0	78.0	100.0	60.0	1.00
25.00	1.96	6.00	25.5	40.0	59.5	75.0	45.0	0.50
30.00	1.79	5.00	20.0	32.0	47.0	50.0	30.0	0.25

f_0	Radar frequency
P_{Bragg}	Period of Bragg scattering ocean wave
λ_{Bragg}	Length of Bragg scattering ocean wave
R_8	Working range at 8 PSU salinity (propagation model)
R_{16}	Working range at 16 PSU salinity (propagation model)
R_{35}	Working range at 35 PSU salinity (propagation model)
R_{opt}	Working range at 35 PSU salinity, optimum sea state (Shearman 1987)
R_{min}	Working range at 35 PSU salinity, high sea state (Shearman 1987)
R_{resol}	Highest range resolution possible

HF radar range resolution can be done by pulses, coded pulses, and linear frequency chirps. To achieve a high range resolution by pulses, they must be short (1.2 km ↔ 8 µs). To keep an average transmit power of 50 W at 128 range cells, a peak power of 128×50 W = 6.4 kW is required. Coded pulses help to improve the Transmit-to-Receive (T/R) ratio and reduce the peak power. In the case of Frequency Modulated Continuous Wave (FMCW) modulation, the transmitter is run continuously and range resolution is achieved by a frequency chirp (Gurgel et al. 1999a). Some systems use Interrupted FMCW (FMICW) to reduce the system's dynamic range requirements.

A low transmit power is easier to generate, increases the chance to get a transmit license, and allows for shorter safety distances from the transmit antenna to the public.

Azimuthal resolution can be done by Beam Forming (BF) and Direction Finding (DF). BF requires a linear antenna array at $\leq \lambda_{Bragg}$ spacing. At low radar frequencies, this leads to a long array size (cf. Table 2). The advantage of BF is, that access the complete backscatter spectrum as shown in Figure 1 is possible. Figure 5 shows a linear antenna array installed in Garchine, near Brest, France. DF implements azimuthal resolution in frequency domain by exploiting the amplitudes and phases of the backscatter Doppler spectrum. Each spectral line is mapped to one, two or more incident angles (Gurgel *et al.* 1999a). Access to the second-order sidebands is masked by the strong first-order Bragg peaks from other directions.

Fig. 5. The linear array of a WERA deployment in Garchine, France.

Table 2 summarizes the characteristics of some available HF radar systems: The Raytheon system (Ponsford *et al.* 2003) is used by the Canadian Navy for surveillance of their EEZ. This is a large system and ship tracking is one of its main applications.

The 5 MHz SeaSonde is a small and compact DF design to give average surface current fields every three hours at long ranges.

The PISCES radar (Shearman *et al.* 1988) is based on FMICW modulation for range resolution and BF for azimuthal resolution. BF is implemented in hardware by using switchable cables as phase shifters. PISCES is currently deployed at the Irish Sea, UK.

The WERA system (Gurgel *et al.* 1999b) can be operated at different frequencies. The 8 MHz system is currently tested for current and wave measurements, as well as ship tracking. The 12.5 MHz system is a component of SHOM's monitoring system implemented in the area off Brest. The 27–30 MHz WERA can be operated in a high-resolution mode, allowing for a range resolution down to 300 m.

Table 2. Operating frequency, max. working range, peak transmit power, modulation for range resolution: Pulse (P), Coded Pulse (CP), Frequency Modulated Continuous Wave (FMCW), Interrupted FMCW (FMICW), azimuthal resolution: Beam Forming (BF), Direction Finding (DF), and antenna array size of existing HF radar systems ordered by operating frequency.

System	Operating frequency	Working range	Transmit power	Modulation	Range resolution	Azimuth resolution	Antenna array size
Ray-theon	3–5 MHz	360 km	16 kW	CP	7.5 km	BF	495 m
Sea-Sonde	5 MHz	200 km	100 W	FMICW	7.5 km	DF	20 m
PISCES	7.0 MHz	200 km	3 kW	FMICW	7.5 km	BF	200 m
WERA	8.0 MHz	250 km	50 W	FMCW	1.5 km	BF,DF	253 m
WERA	12.5 MHz	180 km	30 W	FMCW	1.5 km	BF,DF	162 m
WERA	16.0 MHz	110 km	30 W	FMCW	1.0 km	BF,DF	127 m
Sea-Sonde	25.0 MHz	50 km	100 W	FMICW	1.2 km	DF	10 m
WERA	27.6 MHz	40 km	30 W	FMCW	0.3 km	BF,DF	74 m
COSRA	30.0 MHz	45 km	1 kW	P	2.0 km	BF	68 m
COSME	40.0 MHz	40 km	1 kW	P	0.6 km	BF	56 m

The SeaSonde (Paduan *et al.* 1996) is a very small portable DF system using FMICW for range resolution. Azimuthal resolution is provided by a very small loop antenna using a special DF algorithm based on MUSIC (MUltiple SIgnal Classification).

The Australian COSRAD radar (Coastal Ocean Surface Radar), developed at James Cook University (Heron *et al.* 1985), is operated at 30 MHz. In contrast to all the other systems mentioned, a common linear antenna array for transmit and receive is used to perform BF. This results in a narrower beam and increases the signal-to-noise ratio. However, as the different directions are scanned step by step, the azimuthal surveillance is slow.

The COSMER (Broche *et al.* 1987) has been developed at the University of Toulon, France, and is operated at about 40 MHz. This high frequency avoids problems arising from RFI and enables high spatial resolution.

4. Monitoring the coastal zone by an integrated remote sensing - model system

Currently, there is an increasing demand for monitoring systems in coastal regions for coastal management and marine safety. One of the European initiatives to build up and integrate these systems is EuroGOOS, the European branch of the Global Ocean Observing System (GOOS). Monitoring systems consist of two major components, the measurement instruments and numerical models. Together they deliver now- and forecasts of high accuracy: The measurements keep the model 'close to nature' and the model provides forecasts.

A monitoring system based on HF radar current and wave measurements has been demonstrated within the European EuroROSE project. The actual application was to help crude oil tankers to navigate through a narrow entry in presence of strong ocean currents perpendicular to their course. Forecasts up to 6 hours were provided to the pilots. A description of the system is given by Günther *et al.* (1998). Larger areas can be covered by installing HF radar networks (Gurgel *et al.* 2003). Currently, the French Hydrographic Office (SHOM) is setting up a monitoring system off Brest. Because of the highly dense ship traffic, there are also plans for monitoring systems in the Bosphorus (Sevgi 2003).

Acknowledgments

The data shown in this article have been collected during the SURLITOP campaign August to November 2005 which aimed to demonstrate the operational capabilities of HF radar applications to the French Hydrographic Office (SHOM). SURLITOP was funded by the French Ministry of Research (50%) and by the French companies Actimar and Boost Technologies. We wish to thank Vincent Mariette, Nicolas Thomas, and Veronique Cochin at Actimar, Fabrice Collard at BOOST, and Yves Barbin at LSEET, University of Toulon, for the excellent support. We also wish to thank Monika Hamann and Iris Ehlert at the Remote Sensing Group, University of Hamburg.

References

Barrick DE (1971) Theory of HF and VHF propagation across the rough sea, 2, Application to HF and VHF propagation above the sea. Radio Science 6: 527– 533

Barrick DE, Evans MW, Weber BL (1977) Ocean surface current mapped by radar. Science 198: 138–144

Broche P, Crochet JC, de Maistre JL, Forget P (1987) VHF radar for ocean surface current and sea state remote sensing. Radio Science 22: 69–75

Crombie DD (1955) Doppler spectrum of sea echo at 13.56 Mc/s. Nature 175: 681–682

Essen HH, Gurgel KW, Schirmer F, Sirkes Z (1995) Horizontal variability of surface currents in the Dead Sea. Oceanologica Acta 18: 455–467

Essen HH, Gurgel KW, Schlick T (2000) On the accuracy of current measurements by means of HF radar. IEEE J. Oceanic Engineering 25 (4): 472–480

Essen HH, Breivik Ø, Günther H, Gurgel KW, Johannessen J, Klein H, Schlick T, Stawarz M (2003) Comparison of remotely measured and modelled currents in coastal areas of Norway and Spain. The Global Atmosphere and Ocean System (ISSN 1023–6732) 9 (1–2): 38–64

Georges TM, Harlan JA, Lee TN, Leben RR (1998) Observations of the Florida Current with two over-the-horizon radars. Radio Science 33(4):1227–1239

Graber HC, Haus BK, Chapman RD, Shay LK (1997) HF radar comparisons with moored estimates of current speed and direction: Expected differences and implications. J Geophys Res 102 (C8): 18,749–18,766

Günther H, Gurgel KW, Evensen G, Wyatt LR, Guddal J, Nieto Borge JC, Reichert K, Rosenthal W (1998) EuroROSE - European Radar Ocean Sensing. Proc COST Conf "Provision and Engineering/Operational Application of Wave Spectra", 21–25 September 1998, Paris, France

Gurgel KW (1995) The variability of surface current fields in the Pomeranian Bay as measured by Decameterwave Radar. Polish Academy of Sciences, Institute of Hydroengineering, Hydrotechnical Transactions 60: 53–64

Gurgel KW, Essen HH, Kingsley SP (1999a) HF radars: Physical limitations and recent developments. Coastal Engineering 37 (3–4): 201–218

Gurgel KW, Antonischki A, Essen HH, Schlick T (1999b) Wellen Radar (WERA), a new ground-wave based HF radar for ocean remote sensing. Coastal Engineering 3 7(3–4): 219–234

Gurgel KW, Essen HH, Schlick T (2003) The use of HF radar networks within operational forecasting systems of coastal regions. In: Dahlin H, Flemming NC, Nittis K, Petersson SE (eds) Building the European Capacity in Operational Oceanography. Elsevier, pp 245–250

Gurgel KW, Schlick T (2005) HF Radar Wave Measurements in the Presence of Ship Echoes - Problems and Solutions. Proc IEEE Oceans Europe Conference, vol 2, pp 937–941

Gurgel KW, Essen HH, Schlick T (2006) An empirical method to derive ocean waves from second-order Bragg scattering - prospects and limitations. IEEE J. Oceanic Engineering 31: 804–811

Heron ML, Dexter PE, McGann BT (1985) Parameters of the air-sea interface by high-frequency ground-wave HF Doppler radar. Australian J Marine Freshwater Re. 36: 655–670

Paduan JD, Rosenfeld LK (1996) Remotely sensed surface currents in Monterey Bay from shore-based HF radar (Coastal Ocean Dynamics Application Radar). J Geophys Res 101: 20669–20686

Ponsford AM, Dizaji RM, McKerracher R (2003) HF Surface Wave Radar Operation in Adverse Conditions. Proc CSSIP/IEEE Radar 2003 Conf

Rohling H (1983) Radar CFAR thresholding in clutter and multiple target situations. IEEE Trans. Aerospace Electron Syst 19: 608–621

Sevgi L (2003) Stochastic Modelling and Simulation Studies for Surface Wave High Frequency Radars: Problems and Challenges. Proc CSSIP/IEEE Radar 2003 Conf

Shearman EDR (1987) Chapter 5: Over-the-horizon Radar. In: Scanlan MJB (ed) Modern Radar Techniques. Collins, London, pp 200–240

Shearman EDR, Moorhead MD (1988) PISCES: A Coastal Ground-wave HF radar for Current, Wind and Wave Mapping to 200 km Ranges. Proc IGARSS'88, pp 773–776

Wyatt LR (2000) Limits to the inversion of HF radar backscatter for ocean wave measurement. J Atmospheric Oceanic Technology 17: 1651–1665

Wyatt LR, Green JJ, Gurgel KW, Nieto Borge JC , Reichert K, Hessner K, Günther H, Rosenthal W, Sætra Ø, Reistad M (2003) Validation and intercomparions of wave measurements and models during the EuroROSE experiments. Coastal Engineering 48: 1–28

Section 4:

Multi-Sensor Techniques

Multi-Sensor Observations of Meso-Scale Features in European Coastal Waters

Olga Lavrova[1], Marina Mityagina[1], Tatiana Bocharova[1], and Martin Gade[2]

[1] Space Research Institute, Russian Academy of Sciences, Moscow, Russia
[2] Institut für Meereskunde, Universität Hamburg, Hamburg, Germany

Abstract. Results of long-term multi-sensor observations of coastal zones in the Baltic and Black Seas are discussed. The study is based on remote sensing data acquired over these regions by the ERS-2 SAR, Envisat ASAR, Terra and Aqua MODIS and NOAA AVHRR. The data were analysed to investigate coastal water circulations, in particular the occurrence, evolution and drift of small- and meso-scale vortex structures, because their understanding is crucial for the knowledge of the mechanisms that determine mixing and circulation processes in the coastal zone. To a large extent, these mechanisms determine the coastal zone's ecological, hydrodynamic and meteorological state, whose constant monitoring is vital for these densely populated regions with their well-developed industry and agriculture and a rapidly growing tourist sector. SST fields derived from AVHRR data were used to observe meso-scale water dynamics. Using MODIS SST and ocean colour data we were able to highlight various meso- and small-scale water dynamics features, such as currents, eddies, dipoles, jets, filaments, and river plumes. Surfactants of natural and artificial origin are often encountered in coastal waters, and their signatures on SAR images help the detection of small- and meso-scale surface currents and vortex structures.

1. Introduction

Investigation of eddies and meso-scale features in coastal zones is important for understanding local mechanisms of mixing and circulation processes. To a large extent, these mechanisms determine the ecological, hydrodynamic and meteorological state of coastal zones, whose constant monitoring is vital for densely populated regions.

Both the Black Sea and the Baltic Sea are semi-enclosed seas with narrow and shallow straits connecting them to the ocean. Water replacement in

V. Barale, M. Gade (eds.), *Remote Sensing of the European Seas.*

them takes several decades, which is the main reason for the high vulnerability of their complex aquatic environment. The eight European countries surrounding the Baltic Sea and six European countries surrounding the Black Sea have large ports, developed fishing industries, vast cultivated terrains and ever growing residential construction along the shorelines. The seas carry intense marine traffic, including oil tanker transport. All in all, both seas and their coastal areas are permanently exposed to massive anthropogenic pressure.

The south-eastern zone of the Baltic Sea and the north-eastern coastal zone of the Black Sea are famous tourist regions. Moreover, the south-eastern zone of the Baltic Sea is a site of increasing offshore oil production, while the Blue Flow subsea gas pipeline to Turkey starts in the north-eastern coastal zone of the Black Sea. It is therefore obvious that the understanding of ocean circulation processes and the evolution of inevitable pollution in these particular regions is of vital importance.

Coastal waters are affected by vortex structures of all scales: synoptic, meso-scale and small-scale circulations. Synoptic eddies have diameters of hundreds of kilometres, meso-scale eddies tens of kilometres, and small-scale eddies have diameters of a few kilometres and less. In coastal zones, meso-scale and small-scale cyclonic and anti-cyclonic eddies are found.

Data obtained by the Advanced Very High Resolution Radiometer (AVHRR) on board the National Oceanic and Atmospheric Administration (NOAA) satellites allow the derivation of Sea Surface Temperature (SST) fields for the analysis of meso-scale water dynamics. The Moderate Resolution Imaging Spectroradiometer (MODIS) instruments aboard the Terra and Aqua satellites provide SST, ocean colour and other optical properties measurements. Over short time scales of a few days, the water scatterers can be viewed as passive tracers of surface currents and frontal zones, with their distribution corresponding to flow paths. Hence, the analysis of optical images makes it possible to highlight the main meso- and small-scale water dynamics features, such as currents, eddies, dipoles, jets, filaments and river plumes. In Synthetic Aperture Radar (SAR) images, eddies may be visualized due to numerous bands of slicks produced by surfactants of natural and artificial origin. Surfactant films get entrained in the eddy motion, and since they reduce the backscattered radar power under low to moderate wind conditions (Alpers and Hühnerfuss 1989) they may be visible on SAR images.

Taking advantage of the fact that meso-scale features can be delineated on satellite imagery acquired by different remote sensing sensors we have performed a comprehensive multi-sensor analysis of coastal dynamics in both the south-eastern Baltic Sea and the north-eastern Black Sea. Some of our results are presented herein.

2. The north-eastern coastal zone of the Black Sea

2.1 General circulation features

The circulation in the Black Sea is characterized by a strong basin-wide cyclonic current along the shore, the Rim Current. The current embraces the entire sea along its periphery and is characterized by high hydrodynamic instability (Zatsepin *et al.* 2002). Usually the Rim Current can be inferred from satellite images (Figure 1). The observed meso-scale eddy variability consists of meanders, anti-cyclonic and cyclonic eddies, pinched off eddies, vortex dipoles, filaments and jets.

The combination of the cyclonic Rim Current and anti-cyclonic near shore eddies plays an important role in the hydrodynamics and ecology of the coastal zone. In some cases, like in the north-eastern Black Sea, meso-scale eddies can interrupt the Rim Current locally, thus resulting in substantial cross-shelf transport (Poulain *et al.* 2005).

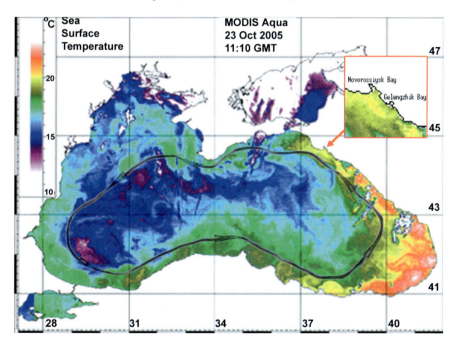

Fig. 1. SST map derived from MODIS data of 23 October 2005. Dark line shows schematically the mean position of the Rim Current. Note the anti-cyclonic near-shore eddies along the southern coast. The region of interest near the north-eastern coast is enlarged in the box (upper right corner).

2.2 Eddy dipoles

There is a particular interest in so-called mushroom flows, which are quasi-dipolar eddies combining a narrow jet with a pair of vortices of opposite signs at its end. The jet length and the size of the vortices are, in most cases, of the same order, while the jet width does not exceed 10–25% of its length. The structure resembles a cross-section of a mushroom. Observations of such eddy pairs reported in literature highlight dipoles as large as 80–100 km in size (Ginzburg *et al.* 2002; Afanasyev *et al.* 2002).

Mushroom flows occur due to local, short-duration momentum actions onto the sea surface or in the near-surface layer and can represent a very effective mechanism for horizontal mixing in the ocean. Because of their self-propelling motion, mushroom flows can transport scalar properties, such as salt, heat and biological constituents, over large distances, and hence they can play an important role in the exchange between shelf and deep-sea waters. On remote sensing images, mushroom structures are often visible through natural tracers of some kind on, or near, the surface.

Our observations over many years indicate that the use of SAR data considerably extends the possibilities of remote sensing detection and examination of such structures and, coupled with data from other sensors, raises the reliability of data interpretation and retrieval of coastal circulation patterns. We will illustrate this herein.

An Envisat Advanced SAR (ASAR) image acquired on 15 May 2006 at 1910 UTC (Figure 2a) reveals an early formation stage of a dipole composed of a jet and a pair of cyclonic and anti-cyclonic eddies. At the time of image acquisition a weak south-westerly wind of 2–3 m/s and weak surface waves were reported. The homogeneously low wind field coupled with the presence of biogenic films on the sea surface favoured clear visibility of flow signatures on the ASAR image. The eddy dipole appears also on MODIS data acquired on the next day, 16 May, at 1039 UTC (Figure 2 b,c,d).

The mushroom flow can easily be delineated in all panels of Figure 2. For a considerable time, the dipole retained its shape and position, slowly being driven into a northwest direction by the Rim Current. Dynamic vortex structures of this kind are regularly observed in this region of the Black Sea and are known to induce not only horizontal, but also vertical mixing of the water. They contribute to hydrodynamic instability of the alongshore current and intensify the transport between coastal waters and the open sea (Zatsepin *et al.* 2003).

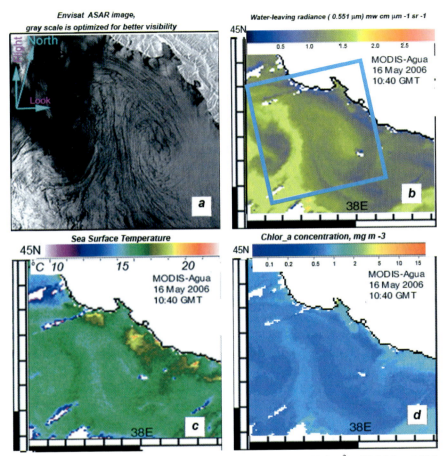

Fig. 2. (a) Envisat ASAR VV-polarization image (100×100 km^2) of 15 May 2006, 19:10 GMT showing an eddy dipole ("mushroom flow") in the north-eastern Black Sea, manifested through slick bands (© ESA, 2006); (b) water-leaving radiance map derived from MODIS data. The position of the ASAR image is marked by the rectangle; (c) SST maps of 16 May; (d) chlorophyll-*a* concentration map derived from MODIS data of 16 May.

2.3 Small-scale nearshore eddies

Previously, satellite observations of vortex structures in the north-eastern part of the Black Sea have been performed using IR or optical data, together with *in-situ* measurements (Zatsepin *et al.* 2002). The spatial resolution of such images of about 1 km makes it possible to study vortex structures of 40 km or larger in size, but only under cloudless conditions. Our results show that the observation of vortex structures of smaller size (less than 30 km) at short time scales (days to weeks) can be supplemented by SAR imagery of higher spatial resolution. The use of SAR data has allowed discovery of intense small-scale vortex activity, which researchers were unaware of before.

Regular satellite monitoring of the Black Sea coastal zone in the region of Novorossiysk - Gelendzhik has been conducted in the summer-autumn seasons since 1998 (Bulatov *et al.* 2003; Lavrova *et al.* 2003). The analysis of the obtained data has revealed that eddies of a few kilometres to some tens of kilometres in size constantly occur up to the distance of 20–30 km from the shore and move to the northwest (Lavrova 2005). Most of the reported small scale eddies are to be anti-cyclonic, because they are generated mainly through baroclinic instability of the Rim Current (Zatsepin *et al.* 2003). However, our SAR data analysis also revealed many instances of cyclonic eddies of less than 10 km in size. As a rule, they are situated closer to the shoreline compared to anti-cyclonic eddies. Figure 3 shows a SAR image of a cyclonic eddy of 5 km in diameter.

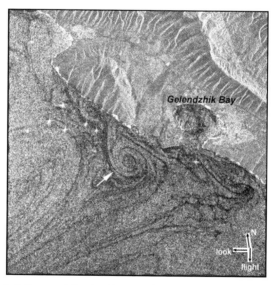

Fig. 3. Fragment (25km × 25km) of an Envisat ASAR VV-polarization image of 03 August 2006 showing imprints of a cyclonic eddy, with a diameter of 5 km, north of Gelendzhik Bay (© ESA, 2006).

Understanding the circulation in semi-enclosed waters such as bays and gulfs is an important task within coastal studies. Local winds are weaker there and the sea surface is often covered with surfactants. These conditions are favourable for the study of vortices outlined by sea surface slicks.

The Black Sea test region also included Gelendzhik Bay, which is an enclosed area with a mouth width of less than 4 km. It is one of the largest Russian resorts on the Black Sea. A large amount of water containing surfactants of biogenic and anthropogenic origin is always present in the Bay. The water replacement process in the Bay may take from 1 to 10 days depending on the wind. Vortices entering the Bay during south-eastern winds play the main role in water circulation here. They are the main water cleaning factor as well, because they carry relatively clean water from the open sea.

3. The south-eastern coastal zone of the Baltic Sea

3.1 Meso-scale features in the Bay of Gdańsk

The Baltic Sea is one of the best observed seas on Earth and hence can be seen as a laboratory for various studies. Up to date, different kinds of (process-oriented, one-, two-, or three-dimensional) models have been applied to study the circulation in the Baltic Sea (Lehmann 1995). However, there still remain many open questions. One of them concerns the role of wind in the formation of sub-surface eddies.

Over the period from July 2004 to November 2005, operational satellite monitoring of the southeast coastal zone of the Baltic Sea was conducted jointly by the Space Radar Laboratory of IKI-RAS, the P.P. Shirshov Institute of Oceanology (Moscow), the Geophysical Center (Moscow), and the Marine Hydrophysical Institute (MHI, Sevastopol, Ukraine) (Kostianoy *et al.* 2006). Observations were based on remote sensing data from the Envisat and ERS-2 SARs, the SAR aboard the Canadian RADARSAT, as well as from Aqua and Terra MODIS. The aim of the project was a daily surveillance of anthropogenic pollution, especially of oil pollution. Moreover, it provided a unique opportunity to gather a huge amount of information on vortex processes and various meso-scale features in the coastal zone of the south-eastern Baltic Sea.

In late July/early August 2004, an intense algae bloom was observed in the Central Baltic Proper. High concentrations of blue-green algae were used as passive tracers, revealing the locations of convergence - divergence zones and, consequently, of current field structures. Figure 4 presents a series of colour composites of Terra/Aqua MODIS images acquired from 28 July to 11 August 2004 over the south-eastern Baltic Sea and the Bay of Gdańsk. A newly developing mushroom current manifests in all spectral bands, and hence it is also visible in MODIS SST and chlorophyll-a maps (not shown herein).

The analysis of the optical image series allowed us to follow the development of the meso-scale features. Although the eddies and dipoles observed in the optical images underwent slight transformations, they remained in practically the same locations over two weeks, notwithstanding that on certain days, the wind speed reached 10 m/s.

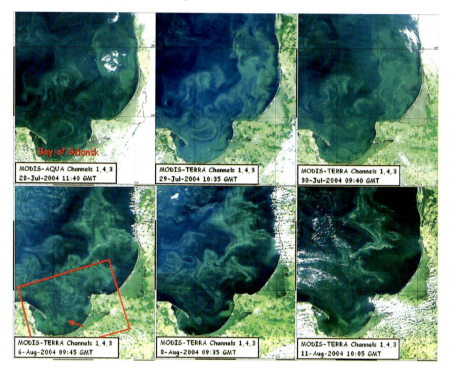

Fig. 4. Series of colour composites of MODIS images (channels 1, 4, and 3, corresponding to red, green, and blue, respectively) acquired on (upper row) 28, 29, 30 July and (bottom row) 6, 8, 11 August 2004. The position of the area shown in Figure 5 is marked by the red rectangle. The red arrow points at the cyclonic eddy seen in Envisat ASAR image (see Figure 5).

According to observations and numerical model simulations (Zhurbas *et al.* 2004), cyclonic eddies can be found in the intermediate layer of the Bay of Gdańsk. These eddies are reported to develop under westerly wind conditions, after the relaxation of the coastal downwelling jet at the tip of the Hel Peninsula in the deep water, when the flow becomes baroclinic. In the case presented here, the stable vortex structures occurred in the presence of a rather weak swell coming from the west and an unstable westerly wind, gradually replaced by a rising northerly wind.

We did not observed those vortex structures on every radar image obtained in the same period over the same area. At higher wind speeds (well above 5 m/s) the surfactant films are disrupted by the strong action of wind and waves, thus making the vortex structures invisible on SAR imagery. However, as an example, Figure 5 shows an Envisat ASAR image acquired on August 5, when the wind speed did not exceed 4 m/s.

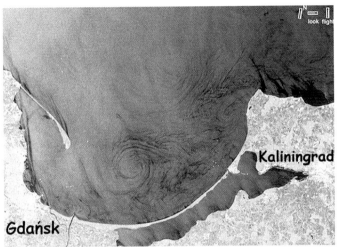

Fig. 5. Fragment (140 km × 100 km) of an Envisat ASAR Wide Swath image, for the Bay of Gdańsk, acquired on 5 August 2004, at 20:19 GMT, showing a cyclonic eddy of diameter 19 km (© ESA, 2004).

The SAR image (Figure 5) does not show all eddies, dipoles and jets that are clearly manifested in the optical images (Figure 4). Apparently, this is not only caused by the local wind speed being too high, but also by the fact that the algae accumulate not at the very water surface, but in some water depth where they can still be detected by optical sensors. Thus, as they are not correlated with surfactants floating on the sea surface, they are invisible to SAR sensors.

3.2 Eddy dipoles near a river mouth

Coastal circulation patterns often feature mushroom or dipole vortices, particularly in the vicinity of river mouths. An example of mushroom flows manifested in both a SAR and a MODIS image is presented in Panels *a* and *b*, respectively, of Figure 6. The ASAR image (Figure 6*a*) shows two eddy dipoles: one, 25 km in size, is formed by the river Vistula outflow, the other, 6 km in size, is formed by fresh water outflow from the

Bay of Kaliningrad. The dipoles are outlined by surfactant slicks accumulated in convergence zones. The Vistula dipole is clearly seen in the optical image due to the different optical properties (high sediment load) of the river water (Figure 6b).

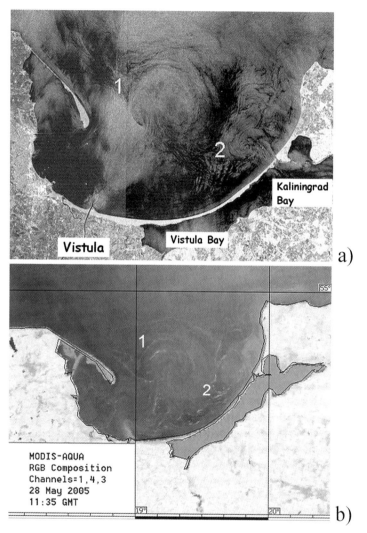

Fig. 6. (a) fragment (115km × 85km) of an Envisat ASAR Wide Swath image acquired on 28 May 2005, at 20:16 GMT (© ESA, 2005); (b) fragment (160 km × 120 km) of a colour composite MODIS image acquired on 28 May 2005, at 11:35 GMT. Manifestations of mushroom vortices in the Bay of Gdańsk are visible on both images. They are caused by (1) the Vistula runoff and (2) the outflow of the Kaliningrad (Vistula) Bay.

The second dipole is less distinct in the optical image and its shape is deformed there. Obviously, the joint analysis of data from different sensors helps to identify certain dynamic features as river plumes, to define their precise shapes and sizes, and to infer roughly (*i.e.* qualitatively) the outflow component compositions.

4. Concluding remarks

Our multi-sensor satellite observations of meso-scale features in the coastal zones of the south-eastern Baltic Sea and north-eastern Black Sea have shown that the combination of remote sensing data from different sensors may lead to a better understanding of the dynamic features observed. Under cloudfree weather conditions, data obtained by optical and remote sensing sensors are well suited to observe and to track meso-scale features in European coastal zones. This is particularly due to the high temporal and spatial coverage of such sensors. However, the dependence on daylight and cloudfree weather conditions may cause data gaps, especially in higher latitudes.

SAR sensors, because of their all-weather capabilities, allow for the detection of eddies and vortices under low to moderate wind speeds and when surfactants are present. In particular, the high resolution of SAR sensors allows observation of small-scale current features, such as flow patterns in the mouth of Gelendzhik Bay. In such a manner SAR imagery has the potential to complement remote sensing data obtained by spaceborne optical and infrared sensors.

In summary, we have shown that the combination of remote sensing data acquired at different electro-magnetic wavelengths gives rise to an improved monitoring of European coastal zones from space, particularly of coastal dynamic features.

Acknowledgements

This work was partly supported by INTAS project 03-51-4987, Black Sea Scientific Network (Contract # 022868) and RFBR grant # 06-05-08072 – ofi . SAR data were obtained under ESA projects C1P.1027, AO3.224 and AO Bear 2775. Aqua/Terra MODIS data were processed and kindly provided by D.M. Soloviev, MHI.

References

Afanasyev YD, Kostianoy AG, Zatsepin AG, Poulain PM (2002) Analysis of velocity field in the eastern Black Sea from satellite data during the Black Sea '99 experiment. J Geophys Res 107 (C8): 3098, doi: 10.1029/2000JC000578.

Alpers W, Hühnerfuss H (1989) The damping of ocean waves by surface films: A new look at an old problem. J Geophys Res (94): 6251–6265

Bulatov MG, Kravtsov YA, Lavrova OY, Litovchenko KT, Mityagina MI, Raev MD, Sabinin KD, Trokhimovskii YG, Churyumov AN, Shugan IV (2003) Physical mechanisms of aerospace radar imaging of the ocean. Physics-Uspekhi 46 (1): 63–79

Ginzburg AI, Kostianoy AG, Krivosheya VG, Nezlin NP, Soloviev DM, Stanichny SV, Yakubenko VG (2002) Meso-scale eddies and related processes in the northeastern Black Sea. J Mar Syst 32: 71–90

Kostianoy AG, Litovchenko KT, Lavrova OY, Mityagina MI, Bocharova TY, Lebedev SA, Stanichny SV, Soloviev DM, Sirota AM, Pichuzhkina OE (2006) Operational Satellite Monitoring of Oil Spill Pollution in the Southeastern Baltic Sea: 18 Months Experience. Environmental research, engineering and management 4 (38): 70–77

Lavrova OY (2005) Slicks as indicators of vorticity in coastal zones. Proc 31st International Symposium on Remote Sensing of Environment, 20–24 June 2005, Saint Petersburg, Russia

Lavrova, OY, Bocharova TY, Mityagina MI (2003) SAR observations of typical phenomena in the Black Sea shore area. Proc International Geoscience and Remote Sensing Symposium (IGARSS'03), Toulouse, France, vol 2 pp 966 – 968

Lehmann A (1995) A three-dimensional baroclinic eddy-resolving model of the Baltic Sea. Tellus A 47: 1013–1031

Poulain PM, Barbanti R, Motyzhev S, and Zatsepin A (2005) Statistical description of the Black Sea near-surface circulation using drifters in 1999–2003. Deep Sea Res. I 52: 2250–2274

Zatsepin AG, Ginzburg AI, Kostyanoy AG, Kremenetskiy VV, Krivosheya VG, Poyarkov SG, Ratner YB, Skirta AY, Soloviev DM, Stanichny SV, Stroganov OY, Sheremet NA, Yakubenko VG (2002) Variability of water dynamics in the northeastern Black Sea and its effect on the water exchange between the near-shore zone and open basin. Oceanology 42 suppl: 1–15

Zatsepin, AG, Ginzburg AI, Kostianoy AG, Kremenetskiy VV, Krivosheya VG, Stanichny SV, and Poulain PM (2003) Observations of Black Sea mesoscale eddies and associated horizontal mixing. J Geophys Res 108 (C8): 3246, doi:10.1029/2002JC001390

Zhurbas VM, Stipa T, Malkki P, Paka V, Golenko N, Hense I, Sklyarov V (2004) Generation of subsurface cyclonic eddies in the southeast Baltic Sea: observations and numerical experiments. J Geophys Res 109 (C05033) doi:10.1029/2003JC002074

Multi-Sensor Remote Sensing of Coastal Discharge Plumes: A Mediterranean Test Site

Martin Gade[1] and Vittorio Barale[2]

[1] Institut für Meereskunde, Universität Hamburg, Hamburg, Germany
[2] Institute for Environment and Sustainability, Joint Research Centre, European Commission, Ispra, Italy

Abstract. Various spaceborne sensors have been used to assess environment features in the north-western Mediterranean Sea, at a test site along the Catalan coast, between the Ebro river delta and the greater Barcelona area. The aim was to demonstrate that the combination of different kinds of data allows for an improved monitoring potential, in particular for applications to coastal zone management. The sample imagery considered for this task was acquired by the ERS-2 Synthetic Aperture Radar (SAR), the Along-Track Scanning Radiometer (ATSR) and the Sea-viewing Wide Field-of-view Sensor (SeaWiFS). By combining different data, it proved possible to overcome the specific drawbacks of each sensor, like insufficient temporal coverage, or dependence on weather and daylight conditions. Within the target area, three main features, visible on many of the analyzed images, were selected as test cases. The Ebro river plume, as seen by SeaWiFS, presents a high load of water constituents, changing seasonally and inter-acting with offshore dynamics. The plume system in the Barcelona area, mostly due to the Llobregat river, exhibits similar traits and high surface-active compounds, so that it is detected by both the SeaWiFS and the SAR, as well as low surface temperature, detected by the ATSR. A smaller plume of cooling water, released from a nuclear power plant, cannot be detected in the optical or thermal images, but causes surface turbulence in the coastal zone, giving rise to signatures detected by the SAR.

1. Introduction

Coastal zones occupy less than 15% of the Earth's land surface, but they accommodate more than 60% of the world's population (EEA, 1999). If the trend continues, this number will rise up to 75% in 2025 (UNCED, 1992), thus causing an increasing pressure on coastal zones by urban and

475

V. Barale, M. Gade (eds.), *Remote Sensing of the European Seas.*
© Springer Science+Business Media B.V. 2008

industrial developments. Land use along European coastlines has already manifested itself, in the past, with a higher level of pollution, be it by direct release of urban or industrial waste water, by indirect release via rivers, or by spillages from ships travelling to or from major harbours.

Passive Remote Sensing (RS) in the visible/infrared spectrum can be used to quantify surface optical properties, which depend on the concentration of water constituents (phytoplankton, suspended sediments, and yellow substance), or surface temperature, which describes selected physical processes in the sea. Sensors working in this spectral range provide high temporal resolution and spatial coverage, but are limited to cloud-free conditions and, moreover, to large-scale assessments, because of their low spatial resolution. Nevertheless, they have proven to be a useful data source for near real-time monitoring and long-term comparisons of environmental dynamics, as *e.g.* in the case of algae blooms (Mobley 1994). Active RS in the microwave spectrum can be used for the evaluation of surface roughness and elevation, which are linked to surface phenomena such as winds, waves, wakes, slicks and to dynamical topography, respectively. These sensors, albeit working with low temporal resolution and spatial coverage, have proven to be a powerful tool to monitor small-scale features such as marine oil pollution (Gade *et al.* 1998).

In the following, a collection of images generated by various spaceborne sensors will be used to appraise the coastal environment of the north-western Mediterranean Sea, along the Catalan coast. The aim is to demonstrate, with some simple examples, that the combination of different kinds of data, generated by complementary RS techniques, allow for an improved monitoring potential for applications to coastal zone management.

2.   Test site and data sets

The test site, chosen to demonstrate the use of multi-sensor RS data for monitoring the coastal environment, is the Catalan near-coastal region. It is located in the north-western Mediterranean Sea, between the Spanish mainland and the Balearic islands, and comprises a stretch of the Costa Dorada, from the Ebro river delta in the south-west, to the greater Barcelona area, in the north-east (Figure 1). Such a target area is well suited for studies of the increasing anthropogenic pressure on European coastal zones, because it suffers from the impact of various potential sources of pollution:

- a large river (Ebro) delta with an extensive agricultural outflow;
- a small river (Llobregat) mouth collecting urban and industrial runoff;
- major urban developments (*e.g.* Barcelona);

- major port facilities (*e.g.* Tarragona) and shipping lanes terminals;
- onshore and offshore industrial facilities (*e.g.* power plants and oil rigs);
- mass tourism facilities (along the Costa Dorada).

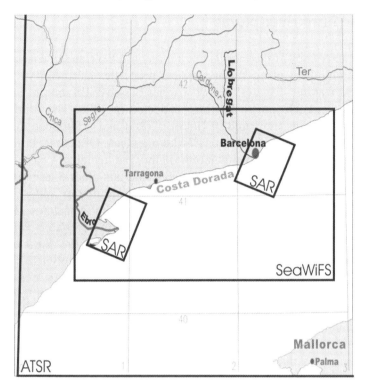

Fig. 1. Map of the north-western Mediterranean region, between 0°E and 3°E, and 39.5°N and 42.5°N. The rectangles show the coverage of the satellite data used (SAR, SeaWiFS and ATSR).

A set of images acquired by the SAR and the ATSR aboard ERS-2, as well as by SeaWiFS aboard OrbView-2 (previously know as SeaStar), was selected to cover the target area above. Particular effort was spent to find data sets not only covering the same geographical district, but also acquired by the various sensors within short time periods. The goal of the analysis and inter-comparison of such data sets was to acquire a better understanding of the coastal and marine processes under investigation and of their imaging by different sensors working in different spectral ranges.

Focusing on the Catalan coastal test site, many of the analysed images revealed a wealth of features that document the anthropogenic pressure put on the local environment, at the interface between drainage basin and open sea. Coastal plumes, in particular, were selected for a more detailed analysis, because of their different origin, visibility, and influence on the local marine ecosystem, and will be presented herein.

3.   The Ebro river plume

The characteristics of the marine environment in the Catalan coastal region can be highlighted using surface optical properties as tracers of its main ecological features. Figure 2 shows a series of SeaWiFS images of the target area (about 200 km × 260 km) collected in 1998. The original data were processed, as indicated in Barale (2005) and references therein, to derive concentrations of chlorophyll-like pigments. Of course, large uncertainties can arise in the computation of this parameter in coastal waters, owing to the presence of optically active materials other than phytoplankton and related pigments (*i.e.* dissolved organic matter and suspended inorganic particles) with partially overlapping spectral signatures. However, by restricting the analysis to patterns and gradients of water constituents in general, rather than considering absolute values of the planktonic pigment concentration, the SeaWiFS imagery can be effectively used to assess coastal anthropogenic loads, runoff areas and pollution sources, even erosion patterns and sediment transport, as well as biological production, monitoring both changes in space and evolution in time (as not possible with the smaller and less frequent coverage of the SAR).

In the series of Figure 2, one image per month was selected to give an indication of the seasonal changes that can occur in the target area. The variability is most noticeable when considering the background water constituents concentration, which appears to be higher in the cold season and lower in the warm one. This is essentially due to the fact that, for the region considered, biological production – the main source of pigments in the open sea – is limited primarily by nutrient availability. Hence, it will be greater when the supply of nutrients from coastal runoff and deep waters, by virtue of vertical mixing, is maintained by the climatic conditions (rain over the continental drainage basins, low temperatures and strong wind regime over marine areas) of the Mediterranean winter (Barale 2005). The same seasonal variations are also displayed by the near-coastal features – the plume originating from the Ebro delta, in particular – superimposed on the background optical signal. Accordingly, the main river plume appears to be larger in winter and smaller in summer, and highly dynamical.

Various factors concur in shaping the dynamical features observed along the Catalan coast, namely a large input of nutrients/sediments of continental origin; several special sites (river mouths, urban centers) where runoff concentrates, generating high concentrations of water constituents; a dynamical current system, interacting with coastal plumes. Sometimes, the structures developing along the coast appear to be confined close to shore and to be separated from the offshore patterns. More often, however, the main plumes and filaments are seen to be entrained in the meso-scale circulation, due to eddies or current meanders, which contribute to the mixing of inshore and offshore waters. As will be seen in the following, the same happens episodically also to the minor plumes of this area.

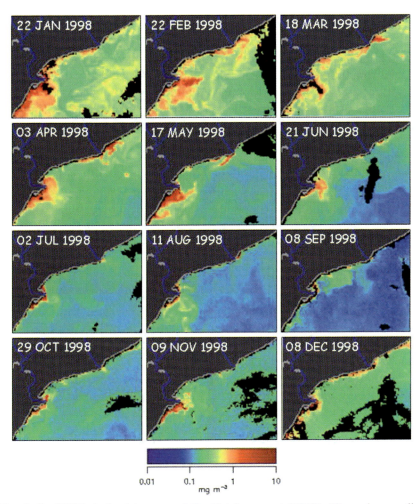

Fig. 2. SeaWiFS-derived images of the Catalan coast (1998). The colour coding indicates chlorophyll-like pigment concentration (co-varying with other water constituents in near-coastal areas). Geographical location in Figure 1.

4. The plume system in the Barcelona area

The mouth of the river Llobregat was chosen as a suitable area for more detailed observations, in order to provide an example of multi-sensor investigations of plume dynamics and impact on the local ecosystem. The river Llobregat originates from the south-eastern Pyrenees and is 170 km long. Its mouth is located south-west of Barcelona, where the original marshy terrain of the Llobregat delta has been transformed into urbanised and industrial areas during the past decades. The growth of population and industrial development in the Llobregat valley has caused an increasing load of pollution within the river outflow. The river plume entering the Mediterranean is a regular feature of the coastal zone near Barcelona.

Data from the ERS-2 SAR, ATSR and SeaWiFS were analysed to detect the surface signature of the Llobregat's plume. Figure 3 shows a series of sub-sections of the SAR images (30 km × 40 km) off Barcelona. The pixel size, 12.5 m × 12.5 m, allows to detect small-scale features, around a few hundred metres in size. All images were acquired in the same year, 1998, but during different seasons and under different atmospheric (wind and temperature) conditions. The city of Barcelona can be seen in the left half of every image, as a light grey area of enhanced radar backscattering, roughly between the river Llobregat in the south and the (much) smaller river Besòs in the north. Different wind speeds in the area caused different sea surface roughness, resulting in the different gray levels of the panels. On Panels (a) and (b), acquired on 11 January and 15 February, respectively, the plumes of both rivers can be seen as dark patches reaching from the river mouths into the open sea. If oily substances within the river out-flow were floating on the water surface, the radar backscattering would be reduced, thus causing dark (irregular) patches. The surface films, in fact, dampen the small-scale surface waves, which in turn result in reduced radar backscattering (Gade et al. 1998).

The Llobregat plume can be observed on every SAR image analysed, whereas the plume of the river Besòs is visible only in winter data. Given its small size, this is possibly due to the fact that it is only in the wet (winter) season that the Besòs presents a discharge large enough to have an impact on the coastal waters. Note that in some cases the plumes are driven by the local current towards south-west (see panels (a) and (f) in Figure 3), and in other cases they remain diffuse and patchy while floating off the coast (panels (b), (c), and (d) in Figure 3). The large dark area in the upper part of panel (f) is likely to be caused by atmospheric effects (low wind speed due to wind shadowing), rather than by any river runoff.

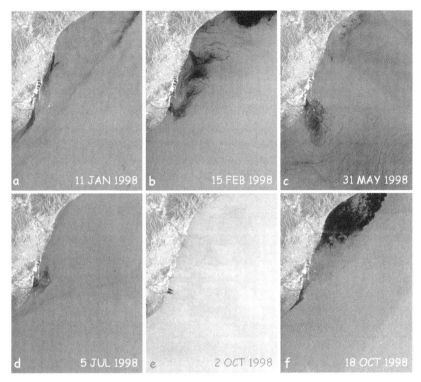

Fig. 3. Subsections (30 km × 40 km) of ERS SAR images of Barcelona and the mouth of the river Llobregat. The Llobregat plume is visible on every image as a dark patchy area. Geographical location in Figure 1.

Surface-active material floating on the river surface, like any kind of oily substance, would be the most probable cause of the permanently visible river plumes. A high load of such material may be a result of municipal or industrial waste water, but may also be caused by high agricultural productivity. Other phenomena such as a significantly reduced sea surface temperature (SST), wind shadowing, or local turbulence can also cause a reduction in the radar backscattering. However, because of the shape of the observed patches, their sharp edges, and their visibility throughout the year, we conclude that the main effect of the river plume, which causes signatures in SAR imagery, is the high load of surfactants (or, more generally, dissolved organic matter).

A multi-sensor approach allows us to gain insight into the influence of river outflow on the local marine ecosystem. It has been seen already that SeaWiFS data can be used to track the evolution of the plumes and their variation in time. In Figure 2, both the Llobregat and the Besòs plumes can be seen in the February, March and possibly April images (again, when the rivers are likely to have a substantial discharge). At other times during the year, only the Llobregat plume is detectable, as it interacts with the coastal current in a number of ways. More details of such interaction, on very short time scales, appear in Figure 4, where two sequences of three (almost) consecutive SeaWiFS images are shown. The images were acquired within only three weeks, on 4, 5, 9 July and on 20, 22, 26 July, 1998.

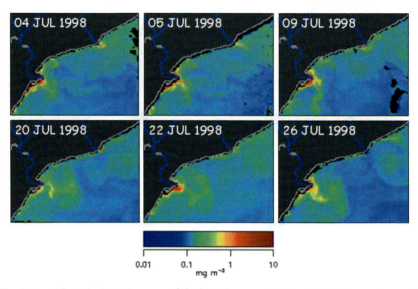

Fig. 4. SeaWiFS-derived images of the Catalan coast (July 1998). The colour coding indicates chlorophyll-like pigment concentration (co-varying with other water constituents in near-coastal areas). Geographical location in Figure 1.

In the earliest image, dated 4 July, the Llobregat plume starts diffusing offshore, possibly as the result of a (rain-driven, as suggested by the thermal image below) outburst of runoff (see also the image dated 2 July, in Figure 2). The spreading continues on the next day, 5 July, but is already gone on 9 July, when a prevailing coastal current from the north-east compresses the plume along the coast, south of the river mouth (displacing even the much larger Ebro river plume toward the south-west). Later on, the 20 July image presents the opposite situation, with a prevailing coastal current flowing from the south-west, compressing the plume along the coast north of the river mouth (and displacing the Ebro plume to the north-east). Interestingly, this situation is maintained also in the following days, 22 and 26 July, even if the costal current is probably changing (as testified by the Ebro plume behavior). The likely reason for this is the offshore anticyclonic eddy, seen off the Barcelona area, which moves closer to shore and causes a northward drift of all coastal waters.

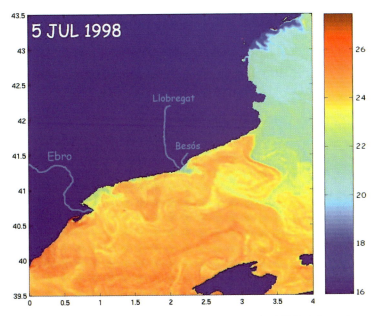

Fig. 5. ATSR-derived image of the Catalan coast (5 July 1998). The colour coding represents SST in °C. Geographical location in Figure 1.

The combination of data from different sensors is of particular interest on 5 July, 1998, when ERS-2 SAR data (acquired at 10:35 UTC) and SeaWiFS data (12:48 UTC) are available (Figures 3 and 4), together with data from the ATSR (21:53 UTC). The Sea Surface Temperature (SST) map derived from ATSR data is shown in Figure 5. Aside from the meso-scale eddies in the southern half of the image and cold upwelling in the north-eastern section, the Llobregat river plume is clearly delineated as an area of lower SST. Note that the plume looks spatially similar in both ATSR and SeaWiFS (Figure 3) images is well correlated, whereas the area of reduced radar backscatter in the SAR image (panel (d) in Figure 2) is much smaller. Therefore, it can be concluded that the observed plume is due to cold, turbid, fresh waters of fluvial origin spreading in a laminar way on top of denser (salty) marine water, whereas an enhanced accumulation of surface films is observed only over a short distance from the river mouth. Note that the Ebro plume is well visible in the SeaWiFS imagery (Figure 4), but it cannot clearly be identified on the SST map (Figure 5), except in the near-coastal area north of the delta. This suggests that the bulk of the Ebro runoff is not coming from a local outburst of rain water (as was suggested for the Llobregat runoff), but rather is composed of "older" water that warmed considerably while flowing through the delta area.

5.   The "Vandellòs II" industrial plume

The area between the Ebro delta, in the south, and l'Hospitalet de l' Infant, in the north, also constitutes an interesting test site where irregular patches can be observed in the SAR images (Figure 6).

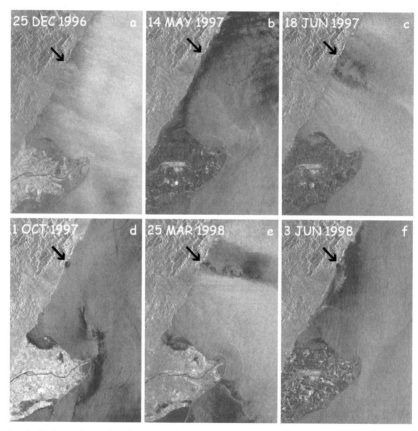

Fig. 6. Subsections (34 km × 51 km) of ERS SAR images of the western Costa Dorada, north of the Ebro estuary. The black arrows denote the location of the nuclear power plant "Vandellòs II". Geographical location Figure 1.

These signatures are typical of local coastal discharges, having a circular shape and being either brighter or darker than the surrounding water surface. Based on information from local sources (J. Redondo, personal communication), it can be safely assumed that the (warm) cooling water of the nuclear power plant "Vandellòs II", located in the area where the observed features reach the land, is producing an elongated outflow plume. Under particular conditions, when the wind is blowing offshore and a considerable amount of cooling water is being released, the plume is visible on SAR imagery, causing the observed features, which may be up to 15 km in length.

Turbulence and/or the upwelling of colder water seem to be causing a reduction in radar backscatter outside of the plume, whereas the plume itself, because of its higher temperature, appears brighter than its surrounding area on the SAR imagery. In the early stage of such a release the plume causes just a small circular dark patch (see *e.g.* panels (a), (b), and (d) of Figure 6), which later changes its appearance into the bright feature described above. The wavy, triangular shape of the plume in panel (c) could be caused by a varying (pulsating) amount of runoff. Although limited to sporadic coverage, the SAR is capable of resolving the power plant runoff, because of its high spatial resolution. The Vandellós plume is visible, in general, neither on the SeaWiFS nor on the ATSR imagery, due to its very limited size, as well as the low contrast in water constituents and in SST, with respect to the surrounding seawater.

6. Conclusions

The multi-sensor satellite imagery presented here provides examples of coastal plumes that can be detected and monitored by at least one of the RS techniques adopted. These plumes may result from river runoff, or from urban or industrial discharges, and manifest themselves in a high occurrence of surface films, an enhanced concentration of water constituents or a change in SST. The simple examples provided for the Catalan coast demonstrate the potential of exploiting complementary RS data for the investigation of runoff impact on coastal zones.

The acquisition of image data by passive sensors working in the visible and infrared bands offers a much higher temporal coverage (up to several images per day) than other techniques. This, coupled with the specificity of the parameters that can be observed and quantified, allows the repeated characterization of a coastal area in terms of its main ecological traits. Basin-wide interactions and seasonal patterns emerge from the analysis of such imagery. On the other hand, the spatial resolution of the data is very low, on the order of 1 km, so that small-scale features become difficult to assess. Moreover, cloud-free weather conditions are needed for monitoring the sea surface, and only daylight data are available using optical sensors.

High-resolution spaceborne SAR proved to be capable of detecting small-scale features, such as minor coastal plumes, but the low temporal coverage of the available SAR data made it difficult to track the single features over time. More recent–as well as future–satellite missions, with wide-swath SAR sensors (*e.g.* ENVISAT, launched in 2002), may be able to fill this gap, even though the resolution of the wide-swath SAR imagery is significantly lower. Finally, the advantage of a SAR sensor as being independent of daylight and cloud conditions is somewhat balanced, on the negative side, by the fact that the visibility of the observed effects, namely of the accumulation of surface-active material on the water surface within a plume, depends very much on the local weather conditions, and predominantly on the local wind speed. These results show that different RS techniques complement each other so that an improved tracking (be it in space or in time) of coastal processes can be achieved.

Acknowledgements

Thanks are due to J. Scholz and F. Melin for the role they had in producing the data sets used here, and to Helen Snaith for providing the ATSR data. Parts of this work were supported by the European Commission under Contract No. ENV4-CT96-0334 (Clean Seas).

References

Barale V (2005) Satellite observations as indicators of the health of the Mediterranean Sea. In: Saliot A (ed) "The Mediterranean Sea", The Handbook of Environmental Chemistry, Volume 5 Water Pollution, Part K. Springer-Verlag, Berlin, Heidelberg, pp 387–408

European Environment Agency (EEA) (1999) Environment in the European Union at the turn of the century, EEA, Copenhagen, pp 446

Gade M, Alpers W, Hühnerfuss H, Masuko H and Kobayashi T (1998) The imaging of biogenic and anthropogenic surface films by a multi-frequency multi-polarization synthetic aperture radar measured during the SIR-C/X-SAR missions. J Geophys Res 103: 18851–18866

Keller WC, Wismann V, Alpers W (1989) Tower-Based Measurements of the Ocean C Band Radar Backscattering Cross Section. J Geophys Res 94: 924–930

Mobley CD (1994) Light and water. Radiative Transfer in natural waters. Academic Press, San Diego

United Nations Conference on Environment and Development (UNCED) (1992) Agenda 21: the Rio Declaration on Environment and Development. Rio de Janeiro, Brazil, 3–14 June 1992

Sea Ice Monitoring in the European Arctic Seas Using a Multi-Sensor Approach

Stein Sandven

Nansen Environmental and Remote Sensing Center, Bergen, Norway

Abstract. Advances in satellite remote sensing of sea ice and icebergs in Arctic regions are described in case studies showing the benefits of using multi-sensor observations. It is demonstrated how Synthetic Aperture Radar (SAR) used in combination with optical images can improve discrimination of open water, nilas, young ice and three gradations of deformed first-year ice. The classification method is based on multi-sensor data fusion and neural network, where *in situ* observations were used for training of the algorithm. Synergetic use of scatterometer and passive microwave (PMW) data is well-established to estimate large scale ice motion, but in straits and marginal seas more detailed ice drift data are needed. In the Fram Strait SAR images from ENVISAT have been used to estimate ice drift and ice area flux since early 2004. It is demonstrated that SAR wide-swath images can provide more accurate and higher-resolution ice drift vectors compared to scatterometer and PMW data. Methods for retrieval of thickness for thin ice are available using thermal infrared, passive microwave and SAR data, but these methods are research-oriented and not used in regular monitoring. Laser and radar altimeter measurements from satellites have shown promising capability to observe sea ice freeboard and thickness for ice thicker than about one meter. Such data used in combination with ice drift and ice types will provide new estimates of ice volume variability and fluxes. SAR and optical data have also been used to observe icebergs in the Barents Sea. The two data types are complementary and can improve iceberg detection if they are used in combination.

1. Introduction

The most mature satellite method for sea ice observation is large-scale ice concentration and ice extent measurements by passive microwave (PMW) radiometer data. A number of retrieval algorithms have been developed and intercompared over the last three decades (*e.g.* Andersen *et al.* 2006)

V. Barale, M. Gade (eds.), *Remote Sensing of the European Seas.*

It is generally agreed that the accuracy in ice concentration and extent estimates from PMW data is in 5–10% except for the melt season. Time series of ice extent and area from these data show that the total ice area in the Arctic has been reduced by 3–4% per decade since 1978, while the multi-year ice area has been reduced by 7% per decade in the same period (Johannessen *et al.* 1999, 2004).

Observation of sea ice has improved significantly by use of satellite Synthetic Aperture Radar (SAR) images, providing detailed information on ice types, ice motion, ice edge, leads, polynyas, ridges, fast-ice boundaries, freezing and melting. In the last decade extensive SAR data sets have been used for ice monitoring in many ice-covered areas (*e.g.* Kwok and Cunningham 2002; Johannessen *et al.* 2007) An overview of remote sensing data used for observation of sea ice variables is given in Table 1.

Table 1. Sea ice variables and remote sensing sensors.

Ice variable	Remote sensing data	Products
Area, extent and concentration	PMW data Scatterometer data optical/IR data	Global maps are available daily from SSMI, AMSR-E, and scatterometer data. Regional ice charts using SAR, optical and IR data
Ice thickness	Radar altimeter/Laser altimeter, IR, SAR	Large-scale maps are provided by ERS and IceSat data. Improved products from CRYOSAT after 2009. Thin ice retrieval from IR and SAR data
Ice drift	PMW data Scatterometer data SAR wideswath and Global Mode data	Large-scale ice drift products are from scatterometer and PMW data for the non-melt season. SAR-based ice drift is available for selected regions.
Ice-snow albedo and surface temp	Optical/IR images	Research activity, pathfinder data sets are produced
Ice type classification	Scatterometer, SAR and PMW	Multi-year and first-year products are available, young and first-year ice can be retrieved from SAR
Ice roughness and deformation	Radar and laser Altimeter, SAR	Ice roughness maps are shown using Icesat data. SAR can provide data on deformation
Icebergs in the Arctic	High resolution optical and SAR images	Maps are produced as part of sea ice services. Detailed mapping with high-resolution images is a research activity

This chapter presents four examples of improved retrieval of sea parameters using the multi-sensor approach: ice classification, ice drift and flux retrieval, ice thickness and iceberg detection.

2. Ice classification by synergy of SAR and optical data

A number of ice classification methods using SAR and other satellite data have been developed in the last two decades, but robust classification methods have not yet been developed and few algorithms are used in operational monitoring systems (*e.g.* Tsatsoulis and Kwok 1998; Onstott and Shuchman 2004; Johannessen *et al.* 2007) In practical ice analysis of SAR images, human interpretation is mostly used for classification of ice types and ice processes. Recent results on SAR ice analysis and classification have been published by *e.g.* Bogdanov *et al.* (2007).

It is demonstrated here that combining two types of SAR images with optical images can improve the classification of five-six ice types (*i.e.* open water, grease ice, nilas, young ice, undeformed and deformed first-year ice) The example shown in Figure 1 is from a typical winter situation in the Kara Sea, where the ice cover consists of various levels of new, young and first-year ice. The images obtained over the same area on the same day are ERS-2 SAR, RADARSAT ScanSAR, and a Meteor 3/5 optical image. The pixels in the RADARSAT image were normalized in range to be comparable to the ERS pixels, compensating for the difference incidence angle of the two SAR images. The study was done during a field experiment with the nuclear icebreaker Sovetsky Soyuz, collecting *in situ* observations of ice types in the study area (Bogdanov *et al.* 2005).

The ice in the study area consisted mostly of medium and thick first-year ice, and coastal polynyas with thin ice (Figure 1) New and young ice has low radiance in visual imagery, which increases as the ice becomes thicker. The dark signature of the SAR images in region A corresponds to grease ice and nilas in a refreezing polynya. The dark area in the lower right corner of the optical image is also a polynya consisting of grey or grey-white ice with thickness of 10–20 cm. The bright signature in the SAR images in region B corresponds to greyish signature in the optical image, indicating that this is young ice with thickness up to 30 cm. The ice floes in area C are undeformed first-year ice with low backscatter in the SAR images and high reflectance in the optical image due to the snow cover on top of the ice. Area D has mainly first-year ice of medium deformation, mixed with some young ice with brighter SAR signature than the first-year ice. Area E contains highly deformed first-year ice with higher backscatter than medium deformed first-year ice. The optical image cannot distinguish between different levels of deformation since the signal depends mainly on the albedo of the snow cover.

The classification method used in this example is based on a multi-sensor data fusion algorithm described by Bogdanov *et al.* (2005) where a number of pre-defined ice classes with training data sets for each of the classes were used as input to the algorithm. The six sea ice classes were: smooth, medium deformed and deformed first-year ice, young ice, nilas, and open water. The results from classification of the two SAR images are shown in Figure 2a while the results from the classification with the optical image included are shown in Figure 2b.

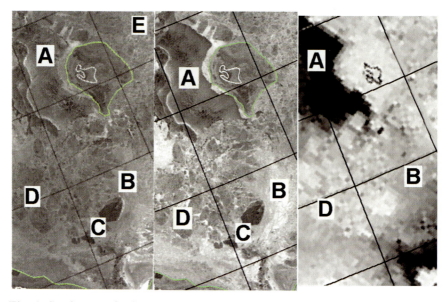

Fig. 1. Sea ice area in the Kara Sea mapped by three types of satellite data on 30 April 1998. Left: ERS-2 SAR; middle: RADARSAT ScanSAR Wide; and right: optical image from Meteor 3/5. The coastline surrounding the island in the upper part of the image and the fast-ice border in the lower part of the image are superimposed on the SAR images. The main ice types in the images are A: nilas, B: young ice, C: first-year level ice, D: first-year medium deformed ice, and E: first-year very deformed ice.

The young ice and first-year ice classes are quite similar in the tworesults, whereas the main improvement is found for the nilas where inclusion of the optical image makes a significant impact. The large nilas in the upper left region (area A in Figure 1) and the smaller nilas in the lower right corner are more correctly classified. The open water class is not found within the study region. To quantify the classification results, a number of test data sets were selected for each of the six classes. The percentage correct classification increased from 75% to 85% by including the optical image. The possibility to improve ice classification further is possible by including infrared radiometer data providing temperature measurements and SAR data with multipolarization and other frequencies.

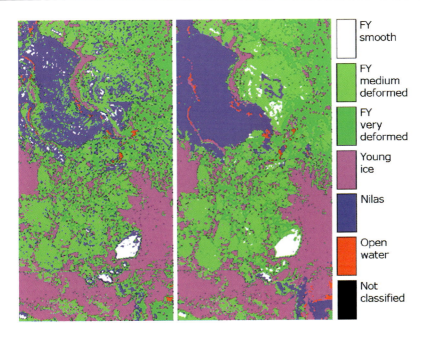

	FY smooth
	FY medium deformed
	FY very deformed
	Young ice
	Nilas
	Open water
	Not classified

Fig. 2. Results of ice classification based on the multilayer neural network method by Bogdanov *et al.* (2005) Left: classification results based on ERS and RADARSAT SAR data; right: classification results after including the optical image from Meteor.

3.   Ice drift in the Fram Strait from active and passive microwave data

Ice drift retrieval from satellite sensors have been developed in the last two decades, using NOAA AVHRR (*e.g.* Emery *et al.* 1991), SSM/I and scatterometer (Gohin *et al.* 1998) as well as high resolution sensors such as SAR (*e.g.* Kwok 1998) The most common algorithms to retrieve ice drift from satellite data are based on the area-correlation principle (*e.g.* Fily and Rothrock 1987), where the correspondence is established by matching image patches using their correlation coefficient.

Ice drift in the Fram Strait has been estimated using PMW data with rather coarse resolution since 1978 (Kwok 2004) In this study, wideswath SAR data have been used to estimate ice drift and ice area flux with repeated observations every three days (Figure 3) From February 2004, ice area flux profiles across 79 N have been calculated using ice drift from

SAR and ice concentration profiles from PMW data. The ice drift time series shows large short-time variability superimposed on a seasonal cycle, with typical mean velocity is $0.20 \ ms^{-1}$ in winter and $0.10 \ ms^{-1}$ in summer.

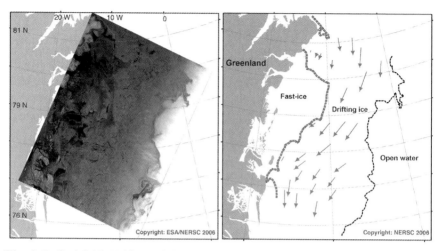

Fig. 3. Left: ASAR Wideswath image from 14 April 2006 covering the sea ice in the Fram Strait; right: Ice drift vectors retrieved from SAR images on 14 and 17 April 2006. The dashed line near the coast of Greenland represent the boundary of the fast-ice, while the dashed line on the right side is the ice edge boundary. Both boundaries are retrieved from the SAR images.

Monthly mean area flux has been estimated from February 2004 to September 2006 as shown in Figure 4. There is a maximum flux in the winter months and a minimum in the summer months, which is in agreement with previous studies (*e.g.* Kwok *et al.* 2004) The annual area flux from the SAR data is estimated to be $0.73 \times 10^{6} \ km^{2}/year$ for 2004 and $0.66 \times 10^{6} \ km^{2}/year$ for 2005. Area fluxes retrieved from PMW data from 1978 to 2002 showed a mean value of $0.86 \times 10^{6} \ km^{2}/year$ (Kwok *et al.* 2004) The SAR-based ice drift was estimated using a correlation algorithm (*e.g.* Fily and Rothrock 1987) as well as manual analysis to obtain maximum accuracy of the ice drift vectors, estimated to be about 10% in this region. The SAR ice drift vectors have been interpolated to a 20 km grid at the 79°N latitude in order to resolve the cross strait profile of the ice drift field. The SAR ice drift retrieval is used as a supplement to large-scale ice drift from the scatterometer and PMW data which are only available for the winter months. SAR-based ice drift will also be used to validate ice drift from the PMW and scatterometer data.

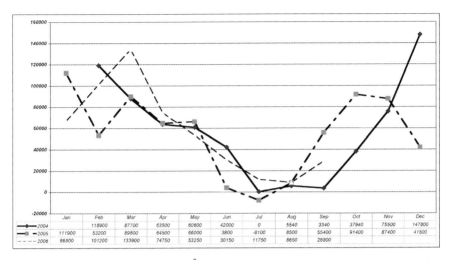

	Jan	Feb	Mar	Apr	May	Jun	Jul	Aug	Sep	Oct	Nov	Dec
2004		118900	87700	63500	60600	42000	0	5540	3340	37940	75500	147800
2005	111900	53200	89800	64500	66000	3800	-8100	8500	55400	91400	87400	41500
2006	66800	101200	133900	74750	53250	30150	11750	8650	28800			

Fig. 4. Monthly ice area flux (km^2 per month) from SAR ice drift and SSMI ice concentration.

4.   Retrieval of thickness and related sea ice parameters

Thickness is the most difficult sea ice parameter to observe from satellites and also the parameter with most severe lack of observations. Available data on ice thickness are mainly obtained by non-space methods using upward looking sonars, electromagnetic induction sensors, ice buoys and other *in situ* measurements techniques.

A method to retrieve ice thickness from thermal infrared (IR) satellite images have been developed for winter conditions for thickness up to about 0.5 m (Drucker *et al.* 2003) The IR-method is not very useful for regular observation of ice thickness because of cloud cover limitations. To study the evolution of ice thickness through the whole freezing season, the possibility to use SAR data has been investigated (Kwok and Cunningham 2002) A SAR algorithm based on the RADARSAT Geophysical Processor System (RGPS) has been developed to derive the distribution of thin ice kinematically. By following a large number of ice trajectories throughout an ice growth season, the RGPS tracks the area changes of each Lagrangian cell (Kwok 1998) The ice thickness retrieval uses assumption about freezing rates and ridging due to the deformation of the ice cover. Ice thickness from SAR images have been compared with IR retrievals in the Beaufort Sea and the Canadian Arctic, showing good agreement for new ice being formed in leads (Yu and Lindsay 2003) However, both methods have their limitations due to assumptions about snow cover, freezing rates and the amount of ridging as a result of ice deformation.

Russian ice scientists have analyzed AVHRR image since 1979 for thin ice thickness retrieval (Bushuev *et al.* 2007) An example of ice thickness map in the eastern Barents Sea is shown in Figure 5 where validation data from coastal stations and ship observations show generally good agreement with the satellite retrievals.

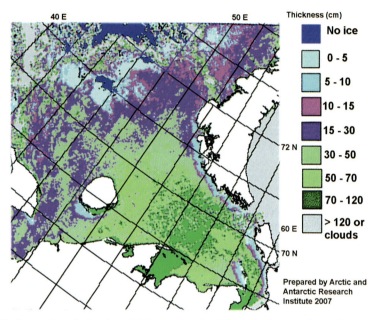

Fig. 5. Example of the ice thickness in the Barents Sea derived from *NOAA* AVHRR IR image on May 4, 2001, 08:20GMT, prepared by Arctic and Antarctic Research Institute (Johannessen *et al.* 2007).

Retrieval of thin ice thickness from SSM/I data, as complementary method to the IR technique, has been demonstrated by Martin et al (2004) in the Bering Sea. These studies showed that 10–20 cm thin ice in re-freezing polynyas could be determined using the vertical to horizontal ratio of the 37 GHz brightness temperature. This method has been adapted to AMSR-E data to improve the resolution of thin ice thickness estimation and heat flux retrieval in polynyas (Martin *et al.* 2005) Use of SAR data has been an important part of recent heat flux and thin ice studies, because SAR can provide accurate definition of polynya areas and estimates of polynya evolution.

A promising new technique to retrieve sea-ice thickness from satellite-borne altimeter data has been developed and tested on ERS-1/2 altimeter data

(Laxon *et al.* 2003) Also laser altimeter data have shown capability to provide synoptic data on ice thickness across the Arctic Ocean (Kwok *et al.* 2006) CryoSat, scheduled for launched in 2009, will carry an improved radar altimeter that has ice thickness measurements as a key objective.

5. Observing icebergs in Arctic ice-covered seas

Until now, satellite data can only observe large icebergs, typically 100 m or more in horizontal extent, under favourable cloud, wind and sea ice conditions. Studies on icebergs detection in satellite images have been conducted in several areas in the Northern hemisphere, such as in eastern Canada (Power et al. 2001), off the coast of Greenland (*e.g.* Gill 2001) and in the northern Barents Sea (Kloster and Spring 1993; Sandven *et al.* 1999)

In this study the iceberg detection capability of ENVISAT ASAR alternating polarization images (AP) and RADARSAT ScanSAR Narrow (SN) mode is compared with Landsat optical images in April 2006. A selection of 15 icebergs of size from 50 to 400 m were identified in a subset of a Landsat panchromatic image (pixel size 15 m) covering the southern part of the Franz Josef Land archipelago (Figure 6a). The subimage covered about 30 by 30 km and overlapped with the SAR AP image (Figure 6b). It should be mentioned that several hundred icebergs could be identified in the whole Landsat image, but only a small subset was selected for further analysis. In optical images icebergs are identified as bright objects against a less bright background combined with a dark shadow due to low sun angle and height of the iceberg. In SAR images icebergs are also identified as bright objects, but detection capability depends on the backscatter of the surrounding sea ice or open water.

The HH image had better signal-to-noise ratio for all 15 icebergs compared to the VV image. Both HH- and VV-image showed many bright spots that were not identified as any object in the Landsat image. A comparison with a RADARSAT ScanSAR Narrow image with pixel size of 25 m and HH polarization was also done. Four of the smaller icebergs could not be identified, suggesting that 25 m pixel size in the RADARSAT image compared to the 12.5 m pixels in the AP image makes a difference.

For iceberg detection in the Barents Sea region, image resolution both in SAR and optical images should be better than 10 m. Further studies are needed to compare iceberg size and shape data from ship and aircraft with satellite data. Since future SAR systems will have different polarization options, it is important to determine which SAR modes should be used for iceberg detection.

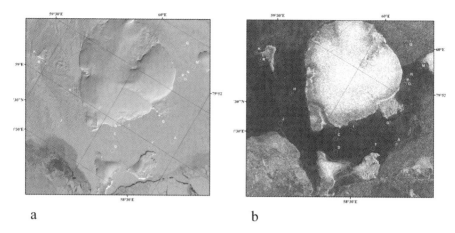

a   b

Fig. 6. (a) Landsat subimage from the southern part of Franz Josef Land region obtained on 14 April 2006. A selection of icebergs with horizontal scale from 50 to 400 m are identified and marked by circles (A – R); (b) ENVISAR ASAR Alternating Polarization image from 12 April 2006, available in both HH- and VV-polarization. The subimage covers the same areas as the Landsat image and the same icebergs are indicated as in (a).

6. Conclusions

Polar orbiting satellites produce an increasing amount of data from several sensors that can be exploited in sea ice observation. In this study, four examples of multi-sensor observations of sea ice and icebergs parameters have been described, demonstrating the benefit of combining data from several sensors to produce higher quality sea ice information. For regional monitoring of ice concentration and ice types, use of high–resolution radar and optical/infrared images can improve the quality of ice classification. For ice drift estimation, synergetic use of scatterometer and PMW data has shown promising results in large-scale mapping, while SAR data are useful for regional mapping of ice drift with higher resolution. Iceberg observations can benefit from the combined use of radar and optical/infrared images, and by increasing image resolution to better than 10 m.

A major task in the coming years will be to exploit new SAR data (RADARSAT-2, ALOS, TerraSAR-X, Sentinel), with multi-polarization capability in combination, altimeter data (IceSat, CryoSat), optical/IR and PMW data, to improve observation of sea ice thickness, motion and fluxes.

Acknowledgements

The studies have been supported by the EU MERSEA Integrated project, the Norwegian SatOcean project, and Hydro Oil and Energy.

References

Andersen S, Tonboe R, Kern S, Schyberg H (2006) Improved retrieval of sea ice total concentration from spaceborne passive microwave observations using numerical weather prediction model fields: An intercomparison of nine algorithms. Remote Sensing of Environment 104: 374–392

Bogdanov A, Sandven S, Johannessen OM, Alexandrov VY, Bobylev LP (2005) Multisensor approach to automated classification using sea ice image data. IEEE Trans Geosci Rem Sens 43 (7): 1648–1664

Bogdanov AV, Sandven S, Johannessen OM, Alexandrov VY, Bobylev LP, Loshchilov VS (2007) Sea ice retrieval algorithms for SAR. In: Johannessen OM *et al.* (eds) Polar Seas Oceanography. Remote Sensing of Sea ice in the Northern Sea Route: Studies and Applications. Praxis Springer 220–243

Bushuev AV, Loshchilov VS, Shcherbahov YA, Paramonov AI (2007) Satellite Remote Sensing of sea ice: Optical and Infrared Imaging. In: Johannessen OM *et al.* (eds) Polar Seas Oceanography. Remote Sensing of Sea ice in the Northern Sea Route: Studies and Applications. Praxis Springer 149–171

Drucker R, Martin S, Moritz R (2003) Observations of ice thickness and frazil ice in the St. Lawrence Island polynya from satellite imagery, upward looking sonar, and salinity/temperature moorings. J Geophys Res 108 (C5): 3149, doi:10.1029/ 2001JC001213

Emery WJ, Fowler CW, Hawkins J, Preller RH (1991) Fram Strait satellite image-derived ice motions. J Geophys Res 96: 4751–4768

Fily M, Rothrock DA (1987) Sea ice tracking by nested correlations. IEEE Trans Geosci Rem Sens GE 25 (5): 570–580

Gill RS (2001) Operational detection of sea ice edges and icebergs using SAR. Canadian J Rem Sens 27 (5): 411–232

Gohin F, Cavanié A, Ezraty R (1998) Evolution of passive and active microwave signatures of a large sea ice feature during its two and half year drift through the Arctic Ocean. J Geophys Res 103 (C4): 8177–8189

Johannessen OM, Shalina EV, Miles MW (1999) Satellite Evidence for an Arctic Sea Ice cover in Transformation. Science 286: 1937–1939

Johannessen OM, Bengtsson L, Miles MW, Kuzmina SI, Semenov VA, Alekseev GV, Nagurnyi AP, Zakharov VF, Bobylev LP, Pettersson LH, Hasselmann K, Cattle HP (2004) Arctic climate change: observed and modeled temperature and sea-ice variability. Tellus, Series A 56A (4): 328–341

Johannessen OM, Alexandrov VY, Frolov IY, Sandven S, Miles M, Bobylev LP, Pettersson LH, Smirnov VG, Mironov EU, eds (2007) Polar Seas Oceanography.

Remote Sensing of Sea ice in the Northern Sea Route: Studies and Applications. Praxis Springer, pp 472

Kloster K, Spring W (1993) Iceberg and glacier mapping using satellite optical imagery during the Barent Sea Ice Data Acquisition Program (IDAP). Proc 12th Int Conf Port and Ocean Engineering under Arctic Conditions, Hamburg. 17–20 August 1993, vol 1, pp 413–424

Kwok R (1998) The RADARSAT Geophysical Processor System. In: Tsatsoulis C, Kwok R (eds) Analysis of SAR data of the Polar Oceans. Springer Verlag, Berlin, pp 235–258

Kwok R, Cunningham GF (2002) Seasonal ice area and volume production of the Arctic Ocean: November 1996 through April 1997. J Geophys Res 107 (C10): 8038, doi:10.1029/2000JC000469

Kwok R, Cunningham GF, Pang SS (2004) Fram Strait sea ice outflow. J Geophys Res 109, C01009, doi:10.1029/2003JC001785

Kwok R, Cunninham GF, Zwally HJ, Yi D (2006) ICESat over Arctic sea ice: interpretation of altimetric and reflectivity profiles. J Geophys Res 111, C06006, doi: 10.1029/2005JC003175

Laxon S, Peacock N, Smith D (2003) High interannual variability of sea ice thickness in the Arctic region. Nature 245: 947–950

Martin S, Drucker R, Kwok R, Holt B (2004) Estimation of the thin ice thickness and heat flux for the Chukchi Sea Alaskan coast polynya from SSM/I data, 1990 – 2001. J Geophys Res,109, C10012, doi:10.1029/2004JC002428

Martin S, Drucker R, Kwok R, Holt B (2005) Improvements in the estimates of ice thickness and production in the Chukchi Sea polynyas derived from AMSR-E. Geophy. Res Lett 32, L05505, doi:10.1029/ 2004GL022013

Onstott RG, Shuchman RA (2004) SAR measurements of sea ice. In: Jackson CR, ApelJR (eds) Synthetic Aperture Radar. Marine User's Manual, National Oceanic and Atmospheric Administration, US Department of Commerce, Washington DC, pp 81–115

Power D, Youden J, Lane K, Randell C, Flett D (2001) Iceberg detection capabilities of RADARSAT Synthetic Aperture Radar. Canadian J Rem Sens 27 (5): 476–486

Sandven S, Johannessen OM, Miles M, Pettersson LH, Kloster K (1999) Barents Sea seasonal ice zone features and processes from ERS-1 SAR. J Geophys Res 104 (C7): 15843–15857

Tsatsoulis C, Kwok R, eds (1998) Analysis of SAR data of the Polar Oceans. Recent Advances. Springer Verlag, Heidelberg

Yu Y, Lindsay RW (2003) Comparison of thin ice thickness distributions derived from RADARSAT Geophysical Processor System and advanced very high resolution radiometer data sets. J Geophys Res 108 (C12), 3387, doi:10.1029/ 2002JC001319, 2003

Acronyms

ABW	Antarctic Bottom Water
ADCP	Acoustic Doppler Current Profiler
ADEOS	Advanced Earth Observing Satellite
AF-FOV	alias-free field-of-view
AIS	Automatic Identification System
ALTICORE	ALTImetry for COastal Regions (project)
AM	Active Microwave(s)
AMSR-E	Advanced Scanning Microwave Radiometer
AOD	Aerosol Optical Depth
ARTIST	Arctic Radiation and Turbulence Interaction Study
ASAR	Advanced Synthetic Aperture Radar
ASCAT	Advanced Scatterometer
ASTER	Advanced Spaceborne Thermal Emission Reflection Radiometer
ATI	Along-Track Interferometric Synthetic Aperture Radar
ATSR	Along-Track Scanning Radiometer
AVHRR	Advanced Very High Resolution Radiometer
AW	Atlantic Waters
BAW	Bundesanstalt für Wasserbau (Federal Waterways Engineering and Research Institute, Germany)
BC	Before Christ
BF	Beam Forming
BSH	Bundesamt für Seeschifffahrt und Hydrographie (Federal Maritime and Hydrographic Agency, Germany)
CASIX	Centre for observation of Air-Sea Interactions and fluxes
CCD	Computer Compatible Disk
CDOM	Chromophoric Dissolved Organic Matter
CDOM	Coloured Dissolved Organic Material
CFAR	Constant False Alarm Rate (detection)
chl	chlorophyll
Chl	Chlorophyll-like pigment concentration
Chl-a	Chlorophyll *a*
CPA	Colour Producing Agents
CPR	Continuous Plankton Recorder
CTD	Conductivity-Temperature-Depth
CZCS	Coastal Zone Color Scanner
DF	Direction Finding
DFT	Discrete Fourier Transform
DLR	Deutsches Zentrum für Luft- und Raumfahrt (German Aerospace Center)

doc	dissolved organic carbon
DOM	Dissolved Organic matter
DWD	Deutscher Wetterdienst (German Weather Service)
E-AF-FOV	extended alias-free field-of-view
EAP	East Atlantic Pattern
EARS	Early Advanced Re-transmission Service
EC	European Commission
ECOOP	EU COastal sea OPerational observing & forecasting
ECMWF	European Centre for Medium-Range Weather Forecast
EEZ	Exclusive Economic Zone
ELF	ENEA LIDAR Fluorosensor
EMT	Eastern Mediterranean Transient
ENEA	Ente per le Nuove tecnologie l'Energia e l'Ambiente (Italian National Agency for New technologies, Energy and the Environment)
ENSO	El Niño / Southern Oscillation
EOF	Empirical Orthogonal Function(s)
EOS	Earth Observation Satellite
ERS	European Remote Sensing Satellite
ESA	European Space Agency
ESRIN	European Space Research Institute
ESTAR	Electronically Steered Thinned Array Radiometer
ESTEC	European Space Research and Technology Centre
EUMETSAT	European Organisation for the Exploitation of Meteorological Satellites
FFT	Fast Fourier Transform
FINO	Forschungsplattformen in Nordsee und Ostsee (German Research Platforms in the North and Baltic Sea)
FLIR	Forward-Looking Infra-Red (camera)
FLS-A	Fluorescent LIDAR System-A
FMCW	Frequency Modulated Continuous Wave
FMICW	Frequency Modulated Interrupted Continuous Wave
FOV	Field-of-View
FYI	First-Year Ice
GAC	Global Area Coverage
GCM	General Circulation Model
GDAC	Global Data Analysis Centres
GDOP	Geometrical Dillution Of Precision
GDR	Geophysical Data Record
GHRSST-PP	GODAE High Resolution Sea Surface Temperature - Pilot Project
GMES	Global Monitoring for Environment and Security
GMF	Geophysical Model Function

GODAE	Global Ocean Data Assimilation Experiment
GOOS	Global Ocean Observing System
GPS	Global Positioning System
GRDC	Global Runoff Data Centre
HAB	Harmful Algal Blooms
HELCOM	HELsinki COMmission
HF	High Frequency
HH	horizontal co-polarisation
InSAR	Interferometric Synthetic Aperture Radar
IOCCG	International Ocean-Colour Coordinating Group
IOP	Inherent Optical Properties
IJRS	International Journal of Remote Sensing
IR	Infra-Red
JPL	Jet Propulsion Laboratory
JRC	Joint Research Centre of the European Commision
K_d	diffuse attenuation coefficient
KNMI	Koninklijk Nederlands Meteorologisch Instituut (Royal Dutch Meteorological Institute)
KSAT	Kongsberg Satellite Services
LAC	Local Area Coverage
LAN	Local Area Network
LASER	Light Amplification by Stimulated Emission of Radiation
LFS-A	Laser Fluoro-Sensor System - A
LIDAR	Light Detection and Ranging
LG	Local Gradient
LIF	LASER Induced Fluorescence
LRIT	Long Range Identification and Tracking
MABL	Marine Atmospheric Boundary Layer
MAST	Marine Science and Technology
MAW	Modified Atlantic Water
MBR	Maximum Band Ratio
MCC	Maximum Cross Correlation
MED	Mediterranean Sea
MERIS	Medium Resolution Imaging Spectrometer
MERSEA	Marine EnviRonment and Security for the European Area
METOP	METeorological OPerational Satellite
MFSTEP	Mediterranean Forecasting System for Environmental Predictions
MHI	Marine Hydrophysical Institute
MIRAS	Microwave Imaging Radiometer by Aperture Synthesis

MLE	Maximum Likelihood Estimation
MMJ	Mid-Mediterranean Jet
MMSEE	Minimum Mean Square Error Estimation
MODIS	Moderate Resolution Imaging Spectometer
MOI	Mediterranean Oscillation Index
MOS	Multi-spectral Optical Sensor
MSS	Multiple Solution Scheme
MW	Mediterranean Waters
MW	Micowave(s)
MWR	Microwave Radiometer
MYI	Multi-Year Ice
NA	North Atlantic
NADW	North Atlantic Deep Water
NAO	North Atlantic Oscillation
NAP	Non-Algae Particles
NASA	National Aeronautics and Space Administration
NCEP	National Centers for Environmental Prediction
NMEA	National Marine Electronics Association (US)
NN	Neural Network
NOAA	National Oceanic and Atmospheric Administration
NP	Non-Pigmented Particulate Material (organic + inorganic)
NRCS	Normalized Radar Cross Section
NSCAT	NASA Scatterometer
NWP	Numerical Weather Prediction
OLS	Oceanographic LIDAR System
OSI	Ocean and Sea Ice
OSPAR	Oslo Paris Commission
OSTIA	Operational SST and Sea Ice Analysis
OTHR	Over-The-Horizon Radar
OW	Open Water(s)
PH	Pigmented Particulate Material (from phytoplankton)
PAR	Photosynthetically Available Radiation
PCA	Principle Component Analysis
PD	Polarization Difference
PDF	Probability Density Function
PM	Passive Microwave(s)
PMT	photomultiplier(s)
PMW	Passive Microwave(s)
POEM	Physical Oceanography of the Eastern Mediterranean
PR	Polarization Ratio
PSMSL	Permanent Service of Mean Sea Level
PSSM	Polynya Signature Simulation Method
psu	practical salinity units
PRF	Pulse Repetition Frequency

QS	QuikSCAT
RADAR	Radio Detection and Ranging
RAR	Real Aperture Radar
RCS	Radar Cross Section
RDAC	Regional Data Assembly Centres
RGB	Red-Green-Blue (colour composite)
RGPS	RADARSAT Geophysical Processor System
R/GTS	(GHRSST-PP) Regional/Global Task Sharing
rms	root mean square
ROV	Remotely Operated Underwater Vehicle
R_{rs}	Remote Sensing Reflectance
RS	Remote Sensor(s)
RWS	Rijkswaterstaat (National Institute for Water Management, The Netherlands)
S	Salinity
SA	South Atlantic
SAF	Satellite Application Facilities
SAG	Southern Adriatic permanent cyclonic Gyre
SAR	Synthetic Aperture Radar
SCAT	Scatteromenter
SeaWiFS	Sea-viewing Wide Field-of-view Sensor
SEVIRI	Spinning Enhanced Visible and Infra-Red Imager
SHOM	Service Hydrographique et Océanographique de la Marine (French Hydrographic Office)
SI	Sea Ice
SIR	Spaceborne Imaging Radar
SLA	Sea Level Anomaly
SLAR	Side-Looking Airborne Radar
SLFMR	Scanning Low Frequency Microwave Radiometer
sm	mineral (suspended) matter
SMMR	Scanning Multichannel Microwave Radiometer
SMOS	Soil Moisture and Ocean Salinity
SNR	signal-to-noise ratio
SRTM	Shuttle Radar Topography Mission
SSH	Sea Surface Height
SSHA	Sea Surface Height Anomaly
SSM/I	Special Sensor Microwave/Imager
SSR	Surface Solar Radiation (SSR)
SSS	Sea Surface Salinity
SST	Sea Surface Temperature
TEC	Total Electron Content
TI	Thin Ice
TM	Thematic Mapper

T/P	Topex/Poseidon
TPJ	Topex/Poseidon and Jason-1
UAV	Unmanned Aerial Vehicle
UHR	Ultra-High Resolution
ULIS	Underwater LIDAR Imaging System
USLA	United States Lifesaving Association
UTC	Universal Time Coordinated
UV	Ultraviolet
VMS	Vessel Monitoring System
VTC	Vessel Traffic Control
VTS	Vessel Traffic Services
VV	vertical co-polarisation
WaMoS II	Wave Monitoring System II
WERA	WEllen RAdar (Wave Radar)
WFD	Water Framework Directive
WiFS	Wide Field-of-view Sensor
WS	White Sea
WS	Wind Speed (from sea surface roughness)
WVC	Wind Vector Cell
XBT	eXpendable Bathy-Thermograph
YS	Yellow Substance

Subject Index

This Volume has been prepared and published with the support of the Joint Research Centre, European Commission.

EUROPEAN COMMISSION

The mission of the JRC is to provide customer-driven scientific and technical support for the conception, development, implementation and monitoring of EU policies. As a service of the European Commission, the JRC functions as a reference centre of science and technology for the Union. Close to the policy-making process, it serves the common interest of the Member States, while being independent of special interests, whether private or national.